ROUTLEDGE LIBRARY EDITIONS:
EVOLUTION

Volume 2

RATES OF EVOLUTION

RATES OF EVOLUTION

Edited by
K.S.W. CAMPBELL AND M.F. DAY

Routledge
Taylor & Francis Group

LONDON AND NEW YORK

First published in 1987 by Allen & Unwin (Publishers) Ltd

This edition first published in 2020
by Routledge
2 Park Square, Milton Park, Abingdon, Oxon OX14 4RN

and by Routledge
52 Vanderbilt Avenue, New York, NY 10017

Routledge is an imprint of the Taylor & Francis Group, an informa business

British Library Cataloguing in Publication Data
A catalogue record for this book is available from the British Library

ISBN: 978-0-367-27938-7 (Set)
ISBN: 978-0-429-31628-9 (Set) (ebk)
ISBN: 978-0-367-26541-0 (Volume 2) (hbk)
ISBN: 978-0-367-26558-8 (Volume 2) (pbk)
ISBN: 978-0-429-29384-9 (Volume 2) (ebk)

Publisher's Note
The publisher has gone to great lengths to ensure the quality of this reprint but points out that some imperfections in the original copies may be apparent.

Disclaimer
The publisher has made every effort to trace copyright holders and would welcome correspondence from those they have been unable to trace.

RATES

—— OF ——

EVOLUTION

Edited by

K. S. W. CAMPBELL

*Department of Geology, Australian
National University, Canberra*

and

M. F. DAY

*Commonwealth Scientific and Industrial
Research Organization, Canberra*

London
ALLEN & UNWIN
Boston   Sydney

Allen & Unwin (Publishers) Ltd,
40 Museum Street, London WC1A 1LU, UK

Allen & Unwin (Publishers) Ltd,
Park Lane, Hemel Hempstead, Herts HP2 4TE, UK

Allen & Unwin, Inc.,
8 Winchester Place, Winchester, Mass. 01890, USA

Allen & Unwin (Australia) Ltd,
8 Napier Street, North Sydney, NSW 2060, Australia

First published in 1987

British Library Cataloguing in Publication Data

Rates of evolution.
1. Evolution   2. Geological time
I. Campbell, K. S. W.   II. Day, M. F.
575   QH371
ISBN 0-04-575030-0

Library of Congress Cataloging-in-Publication Data

Rates of evolution.
Papers delivered at a seminar held at the Geology
Dept., Australian National University, Canberra, Feb.
12–15, 1985.
Bibliography: p.
Includes index.
1. Evolution—Congresses. 2. Paleontology—
Congresses. I. Campbell, K. S. W. (Kenton Stewart Wall)
II. Day, M. F.
QH359.R37   1986   575   86-17373
ISBN 0-04-575030-0 (alk. paper)

Set in 10 on 11 point Sabon by Mathematical Composition Setters Ltd.
and printed in Great Britain by Mackays of Chatham

Preface

This symposium had its origins in discussions about the perennial problem of apparent variation in evolutionary rates during certain periods of the Earth's history or at certain phases of the development of higher taxa. As is obvious from a number of recent attempts to confront aspects of evolutionary history and evolutionary mechanisms, there is at present a greater openness to discussion than at any period since the formulation of the 'Modern Synthesis'. This is largely the result of the revolution in molecular biology which has removed many of the restrictions on discussion of evolutionary mechanisms accepted since the development of that synthesis. But it has also followed on a period of reinterpretation of the fossil record as a source of new historical data, and as a testing ground for mechanisms. The claim of the palaeontologists to exclusive rights to the interpretation of the sequence and dating of evolutionary events has been under siege by advocates of cladistic taxonomy and, more effectively, from proponents of molecular clocks. Finally, the much publicised attack on the logical status of many of the explanatory concepts of Neo-Darwinism has pushed many workers into re-examining possible explanations for evolutionary phenomena that would have had few advocates five years ago. In the resulting turmoil, and it is indeed a turmoil, the once dominant fields of population genetics and developmental biology have provided a stabilising influence, rightly pointing to experimental data on the factors influencing evolutionary processes to counter some of the less well grounded and more speculative views.

It was decided, therefore, that the time was ripe for a symposium that would bring together palaeontologists, geneticists, molecular biologists and developmental biologists to examine some aspects of the problem of evolutionary rates. Support for the proposal was provided by the Australian Academy of Science and subsequently by the Ian Potter Foundation, the Utah Foundation, the 25th International Geological Congress Fund, the Genetics Society of Australia, and the Australian National University. To these organisations we are greatly indebted.

The seminar was held from 12 to 15 February 1985 at the Geology Department, Australian National University, Canberra. Forty workers, including five from overseas, were invited to attend, and 17 of these delivered papers on topics that were loosely defined by the organising committee. Within these guidelines, contributors interpreted their data very broadly, and the papers ranged from discussions of the main features of Precambrian biotic development, to attempts to clarify some of the concepts that form the common currency of debate on evolutionary rates; and from difficulties posed by developmental constraints for hypotheses requiring rapid morphological changes, to proposals for a consideration of evolutionary processes as biological functions that exert an influence on the direction and rate of change.

Inevitably, some topics came up for repeated appraisal. Did large morphological changes really occur rapidly at various times in the geological past, or is the fossil record really too imperfect to be of value in assessing rates of morphological change? What is the measure of 'rapid' change? Is stasis at any taxonomic level established? Is it possible to relate genomic and morphological change? What is the rôle of regulatory and executive genes in controlling evolutionary change? Does the transfer of genetic material between different taxa provide the possibility of increasing evolutionary rates? As might be expected, divergent views on all these and many other topics were expressed.

Surprisingly, other aspects of the problem of rates received relatively little attention. Extinction, which is the subject of considerable research at the present time, was introduced by Raup in the introductory paper, but was not developed by subsequent contributors. Such related topics as the availability of biotic space, diversity equilibria and pervasive global environmental change – in general, the 'externalist' controls on evolutionary rates – were not strongly argued. Clearly, the swing towards 'internalist' explanations has captured the imagination of most workers, and it would be surprising if subsequent symposia did not return to the significance of environmental controls, albeit in somewhat different form.

There was little support for the concept of punctuation at species level, though the case was presented strongly by Williamson. The geneticists present clearly gave a verdict of 'not proven', and they therefore have seen little point in examining possible genetic mechanisms. But stasis may not be a possibility only at the species level; it may be significant over long periods of time and affect whole clades. As long ago as 1967, the geneticist Rendel in his study of canalisation concluded 'that by and large the periods of great evolutionary change will be short lived and will be interspersed by long periods during which nothing much seems to happen'. The significance of this type of concept for the interpretation of palaeontological data needs further exploration. In this connection, the highly contentious work of Schopf (1984) on rates of evolution and living fossils did not receive attention. This probably appeared too late to be assimilated by contributors before the meeting. He concluded that living fossils 'are not a problem in speciation theory but rather a problem for developmental biology'.

There is no doubt that developmental biology has a major new rôle in understanding rates – in fact this area provides one of the most exciting areas for further study. On the one hand there is the extraordinary conservatism of the developmental process, as Anderson so strongly pointed out, and on the other there is the possibility of change in genes that have an especially significant rôle in mediating morphological and structural organisation, and exert a co-ordinating rôle on developmental processes. The isolation of such genes and the elaboration of these effects in a variety of organisms seems to be a priority area for future work.

It is appropriate to conclude this Preface with a comment on the paper by John Campbell that ends the volume. It epitomises the current state of ferment in evolutionary studies in that it attempts to reverse the classical conception of the relation between gene and environment in evolution, viz. that the raw materials for change are genetic but that direction and rate are imposed by the external environment. According to Campbell and co-workers, genes have a dynamic life of their own – no matter how environmentally pampered a species, its genes would evolve. We should be thinking of evolution, therefore, as a process of the unfolding of genetic potential on an environmental stage, the confines of which provide only secondary constraints. Under these circumstances, the analysis of rates of evolutionary change becomes an exceedingly complex affair which 'must begin with a clear description of the sort of change being considered and then proceed to the mechanisms that produce that form of change'.

The comment was recently made in a popular science forum that evolutionary studies had now reached a stage similar to that reached by physics in the early decades of this century. The old paradigms were crumbling on a number of fronts, but new ones were being uncertainly received by many of those actively engaged in the field. Philosophical barriers developed even between the foremost investigators in some areas. Is the analogy with the present state of evolutionary studies too fanciful?

K. S. W. Campbell
M. F. Day

REFERENCES

Rendel, J. M. 1967. *Canalisation and gene control*. London: Logos.

Schopf, T. J. M. 1984. Rates of evolution and the notion of 'living' fossils. *Ann. Rev. Earth Planet. Sci.* **12**, 245–92.

Contents

List of tables

List of contributors

M. Adams   Evolutionary Biology Unit, South Australian Museum, North Terrace, Adelaide, Australia

D. T. Anderson   School of Biological Sciences, University of Sydney, Sydney, NSW 2006, Australia

P. R. Baverstock   Evolutionary Biology Unit, South Australian Museum, North Terrace, Adelaide, Australia

J. H. Campbell   Department of Anatomy, School of Medicine, Center for the Health Sciences, University of California, Los Angeles, California 90024, USA

K. S. W. Campbell   Department of Geology, Australian National University, Canberra, ACT 2601, Australia

H. L. Carson   Department of Genetics, University of Hawaii, Honolulu, Hawaii 96822, USA

I. R. Franklin   Division of Animal Production, Commonwealth Scientific and Industrial Research Organization, PO Box 239, Blacktown, NSW, Australia

B. John   Department of Population Biology, Research School of Biological Sciences, Australian National University, Canberra, ACT 2601, Australia

J. Langridge   Division of Plant Industry, Commonwealth Scientific and Industrial Research Organization, Canberra, Australia

C. R. Marshall   Department of Geophysical Sciences, University of Chicago, Illinois 60637, USA

G. L. G. Miklos   Department of Population Biology, Research School of Biological Sciences, Australian National University, Canberra, ACT 2601, Australia

D. M. Raup   Department of Geophysical Sciences, University of Chicago, Chicago, Illinois 60637, USA

D. C. Reanney   Department of Microbiology, Latrobe University, Bundoora, Victoria 3083, Australia

B. N. Runnegar   Department of Geology and Geophysics, University of New England, Armidale, NSW 2351, Australia

A. R. Templeton   Department of Biology, Washington University, St. Louis, Missouri, 63130, USA

J. A. Thomson   School of Biological Sciences, University of Sydney, Macleay Building A12, Sydney, NSW 2006, Australia

E. M. Truswell   Division of Continental Geology, Bureau of Mineral Resources, Canberra, Australia

M. R. Walter Baas Becking Geobiological Research Laboratory, Bureau of Mineral Resources, PO Box 378, Canberra 2601, Australia

P. G. Williamson Museum of Comparative Zoology, Harvard University, Cambridge, Mass. 02138, USA

1

Major features of the fossil record and their implications for evolutionary rate studies

DAVID M. RAUP

ABSTRACT

The fossil record contains a vast array of data on the distribution of extinct species in space and time. Sampling problems are often severe, however, so that great care must be exercised when using palaeontological data for the analysis of evolutionary rates. But, because the fossil record makes rate data available over geological time spans, information from fossils is indispensable to the complete analysis of rates.

Cladogenetic rates of evolution involve branching and termination of lineages and can be approached through the use of a simple branching model of the birth–death type. This model often provides accurate predictions of the behaviour of the cladogenetic process with generalised, real-world data. In many cases, however, analysis with higher resolution data shows cladogenesis to be much more episodic (punctuated) than would be predicted from the model. The branching model thus provides a time-averaged picture of evolution as well as a null hypothesis with which significant departures from statistical expectations can be uncovered.

It has been suggested from molecular-clock data that rates of genomic change are higher in the early stages of an adaptive radiation, although this view has been challenged. To the extent that clock rates do change predictably and systematically, the cause may lie in the possibility that genomic change is more a function of the frequency of lineage branchings (speciation events) than of total elapsed time. This proposition should be tested with biological groups where genomic divergence data are optimal and where well-dated fossil records are available.

INTRODUCTION

The fossil record contains information on about 250 000 plant and animal species that lived in the geologic past (Raup 1976a). Although recognisable fossils are found in rocks as old as about 3.5×10^9 years (Ga), the vast majority are in rocks of Phanerozoic age (600 million years (Ma) and younger). In spite of the large size of the fossil sample, it represents but a tiny fragment of past life. Mean durations of species vary from one biological group to another but average in the range of 1–10 million years (Valentine 1970, Van Valen 1973, Raup 1978). Thus, the biological world has turned over many times during the Phanerozoic so that the number of species that have lived in the past should be some large multiple of the number living today. Estimates of the total number of species that have lived in the past range as high as 50×10^9 (Simpson 1952). It is for this reason that the 250 000 species known as fossils constitute an extremely small sample of past life.

The fossil record provides a wealth of information on evolutionary rates. Within the limits of the resolution of geological dating, we can document two basically different rates: cladogenetic and morphological. The first has to do with the branching pattern which constitutes the evolutionary tree. Critical cladogenetic elements include rates of lineage branching (speciation) and rates of lineage termination (extinction) and these combine to produce change in taxonomic diversity (richness). Morphological rates refer to change in the phenotype over geological time and this has been of particular interest in the past decade because it bears on the problem of punctuated equilibrium. Is morphological change distributed over time in the evolution of a lineage, or is it concentrated at the branch points in the evolutionary tree?

Cladogenetic and morphological rates of evolution can be studied at various scales. If one defines an evolutionary lineage to be the temporal sequence of populations of a single species, then branching events are true speciation events which serve to split the gene pool into reproductively isolated species. The termination of a lineage is the extinction of a single biological species. The morphological change that may take place during the life of the lineage is caused by population-level mechanisms such as change in allelic frequencies by directional selection or random drift. Although this kind of change (called anagenesis or phyletic transformation) may lead to a new species, the term 'speciation' is here reserved for true lineage branching.

The lineage can also be defined at higher taxonomic levels. Thus, we may define a lineage as all species of a genus or family. If so, the taxon is initiated by a branching event (speciation) but we are not concerned with the details of branching within the genus or family. The taxon becomes extinct when all its species are extinct. In this sort of higher level lineage, morphological change may be dominated by processes such as species selection, wherein trends in morphological change are determined by differential survival of species rather than by changes at the population level (Eldredge & Gould 1972, Stanley 1975).

Although the fossil record provides the only means for observing evolutionary change over long spans of time, recent developments in molecular genetics make it possible to infer rates of change over geological time through the analysis of genetic distance data for pairs of living species. I refer to the use of the molecular-clock logic to reconstruct cladogenetic patterns. Given a few calibration points from the geological record, it is possible to say something about variation in rates of genomic change from one biological group to another or through time. It is thus of special interest to compare evolutionary rates observed in the fossil record with those inferred from the genetics of living organisms.

DATA BASE AND SAMPLING PROBLEMS

The fossilisation and subsequent discovery of an ancient species is a most unlikely event. This follows inevitably from the fact that such a small proportion of past species are known as fossils. It follows that truly unusual circumstances are required for fossilisation, and this means that our sample of past life may be distinctly non-random. As a general rule, dead organisms are consumed by scavenger activity (including bacterial action) shortly after death so that an organism that dies in a biologically active environment has little chance of being preserved. The probability of preservation is greatly enhanced if the dead organism is moved to a biologically inactive environment and this general process is probably responsible for most fossils. A shallow-water marine animal may be thrown up on a beach by a storm or covered *in situ* by a sudden influx of sediment, or a terrestrial animal may be buried suddenly by volcanic or flood debris. These are examples of the large variety of mechanisms that can isolate plants and animals from biologically active environments and thus favour preservation. The study of this process constitutes the important subdiscipline of palaeobiology called taphonomy (see Kidwell & Jablonski 1983, for a recent treatment).

Taphonomic studies have demonstrated that the fossil record is a record of rare accidents, many of which are catastrophic. It is thus not surprising that the record is strongly biased in favour of certain kinds of organisms and certain environments. As a general rule, environments where there is a net accumulation of sediment contain a disproportionate number of preserved plants and animals. This means that preservation of marine organisms is more common than preservation of terrestrial species. Within the terrestrial realm, species living near lakes and rivers are more likely to be preserved than upland forms, and so on. One result of taphonomic differentials such as this is that aquatic (and especially marine) forms provide much the best sample for any analysis of evolutionary rates. Hard-shelled marine animals are thus far more appropriate for rate studies than, for example, insects, birds, or land plants.

Although the fossil record is systematically biased as described above, there are occasional spectacular exceptions. These are the so-called Lagerstätten and include unusual accidents such as the preservation of a complex mammalian fauna in the La Brea tarpits of Pleistocene age in Southern California. Such Lagerstätten provide invaluable windows to the past although they have a strangely disruptive effect on synoptic studies of evolutionary rates. It can be noted, for example, that cladogenetic activity seems to be greatly elevated at Lagerstätten but this is only an artifact of the unusual preservation. Because Lagerstätten often provide the only record of certain species and higher taxa, these taxa will appear to originate and become extinct at a single point in time. Branching rates, extinction rates and total taxonomic diversity increase markedly at these points in the geological record, but this is only an artifact of the absence of preservation of the same taxa above and below this point. It is thus important to remove data from Lagerstätten before analysis of rates. To the extent that Lagerstätten represent extreme cases in a continuum of varying preservation, this variation adds noise to palaeontological data which can never be removed completely.

Another kind of bias in the fossil record is that which has been called 'The Pull of the Recent' (Raup 1979). It is generally true that the probability of occurrence and discovery of fossils increases as one moves toward the Recent (present day). Younger rocks are more widely exposed simply because they are closer to the 'top of the pile' and because they have had less time to be destroyed by metamorphism or removed by erosion. For this reason, the number of fossils found per million years of rock increases toward the Recent, with or without a true increase in biological diversity. Thus, rate of cladogenetic activity gives the appearance of increasing toward the Recent. There are several ways of coping with this time-dependent bias, including normalising for volume or area of exposed sedimentary rock (Raup 1976b). But the problem remains a serious one and it has given rise to considerable debate over basic questions such as whether total taxonomic diversity has increased through time (Bambach 1977, Sepkoski *et al.* 1981).

A related aspect of the 'Pull of the Recent' stems from the fact that modern organisms are far better known (sampled) than fossil organisms. A higher percentage of living species has been found and described than their extinct counterparts and this leads to a systematic bias in certain kinds of evolutionary rate analysis. This can be explained with reference to the observed time ranges of taxonomic groups. Ordinarily, the range of a taxon is simply the time interval between its first and last occurrences in the fossil record. Suppose we have a taxon which has a low probability of preservation and has been found, by chance, at only one horizon, such as in the Cambrian. Suppose further that the organism is not living today. The range of the taxon will thus be recorded as Cambrian. This is acceptable as long as it is understood that the taxon may have had a longer, unrecorded range, in line with the general observation that ranges in the fossil record are often truncated (shortened) by lack of preservation. Alternatively, suppose the taxon is still living today. Because of better sampling in the modern world, we are unlikely to miss its existence today. If the taxon is living today and is found as a fossil only in the Cambrian, the apparent range will be Cambrian to Recent. This general principle has the insidious effect of exaggerating ranges for taxa that are still living today. The only way to eliminate the problem completely is to ignore all living species when establishing geological ranges. In a sense, the modern fauna constitutes the best of all Lagerstätten! The significance of this is most striking if one takes the total record completely at face value: because there are millions of living species with no fossil record, one could (in a most simple-minded fashion) infer that there was a remarkable burst of speciation in the immediate past.

The foregoing discussion has touched on a few highlights of a complex set of issues relating to the nature of fossil data. A particularly excellent analysis of this subject is that of Paul (1982). Any rigorous analysis of evolutionary rates must accommodate the strengths and avoid the weaknesses of the fossil record. In general, the higher taxonomic groups are more completely sampled than are species; also, higher taxa are less prone to the biases and other difficulties described in this section. For this reason, many of the best analyses of evolutionary rates use genera and families rather than species.

RATES OF CLADOGENETIC EVOLUTION

As indicated earlier, the basic unit of cladogenesis is the lineage, whether it be the ancestor–descendent series of populations of a single species or a collection of species having a common ancestor. Let us first consider cladogenesis with respect to species lineages and use an ultrasimple branching model of the type common in particle physics, demography and several other disciplines. This model can be described by the following equation:

$$S_t = S_0\, e^{(p-q)t} \tag{1}$$

where S_0 and S_t are the numbers of coexisting lineages (species) at times zero and t, respectively. The constants p and q are rates of lineage branching (speciation) and lineage termination (extinction) respectively, and are expressed as the number of events per lineage per unit of time. A typical branching rate, for example, is 0.2 speciations per lineage per million years. If $p > q$, the number of coexisting species (diversity) increases exponentially, and if $p < q$, the number decreases exponentially to zero, leading to the total extinction of the group.

Equation 1 is useful in making a variety of predictions and postdictions about the development of an evolutionary tree. For example, if $p > q$, the number of coexisting species will double after a time equal to $\ln 2/(p - q)$ (Stanley 1979). Actual doubling times are highly variable, of course, because speciation and extinction of species do not happen in a deterministic clocklike manner. Thus, Equation 1 is better seen as a probabilistic description such that p and q are probabilities. The outcome in an actual case is dependent on the vagaries of the stochastic process.

Equation 1 can be elaborated and extended almost without limit to include estimates of many aspects of the cladogenetic process (see Raup 1985, for review). Equations have been developed which predict such things as the probability of extinction of a related group of species at or before some time t, given the number of species existing at an arbitrary starting time (S_0). Other equations predict the size frequency distribution of groups and still others predict quantities such as the 'total progeny' (the total number of lineages, surviving and not surviving) over the history of a group. Again, all such predictions are subject to statistical uncertainty and computation of a variance of the estimate is essential with most applications.

In a number of cases with real-world data, Equation 1 and its derivatives have proven accurate and useful. For example, if one is concerned only with extinction, the following variant on Equation 1 can be written:

$$S_t = S_0 \, e^{-qt} \tag{2}$$

This describes the decay of a cohort of lineages present at some time equal to 0 and is, therefore, the equation for a survivorship curve for lineages having an extinction probability of q per lineage per million years. If the value of q is stochastically constant through time and from species to species in the cohort, then the survivorship curve for lineages will be linear when plotted with a logarithmic ordinate (Van Valen 1973). This is illustrated with records from fossil foraminifera in Figure 1.1. Given linear survivorship, the following is true:

$$\text{mean lineage duration} = 1/q \tag{3}$$

and

$$\text{median lineage duration (half-life)} = -(\ln 0.5)/q \tag{4}$$

Contrary to conventional procedure in cohort survivorship analysis, the cohort depicted in Figure 1.1 is made up of all species that were living at a point in time, regardless of how long they had been alive up to that point.

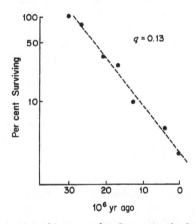

Figure 1.1   Cohort survivorship curve for Cenozoic planktonic foraminifera. The cohort consists of those species in existence 30 Ma ago. The topmost point (at 100%) represents the initial cohort and the other points monitor the decay of the cohort through time extinction of species. The straight line fit to the points has a slope of -0.13 ($= -q$) indicating a mean species duration of 7.7 Ma (Eqn. 3) and a species half-life of 5.3 Ma (Eqn. 4). (Data from Fig. 9C of Hoffman and Kitchell 1984.)

The points in Figure 1.1 are chosen at convenient times about 5 Ma apart in the geological time scale. For this particular plot, the assumption of constant probability of extinction, regardless of species duration, appears to be justified by the loglinear array of points. It is on this basis that a straight line has been fitted to the points.

Serious problems are encountered in cases where it is evident that values of p and/or q are not stochastically constant. This is seen in Figure 1.2 which is based on a fuller version of the dataset that was used for Figure 1.1. There are several differences between the two plots. First, a number of cohorts are analysed instead of just one. Secondly, a much higher resolution time scale is used. And third, the points of known survivorship are connected by straight lines instead of serving just as guides for a generalised fit. As a result, the character of the pattern is changed substantially. The overall impression is one of punctuated extinction with horizontal or gently sloping segments (indicating zero or negligible extinction rate) alternating with vertical or near vertical segments (indicating nearly instantaneous extinction of several species). Some of the horizontal segments are due to widely spaced data points but, in general, their horizontality is real.

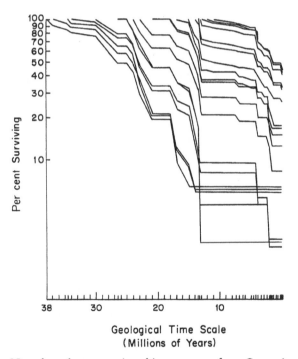

Figure 1.2 Nested cohort survivorship curves for Cenozoic planktonic foraminifera showing punctuated nature of species extinction in this group. A major extinction at about 12 Ma is indicated by the coincidence of sudden drops in the curves for all cohorts. (From Fig. 9C of Hoffman and Kitchell 1984.)

The contrast between Figures 1.1 and 1.2 is only partly due to the method of plotting and curve fitting. The main difference is that when higher time resolution is available, a stair-step pattern can be seen, suggesting a process involving anything but constant rates.

Figure 1.3 shows another departure from the expectations of the simple model. Here, a survivorship curve for genera of marine invertebrates shows a clearly concave-upward shape, indicating a progressive decrease in extinction rate (or probability) during the life of the cohort. In this case, the result can be reconciled with the model by showing (Raup 1978) that if the species that make up genera have a constant extinction rate (q), the genera themselves will show curvilinear survivorship of the type seen in Figure 1.3. The heuristic explanation for this is that if a genus manages to survive, it has probably done so by building up the number of constituent species. To put this another way, the longer a genus survives, the more species it probably has and the less likely it is to become extinct in the next time unit. This leads to a decrease in probability of extinction (in a given time unit) through the lifespan of a higher taxon. This can be expressed mathematically as follows:

$$P_{(0,\,t)} = \frac{q(e^{(p-q)t} - 1)}{p\,e^{(p-q)t} - q} \qquad (5)$$

where $P_{0,t}$ is the probability of extinction of the whole taxon at or before time t. If one is willing to assume that Equation 5 is a valid description of generic extinction as depicted in Figure 1.3, then the equation can be fitted to the curve in Figure 1.3 to obtain estimates of species-level p and q values even though actual species data have not been considered. This is especially

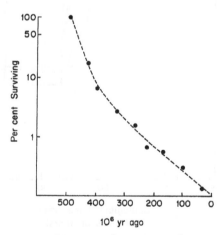

Figure 1.3 Cohort survivorship curve for genera of marine invertebrates that originated in the Ordovician Period (505–438 Ma). The curve is concave upward, as expected for supraspecific taxa, and can be approximated by Equation 5. (Data from Fig. 1 of Raup 1978.)

valuable in the common geological situation in which the vagaries of preservation and sampling preclude direct assessment at the species level.

Again, however, it is necessary to assume that the values of p and q are stochastically constant and as we have seen (Fig. 1.2), this is not the case at some scales. Figure 1.4 shows nested survivorship curves for families of marine animals during the Phanerozoic. Data points have an average spacing of about 7 Ma in contrast to a spacing of about 50 Ma in Figure 1.3. The steplike pattern is similar to that seen at the species level in Figure 1.2. It is clear, therefore, that the smooth curve in Figure 1.3 is a time-averaged version of a more complex process.

We see in the foregoing examples an evolutionary process characterised by gradual, continuous patterns at some scales and episodic, punctuated patterns at other scales. The punctuated character of the cladogenetic process is most striking when we consider mass extinctions and adaptive radiations, representing sudden bursts of extinction and branching, respectively. In Figure 1.4, several of the largest mass extinctions are visible as sharp drops in survivorship, the two most striking being those at the ends of the Palaeozoic and Mesozoic eras.

Adaptive radiations constitute the other common departure from constant cladogenetic rates. It is common that branching rates are substantially higher in the early stages of the evolutionary history of a biological group. This has been recognised by several authors, including Gould *et al.* (1977). Total diversity climbs more rapidly than would be predicted from Equation 1 using mean p values based on the total history of the group. The increase in speciation rate could be caused by the ease of exploitation of a new adaptive zone or by the presence of a vacuum following mass extinction, or a combination of the two. A classic example is the rapid radiation of mammals following the extinction of the dinosaurs. But it must be

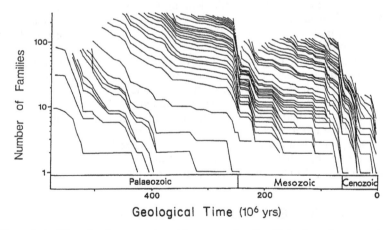

Figure 1.4 Nested cohort survivorship curves for families of marine organisms. The major mass extinctions at the end of the Palaeozoic and at the end of the Mesozoic are shown by sudden drops. (Data from Sepkoski 1982.)

remembered that the same effect on diversity can be produced by a decrease in extinction rate (q). Rigorous tests of this proposition have not been made, but anecdotal data suggest that a typical radiation is accompanied by both elevated p and depressed q.

The most spectacular radiation known from the evolutionary record was the great expansion of metazoans in the late Precambrian and early Palaeozoic: the interval from 700 to 500 Ma ago. This involved not only an extremely rapid rise in diversity, but also unusually rapid morphological divergence. This event will be discussed in more detail below, in the context of molecular phylogenies.

RATES OF MORPHOLOGICAL CHANGE

The theory of punctuated equilibrium is based on the claim by Eldredge and Gould (1972) that most morphological change is concentrated at speciation events (lineage branchings) and that species are in a state of stasis during most of their existence. That is, it is argued that phyletic transformation of species over the duration of a lineage, whether it be by directional selection or genetic drift, is a minor element in evolution. This claim has led to a large number of specialised analyses of closely spaced fossil samples of single species lineages. A few examples of such studies are the works of Arnold (1983), Kellogg (1983), Lohmann and Malmgren (1983), Malmgren and Kennett (1981), Malmgren et al. (1983), and Raup and Crick (1981). In theory, these studies should be straightforward: in geological sections of continuous sedimentation and optimal fossil preservation, it should be a simple matter to perform biometrical analyses to demonstrate the presence or absence of stasis. Unfortunately, these studies have not yet produced unequivocal and consistent results leading to general statements.

Morphological change can also be considered on a larger scale by the analysis of interspecific trends. For example, do organisms become larger or more complex over long periods of time regardless of whether the change is locally punctuated or gradual? If so, do rates of change vary in a definable or predictable manner? Much has been written on these subjects but most of the analyses have been largely anecdotal and thus have not produced rigorous answers. Questions involving complexity have been plagued by the lack of suitable metrics with which to measure complexity although the study by Flessa et al. (1975) represents a bold attempt. On the question of body size, there is a conventional wisdom that size increases through time (Cope's rule) but there have been few rigorous analyses such as that by Stanley (1973). Much remains to be done in this field.

ORGANISMIC VERSUS GENOMIC RATES

The past decade has seen important advances in our ability to chart genomic divergence in evolution. The molecular-clock logic has made it possible to reconstruct evolutionary trees on the basis of matrices of genetic distance data for pairs of living species. In the best cases, the trees that result correspond well to those developed by other methods (e.g. Sibley & Ahlquist 1983). Furthermore, it has been possible to calibrate these trees with divergence times (branch points) from the geological record. These reconstructions are based, at least to some degree, on an assumption of constancy in rates of genomic change.

Goodman (1981 and subsequent papers) has argued, however, that rates of genomic change can vary through time in a systematic fashion. This is based on noting time intervals during which geological dates of divergence times are consistently different from those predicted from molecular genetic data. Goodman finds periods of high genomic rates for vertebrates in the Early Palaeozoic (500–300 Ma ago) and again in the Late Cretaceous and Early Tertiary (90–40 Ma ago), coinciding approximately with major periods of evolutionary diversification. He argues that these bursts of genomic change are associated with the early stages of major adaptive radiations because these are times of strong directional selection. This reading of the data has been challenged by several authors, especially Runnegar (1982). Indeed, the molecular as well as geological data are sufficiently few and uncertain that they are open to quite different interpretations. Nevertheless, the Goodman suggestion of variable rates *vis-à-vis* adaptive radiations is of sufficient importance that it should be tested as thoroughly as possible.

The problem can be illustrated by the example of the timing of the main divergence of the metazoan phyla in the late Precambrian. A literal reading of the fossil record would place this event, or set of events, at no more than about 700 Ma ago, the first occurrence of metazoan fossils. However, a molecular-clock analysis of haemoglobins by Runnegar (1982) indicated a divergence time of 900–1000 Ma ago for the major metazoan phyla. Assuming the genetic distance data to be correct, we have only two alternatives: first, the fossil record is deficient before 700 Ma or secondly, the rate of genomic change was higher near the beginning of the metazoan radiation. Runnegar favours the first alternative, suggesting that the early metazoans were worm-like forms that did not leave a fossil record. The Goodman position would be that the early stages of the radiation were accompanied by a burst of genomic change (faster clock rates).

The molecular-clock logic assumes (often tacitly) that change in the genome takes place continuously throughout evolution: that is, along the length of each lineage. This is in sharp contrast to the claim for organismic evolution made by the punctuated equilibrium theory! This difference has led, among other things, to debate between geneticists and palaeontologists with the former claiming that the genome is in constant change and the latter claiming that most species are in stasis most of the time. This has led some to suggest a fundamental decoupling of genomic and organismic evolution. But are the genetic and morphological results necessarily in conflict? I think not.

We have seen above that branching rates p tend to be higher in the early history of an adaptive radiation, leading to higher than expected diversity at these times. These are also times of most rapid morphological divergence and this would be viewed by the proponents of punctuated equilibrium as one result of the higher frequency of speciation (branching) events (Stanley 1976). If rates of genomic change are also raised at these times, we have the tantalising possibility that genomic change is also concentrated at the branch points of the evolutionary tree. By this interpretation, the molecular clock should be expected to 'keep time' only if branching events are uniformly distributed through time. If the branching events are not uniformly distributed, one should use the actual number of branch points rather than elapsed time as the denominator in rate calculations. This possibility is contrasted schematically with the conventional approach in Figure 1.5.

The foregoing hypothesis is testable, given enough genetic data for biological groups with good fossil records. Branching rate for a given time interval can be estimated by the survivorship techniques described above or simply by raw data on taxonomic diversity, on the assumption that a rise in diversity reflects an increase in speciation rate (as opposed to a decrease in extinction rate). Except for the limited analysis of the late Precambrian diversification of metazoan phyla already referred to, the appropriate data to test the proposition have not yet been assembled.

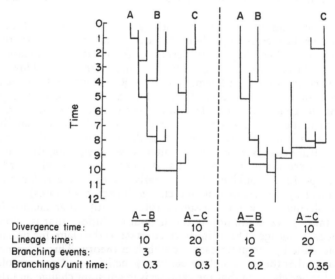

Figure 1.5 Schematic evolutionary trees. Left: lineage branching spread uniformly through the tree. Right: lineage branching concentrated in an adaptive radiation in the 8–10 time range. Both trees have three surviving lineages (with which genetic comparisons could be made) and 12 branching events. If clock rate is dependent on the number of branching events per unit time, conventional molecular measures of divergence time will be in error if branching events are clustered. That is, a clock calibrated on the A–B pair of the tree on the right will yield an exaggerated divergence time when the A–C pair is considered.

REFERENCES

Arnold, A. J. 1983. Phyletic evolution in the *Globorotalia crassaformis* (Galloway and Wissler) lineage: a preliminary report. *Paleobiology* 9, 390–7.

Bambach, R. K. 1977. Species richness in marine benthic habitats through the Phanerozoic. *Paleobiology* 3, 152–67.

Eldredge, N. and S. J. Gould 1972. Punctuated equilibria: an alternative to phyletic gradualism. In *Models in paleobiology*, T. J. M. Schopf (ed.), pp. 82–115. San Francisco: Freeman, Cooper.

Flessa, K. W., K. V. Powers, and J. L. Cisne 1975. Specialization and evolutionary longevity in the Arthropoda. *Paleobiology* 1, 71–81.

Goodman, M. 1981. Decoding the pattern of protein evolution. *Prog. Biophys. Mol. Biol.* 37, 105–64.

Gould, S. J., D. M. Raup, J. J. Sepkoski, T. J. M. Schopf and D. S. Simberloff 1977. The shape of evolution: a comparison of real and random clades. *Paleobiology* 3, 23–40.

Hoffman, A. and J. A. Kitchell 1984. Evolution in a pelagic planktic system: a paleobiologic test of multispecies evolution. *Paleobiology* 10, 9–33.

Kellogg, D. E. 1983. Phenology of morphologic change in radiolarian lineages from deep-sea cores: implications for macroevolution. *Paleobiology* 9, 355–62.

Kidwell, S. M. and D. Jablonski 1983. Taphonomic feedback: ecological consequences of shell accumulation. In *Biotic interactions in recent and fossil benthic communities*, M. J. Tevesz and P. L. McCall, (eds), New York: Plenum Press. pp. 195–248.

Lohmann, G. P. and B. A. Malmgren 1983. Equatorward migration of *Globorotalia truncatulinoides* ecophenotypes through the Late Pleistocene: Gradual evolution or ocean change? *Paleobiology* 9, 414–21.

Malmgren, B. A. and J. R. Kennett, 1981. Phyletic gradualism in a Late Cenozoic planktonic foraminiferal lineage; DSDP Site 284, southwest Pacific. *Paleobiology* 7, 230–40.

Malmgren, B. A., W. A. Berggren and G. P. Lohmann 1983. Evidence for punctuated gradualism in the Late Neogene *Globorotalia tumida* lineage of planktonic foraminifera. *Paleobiology* 9, 377–89.

Paul, C. R. C. 1982. The adequacy of the fossil record. *Syst. Assoc. Spec. Vol.* 21, 75–117.

Raup, D. M. 1976a. Species diversity in the Phanerozoic: a tabulation. *Paleobiology* 2, 279–88.

Raup, D. M. 1976b. Species diversity in the Phanerozoic: an interpretation. *Paleobiology* 2, 289–97.

Raup, D. M. 1978. Cohort analysis of generic survivorship. *Paleobiology* 4, 1–15.

Raup, D. M. 1979. Biases in the fossil record of species and genera. *Carnegie Mus. Nat. Hist., Bull.* 13, 85–91.

Raup, D. M. 1985. Mathematical models of cladogenesis. *Paleobiology* 11, 42–52.

Raup, D. M. and Crick, R. E. 1981. Evolution of single characters in the Jurassic ammonite *Kosmoceras*. *Paleobiology* 7, 200–15.

Runnegar, B. 1982. A molecular-clock date for the origin of the animal phyla. *Lethaia* 15, 199–205.

Sepkoski, J. J. 1982. A compendium of fossil marine families. *Contrib. Milwaukee Pub. Mus.* 51, 1–125.

Sepkoski, J. J., R. K. Bambach, D. M. Raup and J. W. Valentine 1981. Phanerozoic marine diversity and the fossil record. *Nature* 293, 435–7.

Sibley, C. G. and J. E. Ahlquist 1983. The phylogeny and classification of birds, based on the data of DNA–DNA hybridization. In *Current ornithology*, R. F. Johnston (ed.), Vol. 1. New York: Plenum Press.

Simpson, G. G. 1952. How many species? *Evolution* 6, 342.

Stanley, S. M. 1973. An explanation for Cope's Rule. *Evolution* 27, 1–26.

Stanley, S. M. 1975. A theory of evolution above the species level. *Proc. Natl Acad. Sci. USA* 72, 646–50.

Stanley, S. M. 1976. Ideas on the timing of metazoan diversification. *Paleobiology* 2, 209–19.

Stanley, S. M. 1979. *Macroevolution*. San Francisco: New York.

Valentine, J. W. 1970. How many marine invertebrate fossil species? *J. Paleontol.* 44, 410–15.

Van Valen, L. 1973. A new evolutionary law. *Evol. Theory* 1, 1–30.

2

The timing of major evolutionary innovations from the origin of life to the origins of the Metaphyta and Metazoa: the geological evidence

M. R. WALTER

ABSTRACT

The geological record of life from 3.5×10^9 (Ga) years ago to 0.57 Ga ago is considered within the framework of published phylogenies derived from molecular biology. The earliest steps in evolution are not recorded in any known rocks. The record starts 3.5 Ga ago, with coccoid and filamentous benthic anaerobic photoautotrophs (Eubacteria). Evidence for methanogens (Archaebacteria) is found in rocks 2.8 Ga old, but because of the paucity of ancient rocks, of relevant research, and of means for recognising these organisms in the geological record, this is only a minimum estimate of the age of this group. The oldest eukaryotes seem to be about 1.4 Ga old; megascopic algae have a record back to 1.3 Ga, metazoans tentatively to about 1.0 Ga and protists back to 0.8 Ga. No Fungi are known from the Archaean or Proterozoic. The geological record is consistent with, but at present cannot test, phylogenies derived from the sequencing of proteins and nucleic acids. Nevertheless, it does provide a time calibration, and is consistent with at least one major prediction made using the 'molecular clock' approach: that the initial divergence of the animal phyla began about 1.0 Ga ago.

INTRODUCTION

In recent years there have been quite remarkable changes in our perception of the diversity and phylogenetic relationships within the prokaryotes. Steady progress was being made in the determination of biochemical pathways, in ultrastructural studies, and in the chemical characterisation of cell components, all of which provide insights into evolutionary relationships amongst the prokaryotes (Broda 1978), when the techniques of protein and nucleic acid sequencing were introduced. In just a few years microbiologists have progressed from debating whether 'blue-green algae' are really bacteria (which they are, hence the new name Cyanobacteria) to discovering amongst the 'bacteria' a whole new kingdom of organisms, the Archaebacteria (Woese & Fox 1977, Fox et al. 1980, Woese 1982; Fig. 2.1). Now there are hints of yet another kingdom, the 'Eocyta' (Lake et al. 1984). In retrospect, we should not have been too surprised by these discoveries that life is even more diverse than we had thought, because during the past two decades or so it has been established by geologists and palaeobiologists that the history of life extends back at least 3.5×10^9 (Ga) years (Schopf 1983), which seems more than enough time to account for almost any degree of diversity.

Three major groups ('urkingdoms') of organisms are now recognised by some biologists (e.g. Fox et al. 1980): Archaebacteria, Eubacteria and eukaryotes. The cytoplasm of eukaryotic cells may have an evolutionary history as long as that of the Archaebacteria and Eubacteria, and the name

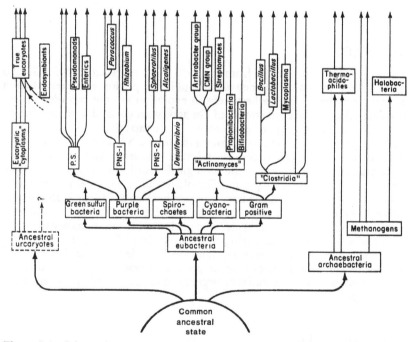

Figure 2.1   Schematic representation of the major lines of prokaryotic descent, derived mainly from molecular sequencing (from Fox et al. 1980).

'Urcaryote' has been suggested for that lineage (Woese & Fox 1977, Fox *et al.* 1980). The eukaryotes can be divided into four kingdoms, the Protista, Plantae, Fungi and Animalia (Whittaker 1969; Figs. 2.2, 2.3). All these major groups have a geological record, enabling some constraints to be applied to phylogenetic schemes derived from neobiology. At least five of the six kingdoms were differentiated during the Archaean or Proterozoic, whereas the history of the skeletonised invertebrates and vertebrates, and of land plants, which together provide the stuff of the more familiar fossil record, began just before the end of the Proterozoic (in the case of the animals) and in the Silurian (in the case of the vascular plants); their record is that of the Phanerozoic, and is not discussed here. The geological record has the potential to allow the testing of phylogenetic interpretations, and to provide the time calibration that at present biology cannot give.

SOURCES OF INFORMATION AND THEIR LIMITATIONS

The fossil record is fairly abundant and well preserved for times less than 2.5 Ga ago (Table 2.1), is very limited from 2.5–3.0 Ga ago, is rare and intermittent from 3.0 to 3.9 Ga, and is essentially non-existent for times

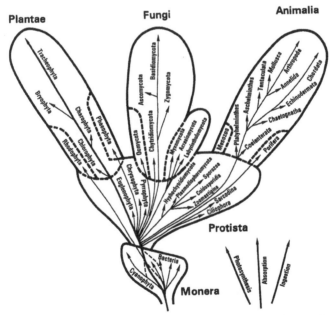

Figure 2.2 A five-kingdom classification of life based on three levels of organisation, prokaryotic (kingdom Monera), unicellular (Protista) and eukaryotic multicellular and multinucleate (from Whittaker 1969). The 'Monera' can now be subdivided as in Figure 2.1.

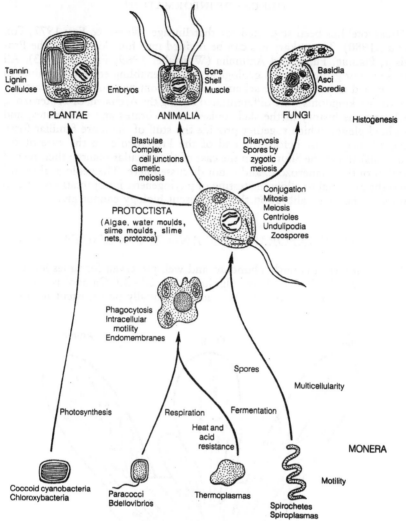

Figure 2.3 Model for the origin of eukaryotic cells by symbiosis (from Margulis 1981).

Table 2.1 Nomenclature of major intervals of Earth history referred to herein. A discussion of this and other variants can be found in Harland *et al.* (1982). Ediacarian is a new name defined by Cloud and Glaessner (1982).

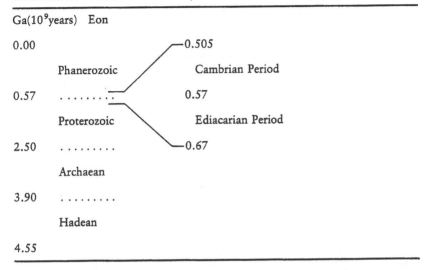

Ga(10^9years) Eon

0.00

Phanerozoic

0.57 . . . .

Proterozoic

2.50 . . . .

Archaean

3.90 . . . .

Hadean

4.55

0.505

Cambrian Period

0.57

Ediacarian Period

0.67

greater than 3.9 Ga ago (Ernst 1983). The record of life is preserved within the sediments of oceans, lakes and other environments, but those sediments are subjected to later destruction by metamorphism and erosion. As a result, the fraction of sediment remaining to give us information on any particular period of evolutionary history is approximately inversely proportional to the age of that period (Veizer 1976, in press). Furthermore, although sedimentary rocks younger than 2.5 Ga are relatively abundant, they do not uniformly record the full spectrum of environments: deep oceanic sediments seem to be rare in the record older than 1 Ga. That is not because such environments did not exist, but simply because the oceanic crust is tectonically subducted and destroyed more rapidly and more thoroughly than is continental crust.

The most direct and obvious evidence of life is of course morphologically preserved organisms, fossils. Macroscopic fossils are abundant and readily interpretable in rocks as far back as 0.57 Ga ago (the base of the Cambrian and the Phanerozoic), but all of the events that concern us in this review predate that time. Some older macrofossils are clearly metazoan, but others with simple shapes may be either animals or algae, as is discussed below. The problems of interpreting morphology are particularly severe with respect to the prokaryotes and eukaryotic protists. Evolution from the earliest prokaryotes to the first macroscopic organisms was predominantly intracellular, with only very limited morphological expression, and so the fossil record is inherently limited (see, for instance, the discussion by Schopf and Walter (1983)). To complicate the issue further, some simple objects that have been interpreted as microfossils have proved, on closer examination, to be inorganic structures. Information on cell shapes still makes a significant contribution, as is indicated in later sections of this chapter, but it is necessary also to seek other sources of evidence.

As the evolutionary events we wish to recognise are biochemical, we can consider whether useful evidence can be found in organic molecules preserved in sediments. These could be, for instance, key 'information-rich' molecules such as light-harvesting pigments and cell-wall components. This approach was tried, using Proterozoic and Archaean rocks, up to 30 years ago, and amino acids were found. Only later, when new techniques were devised, was it discovered that the amino acids were still optically active and must have been much younger than their enclosing rock, as complete racemisation should occur in less than one million years. They apparently were contaminants from modern organisms, such as might be found living on and in rocks, and in soils. Furthermore, the very process of fossilisation, involving burial to a considerable depth in sediments, leads to heating for long periods of time, and the molecules of interest are all thermally labile (for a discussion of these and other related issues see Hayes *et al.* 1983). For these reasons this organic geochemical approach of searching for 'chemical fossils' was largely abandoned, but recently some extraordinarily well preserved Proterozoic and Cambrian sedimentary rocks have been found, and these contain relatively large quantities of hydrogen-rich organic compounds such as steranes and triterpanes. This has revived interest, and research has begun anew (McKirdy & Hahn 1982, Hahn 1982), though it is too early for the results to have had any significant impact on the issues considered in this review. It is worth noting that the oldest rocks now known to contain probably indigenous 'biomarker' molecules are 1.6–1.7 Ga old.

A second, less direct, approach to obtaining biochemical information is based on the fact that some biochemical processes fractionate the stable isotopes of carbon and sulphur to a greater extent than non-biological processes under equivalent conditions. The fractionation products are preserved in the organic matter of degraded cells (kerogen) and in sedimentary minerals such as calcite ($CaCO_3$) and pyrite (FeS_2). Carbon isotopes are fractionated during the autotrophic fixation of CO_2 so that ^{12}C is concentrated in the cell materials leaving ^{13}C to accumulate in the hydrosphere (from which it is sampled by sedimentary carbonate minerals). The predominant biochemical process that fractionates sulphur isotopes is dissimilatory sulphate reduction, an energy-yielding process that releases H_2S which frequently reacts with iron to produce pyrite (Chambers & Trudinger 1979). A full discussion of these processes and their geological products can be found in Schidlowski *et al.* (1983). Bacterial sulphate reduction in sediments utilises buried organic matter as a carbon source, with the result that most sediments have a crudely uniform C : S ratio after the metabolically usable proportion of the carbon has been oxidised and the sulphide produced has reacted with iron. This same C : S ratio is found in sedimentary rocks back through time and has been used to argue that bacterial sulphate reduction was already present by late Archaean times (see discussion in Holland, 1984, pp. 352–5).

Additional physiological information is encoded in the shapes of stromatolites: these are sedimentary structures in the shape of sheets, domes and columns from millimetres to tens of metres in size. They are constructed by benthic mats of bacteria, cyanobacteria and microalgae, and are abundant in late Archaean and Proterozoic rocks (Walter 1976). An approach to the interpretation of stromatolites has been outlined by Walter (1983):

> The construction of a stromatolite with a uniform [internal] fabric requires a consistent and repeatable set of behavioral responses from a particular, relatively invariant, microbial community... Examples of such responses include filament orientation, gliding, positive phototaxis, mucus production, and sheath pigmentation. All these factors have recognizable results in the stromatolite fabric, even though no cells might be preserved. Each single response is a product of one or more biochemical systems and the total behavioral syndrome may be a consequence of several interrelated systems. Thus, a stromatolite fabric can be regarded as recording the former presence of a set of interrelated biochemical systems.

This type of analysis has the potential to provide much information because stromatolites are diverse and abundant in the Proterozoic, and to a lesser extent in the Archaean, but it is limited currently by our ignorance of the factors that control the morphology of extant microbial mats. This is a subject of little interest to most microbiologists, but it is basic to the interpretation of ancient stromatolites. An example of such research and its application is given by Walter *et al.* (1976a) and Walter (1977). An additional complication is that some stromatolite-like structures are abiogenic (Buick *et al.* 1981, Walter 1983).

Finally, there is an approach that is yet more indirect and, as a result, more uncertain. Before the advent of oxygenic photoautotrophy, inorganic processes would have produced only minor amounts of free oxygen. As a result, the Earth would have been anaerobic (i.e. could not have supported aerobic respiration) and perhaps also anoxic (in the sense that oxidation of weathering products would have been very limited). So a transition from unoxidised to oxidised sediments could be equated with the rise of oxygenic photoautotrophy. As Veizer (1983) shows, a problem with this approach is that the rise of oxygen concentrations in the atmosphere and hydrosphere is governed not only by oxygen production, but also by the availability of oxygen sinks (reduced minerals and gases from the Earth's mantle). Nevertheless, useful information can be derived from this approach and, in particular, the distribution of ferric iron and of readily oxidisable uranium minerals is used to trace the progressive oxygenation of the atmosphere and hydrosphere (see discussions in Schopf 1983, Holland 1984).

THE GEOLOGICAL RECORD

The biosphere 3.5 Ga ago

Our information is derived from studies of two areas of sedimentary rocks that are only slightly metamorphosed, the Barberton Mountain Land in South Africa, and the Pilbara region of Western Australia (Lowe 1980). In both examples the ages of the palaeobiologically significant rocks are well established at 3.45–3.56 Ga. The Pilbara area was a platform (Hickman 1983) or micro-continent characterised by extensive volcanism that was both subaerial and subaqueous (Barley *et al.* 1979, Groves *et al.* 1981). Calcareous and siliceous sediments were deposited in shallow water bodies that in some places evaporated to produce gypsum ($CaSO_4.2H_2O$) (Groves *et al.* 1981). The physiographic setting may have been very similar to that of the shallow sea and lagoons of the southern margin of the Persian Gulf, but there is no unequivocal evidence of marine conditions (Lowe 1983). Similar conditions pertained in the Barberton Mountain Land (Lowe & Knauth 1977).

There is no convincing evidence for the existence of planktonic organisms at this time. On the other hand, several occurrences of stromatolites are now known (Walter 1983, Lowe 1983), indicating the presence of benthic mat-forming microbial communities. It can be deduced from microstructural features of the stromatolites that the constructing organisms were filamentous and most probably moved in response to gradients in light intensity. Indeed, filamentous microfossils occur in some of the stromatolites (Awramik *et al.* 1983, Schopf & Walter 1983, Walter 1983). These include long, sinuous narrow threads 0.3–0.7 μm wide, possibly septate filaments about 1 μm wide, tubular sheaths 3.0–9.5 μm wide, and septate filaments up to 7 μm wide. The kerogen in the microfossiliferous cherts is enriched in ^{12}C ($\delta^{13}C = -34$) suggesting that the microbiota included autotrophs (Hayes *et al.* 1983). Some of the stromatolites seemed to have formed in environments where gypsum was precipitating and which were intermittently desiccated, suggesting that the micro-organisms were able to survive the stresses of hypersalinity, low water potential and high light intensity. Geochemical evidence points to an anaerobic environment at this time (Walker *et al.* 1983). Putting all the evidence together, it seems likely that the stromatolite builders were anaerobic photoautotrophic filamentous bacteria with many of the capabilities of modern members of that group. There are, in addition, occurrences of possible coccoid microfossils in both the South African and Australian sequences (reviewed by Schopf & Walter 1983) but it is not possible at present to demonstrate the biogenicity of these objects convincingly. Hayes (1983) reminds us that although the carbon isotopic evidence indicates the presence of autotrophs, these need not have been autotrophs using the Calvin cycle: they could, for instance, have been methanogens.

The sulphates and sulphides of these early Archaean rocks lack the typical isotopic signature of bacterial sulphate reduction. The sulphides average $-0.9‰$ $\delta^{34}S$ and the sulphates average about $+4.0‰$. This small isotopic difference may be the result of oxidation of reduced sulphur by anaerobic photoautotrophic bacteria (Lambert *et al.* 1978, Schidlowski *et al.* 1983).

Records of the earliest stages of life must be sought in rocks older than 3.5 Ga. These are extremely rare, and those few older rock sequences that are known (for instance the 3.8 Ga old rocks at Isua in Greenland) have been metamorphosed repeatedly. As a result, and despite many statements to the contrary, these rocks have yielded no convincing evidence for or against the presence of life on Earth prior to 3.5 Ga ago (Schopf *et al.* 1983). They do at least show that there was liquid water at the Earth's surface at that time, and that sediments were being deposited. An appropriate environment for life may well have existed.

Archaebacteria

This group includes the methanogens, some acidophiles and the extreme halophiles. There are two means of recognising them in the geological record: first, from the degradation products of their distinctive lipids and secondly from the extreme depletion of ^{13}C in the methane produced by the methanogens (and subsequently incorporated in sedimentary carbon, kerogen). Methanogens contain abundant acyclic isoprenoids in the range C_{15} to C_{30} (Holzer *et al.* 1979). The extreme halophiles contain appreciable concentrations of squalene (Tornabene 1978). Some of these compounds seem to occur only in Archaebacteria, or are abundant in Archaebacteria and only rare in other organisms (Langworthy *et al.* 1978). As a result, the discovery of high concentrations of sesterterpanes (C_{25} acyclic isoprenoids) and squalane in Cambrian oils and their source rocks of the Officer Basin, Australia, is regarded as firm evidence for the presence of Archaebacteria in the Cambrian alkaline lake in which these sediments were deposited (McKirdy *et al.* 1981, 1984). No such biomarkers are yet known from older rocks. Proterozoic oils and petroleum source rocks are relatively rare and few have been studied by organic geochemists. Currently the oldest known rocks sufficiently well preserved for such research are in the 1.6–1.7 Ga old McArthur Basin of Australia. These and younger Proterozoic organic-rich sediments in Australia are the subjects of an active research programme, and new results can be expected in the near future.

Some late Archaean kerogens from Australia, South Africa (Hayes *et al.* 1983) and Canada (Schoell & Wellmer 1981) are extraordinarily enriched in ^{12}C, the lightest measured value being δ ^{13}C = -51.9 (contrasting with values of -20 to -35 for most Proterozoic and Archaean kerogens). At least the South African and Australian examples occur with carbonate with the usual marine isotopic composition of near zero. This constancy of the isotopic composition of carbonates coupled with major variations in that of kerogens seems to be best interpreted as resulting from 'a carbon cycle in which the dominant means of primary production ... is unchanged from earlier times but in which the isotopic record (in organic material) of that constancy has been scrambled by reprocessing the primarily produced organic matter' (Hayes 1983). Schoell and Wellmer (1981) and Hayes (1983) conclude that the only known form of reprocessing that could produce the observed result is the production of ^{13}C-depleted methane (CH_4) by methanogens, and then its oxidation or utilisation by methylotrophs. Hayes (1983) reasons that oxygen-utilising methylotrophy must have been involved, and therefore that these observations also establish a minimum age for the advent of oxygenic photoautotrophy.

If the interpretation is correct, then these observations establish a minimum age (late Archaean) for the origin of methanogens, and therefore of Archaebacteria, but we should note that it is possible that the autotrophy recognised by the carbon isotopic signatures of older rocks (discussed in an earlier section) was that of methanogenic bacteria rather than of photoautotrophs (Hayes 1983).

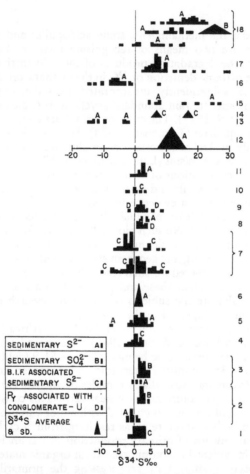

Figure 2.4 Sulphur isotopic distributions in sulphide and sulphate minerals in Archaean and Proterozoic sedimentary rocks (from Skyring & Donnelly 1982). Older rocks are at the base, younger at the top. Note that in the lower two-thirds of the diagram the analyses cluster near zero (the composition of sulphur in the Earth's mantle), whereas in the upper third there are major deviations, which are considered to be caused by bacterial sulphate reduction. The change occurs at about 2.3 Ga ago. For the locality and age details the reader is referred to Skyring and Donnelly (1982).

Sulphate-reducing bacteria and other anaerobic respirers

The sulphur isotope fractionation characteristic of bacterial sulphate reduction can be found in sedimentary rocks of all ages back to approximately 2.3 Ga ago (Fig. 2.4). Sedimentary sulphides and sulphates older than that on average have approximately the same isotopic composition as each other and as sulphur in the Earth's mantle, from which the sedimentary sulphur is derived. For this and other reasons, Cameron and Garrels (1980), Cameron (1982), Skyring and Donnelly (1982) and other authors have concluded that the Archaean oceans contained little sulphate, that sulphate-reducing bacteria had not evolved, or both. Thode and Goodwin (1983) have shown that some 2.7 Ga old sedimentary rocks of Canada have sulphur isotope distributions somewhat like younger bacterially produced distributions, but still centred on the mantle sulphur value, unlike the distributions in younger rocks. Although Thode and Goodwin (1983) interpret these distributions as bacterial, that is unlikely because it is inconsistent with results from other areas, and because the distributions are closely comparable to those from younger seafloor abiogenic hydrothermal deposits (Skyring & Donnelly 1982). In addition, bacterial sulphate reduction produces a characteristic C : S ratio in modern sediments (Holland 1984, p. 475), and that ratio is found in sedimentary rocks of all ages, except Archaean. Comparable analyses are available only from the late Archaean, and there the ratio is different (Cameron & Garrels 1980, but see Holland 1984, for a different interpretation of the same data). However, this is not a strong point, as in fact the ratio in modern sediments is very variable, and as yet relatively few analyses are available for the Archaean and Proterozoic. Nonetheless, there is a growing and consistent body of evidence of this type indicating that there was little sulphate in the Archaean oceans. If this is correct, then bacterial sulphate reducers could have been present in the Archaean but they would have left no clear isotopic evidence of their presence, as they would probably have reduced all of the small amount of sulphate available to them and thus not have caused any net fractionation of sulphur isotopes. So the geological evidence indicates that bacterial sulphate reduction had evolved by 2.3 Ga, but allows the possibility that it evolved much earlier.

Some further constraints are available. Sulphate *was* abundant at least in some local environments 3.5 Ga ago, as evidenced in Western Australia and South Africa and mentioned above. However, the sulphides and sulphates preserved in those rocks lack the isotopic fractionation characteristic of bacterial sulphate reduction (e.g. Lambert *et al.* 1978). As it seems reasonable to assume that once sulphate-reducers had evolved they would have been widely distributed, we can conclude that this group of bacteria had not evolved 3.5 Ga ago. Skyring and Donnelly (1982) discuss the possibility that sulphate reduction might have been preceded by sulphite reduction as a widespread process in the Archaean oceans (but not until after 3.5 Ga ago).

A recent study of the isotopic composition of carbonate minerals in 2.5 Ga old banded iron formations has revealed carbon-isotope distribution patterns that seem to relate to the microbial ecology of the original sediments. Research on this topic is far from complete, but Baur *et al.* (1985) conclude that, given certain preconditions, the carbonate minerals could have been formed by 'a variety of processes, probably including fermentation and possibly also sulfate reduction or anaerobic respiratory processes in which Fe^{3+} served as an electron acceptor'. Their study indicates that a rich isotopic record of microbial processes exists, and that we can expect further research to be fruitful.

Cyanobacteria

The presence of cyanobacteria in the Late (e.g. Schopf 1968) Middle (e.g. Oehler 1978) and Early (e.g. Golubic & Hofmann 1976) Proterozoic has been established by discoveries of distinctive microfossils. The Early Proterozoic occurrences are the subject of an encyclopaedic review by Hofmann and Schopf (1983). Those authors consider all the available evidence for the biological affinities of the microfossils, and though they recognise the difficulties of working primarily with cell morphology, they are able to develop a convincing case for the cyanobacterial affinities of some of the reported microfossils. That is particularly so for the coccoid microfossils, many of which are substantially larger than nearly all extant non-cyanobacterial prokaryotes. Some also occur within fossil sheaths and form specific types of stromatolites that permit very close comparison with extant cyanobacteria such as *Entophysalis*. There remains a significant uncertainty, and that is the age of the key microfossiliferous formations. None of them is well dated, and it is quite possible that none is older than about 2.0 Ga. As a result, we have to conclude that, in the evidence we have just reviewed, there is no confirmation of the view that cyanobacteria existed before 2.0 Ga ago.

As indicated in the previous section, interpretation of the carbon isotopic composition of late Archaean kerogens seems to require the assumption that oxygenic photoautotrophy operated at least 2.8 Ga ago. Support for this assumption comes from studies of stromatolites (Walter 1983) and microfossils (Schopf & Walter 1983) from the 2.7–2.8 Ga old Fortescue Group of Western Australia. The Fortescue stromatolites are extraordinarily well preserved, and as a result it is possible to reconstruct many of the features of the microbial mats that built them. To date, few microfossils have been found in these stromatolites (though further work is likely to be productive), but those that occur are consistent with inferences drawn from analysis of the fabric of the stromatolites. The stromatolites are products of at least three distinct microbial communities. One is not yet well known; there is evidence that the second was produced by finely filamentous photoautotrophs comparable to oscillatoriacean cyanobacteria; and the third, the most abundant and best known, is very similar to stromatolites constructed now by oscillatoriacean and nostocalean cyanobacteria. Because of the apparent affinity of a single microfossil that occurs in one of these stromatolites, it is suggested that the constructing organisms were *Lyngbya*-like oscillatoriacean cyanobacteria.

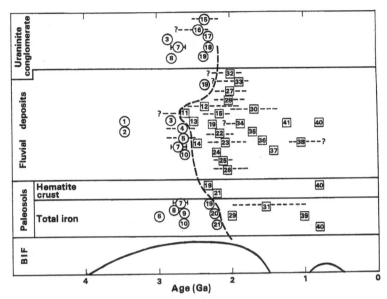

Figure 2.5 Sedimentary indicators of reducing (circle) and oxidising (square) state of the Archaean and Proterozoic atmosphere and hydrosphere (from Walker *et al.* 1983; reprinted with permission of Princeton University Press). BIF is an abbreviation for banded iron formation, and the lines indicate the frequency of occurrence of such rocks; paleosols are ancient soils; uraninite is an easily oxidised uranium mineral. The vertical dashed line is merely a guide to the eye, and separates indicators of reducing and oxidising conditions. The distribution of banded iron formations was controlled by the oxidation state of the hydrosphere, whereas the other indicators were controlled by the atmosphere. For further details the reader is referred to Walker *et al.* (1983).

Some cyanobacteria are facultative anaerobic photoautotrophs, but as far as is known all are capable of aerobic photoautotrophy (Padan & Cohen 1982). No other prokaryotes have this capability. As a result, and because there are no known major inorganic sources of atmospheric oxygen, the rise of atmospheric oxygen concentrations can be related to the advent of cyanobacteria. Evidence for the change from an anaerobic to an aerobic Earth comes from the geochemistry of sedimentary rocks. This work need not be reviewed here, but the approach is outlined in Figure 2.5. The evidence is that the Earth's atmosphere changed from a predominantly reducing to a more oxidising condition early in the Early Proterozoic, perhaps about 2.3 Ga ago, but that the deeper hydrosphere was not oxygenated until 1.7 Ga ago (Walker *et al.* 1983). The evidence on which such interpretations is based is discussed in detail by Holland (1984).

Microscopic eukaryotes

It is likely that the eukaryotes are a polyphyletic group that arose through a series of endosymbiotic relationships between populations of host cells and other bacteria they engulfed. The identity of the original host cells is not clear, although they may have been members of a separate lineage as ancient and as distinctive as the Eubacteria and the Archaebacteria; Woese (1982) calls this hypothetical lineage the 'Urcaryota'. The first significant endosymbiosis apparently gave rise to the mitochrondrion: two extant species of bacteria have been identified as having close similarities to mitochondria, the photoautotroph *Rhodopseudomonas*, and the chemoautotroph *Paracoccus*. The plastids of red algae were probably cyanobacteria, but the origins of the plastids of the other algae are less clear. The origin of the defining feature of eukaryotic cells, the nucleus, is not understood; perhaps the cell-membrane flexibility that allowed phagocytosis also allowed invagination and the formation of the nuclear membrane (Cavalier-Smith 1978). The extensive literature on these subjects is summarised by Knoll (1983), who also reviews the relevant fossil record.

Ideally, one would want to recognise in the fossil record the direct results of endosymbiosis. The quality of preservation of cells is such that this is unlikely to be achieved, but there is one well-documented example of preserved cell contents possibly including organelles. Spheroidal unicells from the approximately 0.85 Ga old Bitter Springs Formation of central Australia contain dark spots interpreted as pyrenoids (photosynthate storage granules found only in eukaryotes) and membranous structures that may be chloroplasts (Oehler 1976, 1977). This interpretation is vigorously disputed (e.g. Francis *et al.* 1978a,b) because similar structures result from lysis of prokaryotic cells. However, Oehler's interpretation cannot be dismissed; there has been no comparable detailed study of lysed cells and, as a result, it is not known whether the structures in these only superficially resemble those found by Oehler.

Although the internal structures of cells are rarely, if ever, preserved in a recognisable way, many cells have robust walls that do preserve well. The walls of prokaryotes are simple and unadorned, whereas those of some eukaryotes are complex, with spines, pores, ribs and other structures. Fossil cells with such complex morphologies are well known and abundant in Phanerozoic rocks, and in recent years have been discovered also in the late Proterozoic. These fossils (called acritarchs) with sculptured walls are well known from rocks as old as about 0.8 Ga (Vidal & Knoll 1983). In addition, there are two known occurrences of 1.4 Ga old, large robust-walled organic spheroids (Peat *et al.* 1978, Horodyski 1980) that Vidal and Knoll (1983) and Knoll (1983) regard as probable eukaryotes. Vidal and Knoll (1983) dismissed one of these examples (from the Roper Group, northern Australia) as being inadequately dated, but recent isotopic dating (Kralik 1982) supports earlier work and increases the reliability of the indicated age. The other occurrence is in the Belt Supergroup of Montana, USA. Spheroidal microfossils show a marked increase in both mean diameter and range of diameters at about 1.4 Ga. It may well be that this marks the advent of eukaryotes, as these generally have larger cells than those of prokaryotes (Schopf 1977). This date is consistent with geochemical evidence

(Fig. 2.5) that both the atmosphere and the hydrosphere had oxygenated by 1.7 Ga ago, thus allowing the development of aerobic respiration (that is characteristic of eukaryotes). Morphologically complex microfossils are known from rocks as old as about 2.0 Ga, and have been considered as possible eukaryotes, but all the interpretations have been challenged (Awramik & Barghoorn 1977, Hofmann & Schopf 1983). Currently the weight of evidence indicates that these are unusual prokaryotes, but new evidence could easily change that view. Additional evidence on Proterozoic eukaryotes is reviewed by Knoll (1983).

The only Proterozoic fossils that can confidently be referred to the Protista (in the formal sense of Whittaker (1969), are the 'vase-shaped microfossils' found in rocks up to 0.8 Ga old. They are considered to have been planktonic heterotrophic protists, and they resemble testate protozoans such as the tintinnids (Knoll 1983).

Fungi

Sokolov and Fedonkin (1984) report fossil fungi from the Vendian (latest Proterozoic, 0.68–0.57 Ga) of Europe; they seem to be referring to fossils that elsewhere in their article they describe thus: 'They are simple microscopic, tubular organisms with a wide "head" (sporangium) and a long "tail". They form dense clusters, and some occur on the surfaces of vendotaenid [algal] thalli . . .' They are 'definitely assigned to the oldest Actinomycetes'. The Actinomycetes are bacteria, not fungi; they 'tend to form branching filaments which in some families develop into a mycelium' (D. Gottlieb in Buchanan and Gibbons 1974, p. 657). Some members of the group bear spores on hyphae, and some are pathogenic to plants. 'Members of the Actinomycetales, despite the filamentous growth of some members, are now generally accepted to be bacteria and not fungi. The properties relating them to bacteria are: the absence of a nuclear membrane, small hyphal diameters, sensitivity to lysozyme, chemical nature of the cell wall, sensitivity to antibacterial agents, and flagella, when produced, of the bacterial type' (Gottlieb op. cit.). Assignment to the eubacteria has been confirmed by 16SrRNA sequencing (Fox et al. 1980).

The fossil record of fungi has been reviewed by Tiffney and Barghoorn (1974). There are very few reports of Proterozoic fungi; most refer to nonseptate, hypha-like filaments (Eomycetopsis is an example), the fungal affinities of which are now disputed. These tubular structures have been reinterpreted as cyanobacterial sheaths (Knoll & Golubic 1979). The first report of the 2.0 Ga old microfossils of the Gunflint Iron Formation interpreted some of them as fungi (Tyler & Barghoorn 1954), but that interpretation was later considered unlikely (Barghoorn & Tyler 1965) and is no longer accepted (Hofmann & Schopf 1983). An 'ascus-like' microfossil was reported from the late Proterozoic by Schopf and Barghoorn (1969) but their interpretation was not accepted by Tiffney and Barghoorn (1974). In short, there are no convincing reports of fossil fungi from Proterozoic or older rocks.

Metaphytes

The term metaphyte is used here to mean megascopic, multicellular thallophyte. Macroscopic algae (vendotaenids) are known from many latest Proterozoic (Ediacarian) rock sequences (Gnilovskaya 1971, Cloud & Glaessner 1982, Sokolov & Fedonkin 1984). These are considered to be Phaeophytes (brown algae) but the cell structure is poorly preserved and so the exact biological affinities within the algae are not clear.

Carbonaceous ribbons up to 6 mm wide and 77 mm long, with rounded ends, have been described from the Little Dal Group in Canada, which is probably 0.8–1.1 Ga old. Apparently no cells are preserved within these structures (*Tawuia dalensis*), but they are megascopic organisms, probably algae (perhaps Phaeophytes) though possibly metazoans (Hofmann & Aitken 1979). They are also known from 0.8–0.9 Ga old rocks in Svalbard (Knoll 1983).

The oldest known possible macroscopic algae occur in 1.3 Ga old rocks of the Belt Supergroup in Montana (Walter *et al.* 1976b). These are preserved as carbonaceous ribbons up to 2.0 mm wide and 125 mm long, some of which form spirals. Horodyski (1980) suggested that they may be fragments of microbial mats, but many of them are far too regular for that to be a reasonable explanation. This comment applies particularly to *Grypania spiralis*. Hofmann and Aitken (1979) have tentatively identified one of these fossils as a second species of *Tawuia*.

Metazoans

It is likely that distinct groups of metazoans arose separately from protozoans on several occasions (Anderson 1981). The oldest unequivocal metazoan body fossils constitute what is known as the Ediacara fauna, after the location in South Australia at which it was discovered. This non-skeletal fauna is 0.57–0.64 Ga old and is now known from many localities, world wide. Its composition and evolutionary significance have been comprehensively documented and reviewed by Glaessner (1979, 1984). The biological affinities of some elements of the fauna remain controversial, but it is generally considered that the Cnidaria, Annelida and Arthropoda are represented. Possibly older annelid and other fossils have been briefly reported from China (Wang *et al.* 1983, Yan & Qian 1983) but the age of these has yet to be convincingly determined.

Animals also leave a record in the form of tracks, trails and burrows, collectively referred to as trace fossils. These are abundant and well known in Phanerozoic rocks. They also occur in the Ediacarian. There is a very scanty and vigorously disputed record of possible trace fossils in rocks as old as about 1.0 Ga or even older (Glaessner 1983). These include one associated with *Tawuia* in the Little Dal assemblage (Hofmann & Aitken 1979) mentioned above. None is both convincingly metazoan and convincingly pre-Ediacarian (as Runnegar (1982a) has also concluded). These occurrences must each be critically restudied. We must conclude that no convincing pre-Ediacarian metazoan fossils are known at present.

The hypothesis that the metazoans evolved early, before 1.0 Ga ago, receives some indirect palaeontological support from another source. A major food source available to the early metazoans would have been the benthic microbial mats that were abundant and widespread on the early Earth. These mats of photoautotrophs are expressed in the geological record as stromatolites (Walter 1977). The present distribution of such mats is severely limited as a result of grazing and burrowing by animals, particularly gastropods. It can be postulated, therefore, that the reason for the great abundance of Proterozoic stromatolites was the lack of grazing and burrowing. Following the evolution of animals with those behaviours, the distribution of mats, and therefore stromatolites, would have become progressively more restricted. Recent analysis of the available information (Walter & Heys 1985) on the abundance and diversity of stromatolites through time has revealed three major events. First, those stromatolites that formed in quiet, subtidal environments began to decline in both abundance and diversity about 1.0 Ga ago. Second, this was followed by a general decline of all stromatolites beginning 0.7–0.8 Ga ago. Third, during the Cambrian, there appeared the first thrombolites, stromatolites with a clotted, unlaminated fabric. Walter and Heys (1985) consider that the thrombolites owe their origin to animals and they record the first macroscopic burrowing and possible boring by these animals. This interpretation is consistent with the well known first abundant occurrence of vertical burrows in the earliest Cambrian. The 0.7–0.8 Ga decline can be attributed to the first widespread grazing by animals (an interpretation made earlier by other authors). The decline of subtidal stromatolites which began about 1.0 Ga ago may well record the early, subtle effects of grazing, presumably by small acoelomate animals. It seems reasonable to postulate that such early acoelomates would have been most abundant in equable, subtidal environments rather than in the intertidal zone where many stromatolites formed but where there are extremes of salinity and temperature. At the present state of knowledge this is a speculative hypothesis, but it is amenable to testing.

I conclude, tentatively, that the Metazoa evolved not long before 1.0 Ga ago, but no definite metazoan fossils more than 0.64 Ga old are known. By about 0.64 Ga ago, there was already a diverse metazoan fauna.

DISCUSSION AND CONCLUSIONS

Eubacteria, Archaebacteria and Urcaryota, the three 'urkingdoms', all extend far back into Earth history (Fig. 2.6). Coccoid and filamentous benthic anaerobic photoautotrophs (Eubacteria) existed 3.5 Ga ago. Depending on one's views about the earliest period of evolution, these may already have been fairly distant from the origin of life. The probable record of Archae-

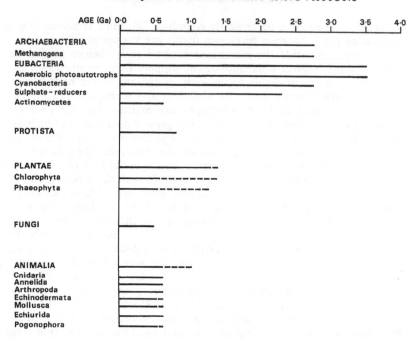

Figure 2.6 The antiquity of major groups of organisms, as interpreted from the geological record. Only those groups known from rocks older than 0.57 Ga (the base of the Phanerozoic) are listed (with the exception of the Fungi, which are only known from the Phanerozoic).

bacteria extends back to 2.8 Ga ago, but they may well be more ancient than that, for three reasons: the Archaean rock record is sparse, little research has been directed to this question, and few means are available for detecting the former presence of these organisms. On the other hand, eukaryotes are recognisable in the rock record, albeit often with difficulty, and their earliest representatives have been sought for many years. Several lines of evidence converge in suggesting that they evolved about 1.4 Ga ago. However, that is a date for nucleated cells with organelles, not for the precursor 'urcaryotes' that, if they existed at all, may have had a much longer history. At present we know of no means of recognising these in the rock record.

Within the eukaryotes, four kingdoms are recognised, the Protista, Plantae, Animalia and Fungi. Fungi are not known from the Archaean or Proterozoic; this may be consistent with the view (Broda 1978) that they evolved from protozoans (Protista), and thus are a relatively recent innovation. Probable protists are known back to 0.8 Ga ago, and a well documented record of megascopic algae (Plantae) and metazoans (Animalia) extends back to about 0.64 Ga ago. There is additional evidence that megascopic algae extend back 1.3 Ga and animals about 1.0 Ga. The phylogenetic schemes of Woese and Fox (1977), Fox *et al.* (1980) and Stackebrandt and Woese (1981) are consistent with the geological record. However, that record is not yet known well enough for it to test such schemes rigorously. Nevertheless, it can calibrate them against time, as is shown in Figure 2.6.

In addition, protein and nucleic acid sequencing can be used to predict the timing of evolutionary events, provided the 'molecular-clock' can be calibrated against the geological record. The prediction that the initial divergence of the animal phyla occurred 0.9–1.0 Ga ago (Runnegar 1982b) is consistent with the geological record as I read it (Walter & Heys 1985), although not everyone would agree. It is thought that the evolution of meiosis (probably some time between 1.0–1.4 Ga ago) led to a rapid increase in rates of evolution. Emphasis is put on the long and demonstrably conservative history of the cyanobacteria, which is contrasted with the relatively rapid diversification of more recently evolved eukaryotes, such as the mammals. However, disguised within the unicellular morphology of the 'bacteria' is a diversity of biochemical systems that rivals or even exceeds that of all the eukaryotes, depending on the measure one chooses to use. Some caution is necessary in any comparison of rates of evolution amongst the prokaryotes with those of the more conspicuously diverse eukaryotes. When our scale is calibrated in hundreds of millions of years, we find that the biological molecular clock has ticked steadily and inexorably for the last thousand million years; it may well have been running at much the same rate for 3.5×10^9 years. That is, on average, rates of mutation of specific molecules may have been more or less constant at this scale. The other underlying mechanism of evolution is genetic recombination, a process that occurs in prokaryotes as well as eukaryotes. Mitosis and meiosis are characteristics of eukaryotes, but it is not at all clear fom the geological record that the advent of those mechanisms some 1.4 Ga ago accelerated evolution. Questions concerning gradualism and punctuated evolution relate to another, much shorter, scale than that we are considering here.

ACKNOWLEDGEMENTS

Drs P. A. Trudinger, G. W. Skyring and E. M. Truswell provided helpful comments on the manuscript of this chapter. Figures 2.1–2.5 are republished with the permission of the authors and publishers. The Baas Becking Geobiological Laboratory is supported by the Bureau of Mineral Resources, the Commonwealth Scientific and Industrial Research Organisation, and the Australian Mineral Industries Research Association. M. R. Walter publishes with the permission of the Director of the Bureau of Mineral Resources.

REFERENCES

Anderson, D. T. 1981. Origins and relationships among the animal phyla. *Proc. Linn. Soc. N.S.W.*, 106, 151–66.

Awramik, S. M. and E. S. Barghoorn 1977. The Gunflint microbiota. *Precambrian Res.* 5, 121–42.

Awramik, S. M., J. W. Schopf and M. R. Walter 1983. Filamentous fossil bacteria from the Archean of Western Australia. *Precambrian Res.* 20, 357–74.

Barghoorn, E. S. and S. A. Tyler 1965. Microorganisms from the Gunflint Chert. *Science* 147, 563–77.

Barley, M. E., J. S. R. Dunlop, J. E. Glover and D. I. Groves, 1979. Sedimentary evidence for an Archaean shallow-water volcanic-sedimentary facies, eastern Pilbara Block, Western Australia. *Earth Planet. Sci. Lett.* 43, 74–84.

Baur, M. E., J. M. Hayes, S. A. Studley and M. R. Walter 1985. Millimeter-scale variations of stable isotopic abundances in carbonates from banded iron formations in the Hamersley Group of Western Australia. *Econ. Geol.* 80, 270–82.

Broda, E. 1978. *The evolution of the bioenergetic processes.* Pergamon Press, Oxford.

Buchanan, R. E. and N. E. Gibbons (eds) 1974. *Bergey's manual of determinative bacteriology*, 8th edn. Baltimore: Williams & Wilkins.

Buick, R., J. S. R. Dunlop and D. I. Groves 1981. Stromatolite recognition in ancient rocks: an appraisal of irregularly laminated structures in an Early Archaean chert-barite unit from North Pole, Western Australia. *Alcheringa* 5, 161–81.

Cameron, E. M. 1982. Sulphate and sulphate reduction in early Precambrian oceans. *Nature* 296, 145–8.

Cameron, E. M. and R. M. Garrels 1980. Geochemical compositions of some Precambrian shales from the Canadian Shield. *Chem. Geol.* 28, 181–97.

Cavalier-Smith, T. 1978. The origin of nuclei and of eukaryotic cells. *Nature* 256, 463–8.

Chambers, L. A. and P. A. Trudinger, 1979. Microbiological fractionation of stable sulfur isotopes: a review and critique. *Geomicrobiol. J.* 1, 249–93.

Cloud, P. E. and M. F. Glaessner 1982. The Ediacarian period and system: metazoa inherit the Earth. *Science* 217, 783–92.

Ernst, W. G. 1983. The early Earth and the Archean rock record. In *Earth's earliest biosphere: its origin and evolution*, J. W. Schopf (ed.), 41–52. Princeton: Princeton University Press.

Fox, G. E., E. Stackebrandt, R. B. Hespell, J. Gibson, J. Maniloff, T. A. Dyer, R. S. Wolfe, W. E. Balch, R. S. Tanner, L. J. Magrum, L. B. Zablen, R. Blakemore, R. Gupta, L. Bonen, B. J. Lewis, D. A. Stahl, K. R. Luehrsen, K. N. Chen and C. R. Woese 1980. The phylogeny of prokaryotes. *Science* 209, 457–63.

Francis, S., L. Margulis and E. S. Barghoorn 1978a. On the experimental silicification of microorganisms II. On the time of appearance of eukaryotic organisms in the fossil record. *Precambrian Res.* 6, 65–100.

Francis, S., E. S. Barghoorn and L. Margulis 1978b. On the experimental silicification of microorganisms III. Implication of the preservation of the green prokaryotic alga *Prochloron* and other coccoids for interpretation of the microbial record. *Precambrian Res.* 7, 377–83.

Glaessner, M. F. 1979. Precambrian. In *Treatise of invertebrate paleontology*, Pt.A., R. A. Robison and C. Teichert (eds), 79–118. Geological Society of America and University of Kansas, Boulder and Lawrence.

Glaessner, M. F. 1983. The emergence of metazoa in the early history of life. *Precambrian Res.* 20, 427–41.

Glaessner, M. F. 1984. *The dawn of animal life: a biohistorical study*. Cambridge: Cambridge University Press.

Gnilovskaya, M. B. 1971. Drevneyshiye vodnyye rasteniya venda Ruskoy platformy (pozdniydokembriy) [The oldest aquatic plants of the Wendian of the Russian Platform (Late Precambrian)]. *Paleont. Zh.*, 3, 101–7 [*Paleont. Jl* 5, 372–8.]

Golubic, S. and H. J. Hofmann 1976. Comparison of Holocene and mid-Precambrian Entophysalidaceae (Cyanophyta) in stromatolitic algal mats: cell division and degradation. *J. Paleont.* 50, 1074–82.

Groves, D. I., J. S. R. Dunlop and R. Buick 1981. An early habitat of life. *Sci. Am.*, October 64–73.

Hahn, J. 1982. Geochemical fossils of a possibly Archaebacterial origin in ancient sediments. *Zbl. Bakt. Hyg. I Abt. Orig.* C3, 40–52.

Harland, W. B., A. V. Cox, P. G. Llewellyn, C. A. G. Pinkton, A. G. Smith and R. Walters 1982. *A geologic time scale*. Cambridge: Cambridge University Press.

Hayes, J. M. 1983. Geochemical evidence bearing on the origin of aerobiosis, a speculative hypothesis. In *Earth's earliest biosphere: its origin and evolution*, J. W. Schopf (ed.), pp. 291–301. Princeton: Princeton University Press.

Hayes, J. M., I. R. Kaplan and K. W. Wedeking 1983. Precambrian organic geochemistry, preservation of the record. In *Earth's earliest biosphere; its origin and evolution*, J. W. Schopf (ed.), pp. 93–134. Princeton: Princeton University Press.

Hickman, A. H. 1983: Geology of the Pilbara Block and its environs. *Bull. Geol. Surv. W. Aust.* 127 (268 pp).

Hofmann, H. J. and J. D. Aitken 1979. Precambrian biota from the Little Dal Group, Mackenzie Mountains, northwestern Canada. *Can. J. Earth Sci.* 16, 150–66.

Hofmann, H. J. and J. W. Schopf 1983. Early Proterozoic microfossils. In *Earth's earliest biosphere: its origin and evolution*, J. W. Schopf, (ed.), pp. 321–60. Princeton: Princeton University Press.

Holland, H. D. 1984. *The chemical evolution of the atmosphere and oceans*. Princeton: Princeton University Press.

Holzer, C., J. Oro and T. G. Tornabene 1979. Gas chromatographic–mass spectrometric analysis of neutral lipids from methanogenic and thermoacidophilic bacteria. *J. Chromatog.* 186, 795–809.

Horodyski, R. J. 1980. Middle Proterozoic shale-facies microbiota from the lower Belt Supergroup, Little Belt Mountains, Montana. *J. Paleont.* 54, 649–63.

Knoll, A. H. 1983. Biological interactions and Precambrian eukaryotes. In *Biotic interactions in recent and fossil benthic communities*, M. J. S. Tevesz and P. L. McCall (eds), pp. 251–83. New York: Plenum Press.

Knoll, A. H. and S. Golubic 1979. Anatomy and taphonomy of a Precambrian algal stromatolite. *Precambrian Res.* 10, 115–51.

Kralik, M. 1982. Rb-Sr age determinations on Precambrian carbonate rocks of the Carpentarian McArthur Basin, Northern Territories, Australia. *Precambrian Res.* 18, 157–70.

Lake, J. A., E. Henderson, M. Oakes and M. W. Clark 1984. Eocytes: a new ribosome structure indicates a kingdom with a close relationship to eukaryotes. *Proc. Natl Acad. Sci. USA.* 81, 3786–90.

Lambert, I. B., T. H. Donnelly, J. S. R. Dunlop and D. I. Groves 1978. Stable isotope studies of early Archaean sulphate deposits of probable evaporitic and volcanogenic origins. *Nature* 276, 808–11.

Langworthy, T. A., T. G. Tornabene and G. Holzer 1978. Lipids of Archae-
bacteria. *Zbl. Bakt. Hyg.*, *I. Abt. Orig.* C3, 228–44.
Lowe, D. R. 1980. Archean sedimentation. *Ann. Rev. Earth Planet. Sci.* 8, 145–67.
Lowe, D. R. 1983. Restricted shallow-water sedimentation of early Archean
stromatolitic and evaporitic strata of the Strelley Pool Chert, Pilbara Block,
Western Australia. *Precambrian Res.* 19, 239–83.
Lowe, D. R. and L. P. Knauth 1977. Sedimentology of the Onverwacht Group (3.4
billion years), Transvaal, South Africa, and its bearing on the characteristics and
evolution of the early Earth. *J. Geol.* 85, 699–723.

Margulis, L. 1981. *Symbiosis in cell evolution.* New York: W. H. Freeman.
McKirdy, D. M. and J. H. Hahn 1982. The composition of kerogen and hydrocar-
bons in Precambrian rocks. In *Mineral deposits and the evolution of the
biosphere*, H. D. Holland and M. Schidlowski (eds), pp. 123–54. Berlin:
Springer-Verlag.
McKirdy, D. M., A. K. Aldridge and P. J. M. Ypma 1981. A geochemical comparison of some crude oils from pre-Ordovician carbonate rocks. In *10th Annual
meeting on organic geochemistry, University of Bergen, Norway, Sept. 14–18,
1981; Advances in organic geochemistry 1981*, M. Bjoroy (ed.), pp. 99–107.
Chichester: John Wiley.
McKirdy, D. M., A. J. Kantsler, J. K. Emmett and A. K. Aldridge 1984. Hydrocar-
bon genesis and organic facies in Cambrian carbonates of the eastern Officer
Basin, South Australia. In Petroleum geochemistry and source rock potential of
carbonate rocks, J. G. Palaes (ed.), *Am. Assoc. Petrolm Geol. Studies in Geology*
18, 13–32.

Oehler, D. Z. 1976. Transmission electron microscopy of organic microfossils from
the late Precambrian Bitter Springs Formation of Australia: techniques and survey
of preserved ultrastructure. *J. Paleont.* 50, 90–106.
Oehler, D. Z. 1977. Pyrenoid-like structures in late Precambrian algae from the Bitter Springs Formation of Australia. *J. Paleont.* 51, 885–901.
Oehler, D. Z. 1978. Microflora of the middle Proterozoic Balbirini Dolomite
(McArthur Group) of Australia. *Alcheringa* 2, 269–309.

Padan, E. and Cohen, Y. 1982. Anoxygenic photosynthesis. In *The biology of
cyanobacteria*, N. G. Carr and B. A. Whitton (eds), 215–35. Oxford: Blackwell
Scientific.
Peat, C. J., M. D. Muir, K. A. Plumb, D. M. McKirdy and M. S. Norvick 1978.
Proterozoic microfossils from the Roper Group, Northern Territory, Australia.
BMR J. Austral. Geol. Geophys. 3, 1–17.

Runnegar, B. 1982a. The Cambrian explosion: animals or fossils. *J. Geol. Soc.
Aust.* 29, 395–411.
Runnegar, B. 1982b. The molecular-clock date for the origin of the animal phyla.
Lethaia 15, 199–205.

Schidlowski, M., J. M. Hayes and I. R. Kaplan 1983. Isotopic inferences of ancient
biochemistries: carbon, sulfur, hydrogen, and nitrogen. In *Earth's earliest
biosphere: its origin and evolution*, J. W. Schopf (ed.), 149–86. Princeton:
Princeton University Press.
Schoell, M. and F. W. Wellmer 1981. Anomalous ^{13}C depletion in early Pre-
cambrian graphites from Superior Province, Canada. *Nature* 290, 696–9.
Schopf, J. W. 1968. Microflora of the Bitter Springs Formation, Late Precambrian,
Central Australia. *J. Paleont.* 42, 651–88.

Schopf, J. W. 1977. Biostratigraphic usefulness of stromatolitic Precambrian microbiotas: a preliminary analysis. *Precambrian Res.* 5, 143–73.

Schopf, J. W. (ed.) 1983. *Earth's earliest biosphere: its origin and evolution*. Princeton: Princeton University Press.

Schopf, J. W. and E. S. Barghoorn 1969. Microorganisms from the late Precambrian of South Australia. *J. Paleont.* 43, 111–18.

Schopf, J. W. and M. R. Walter 1983. Archean microfossils: new evidence of ancient microbes. In *Earth's earliest biosphere: its origin and evolution*, J. W. Schopf (ed.) 214–39. Princeton: Princeton University Press.

Schopf, J. W., J. M. Hayes and M. R. Walter 1983. Evolution of Earth's earliest ecosystems: recent progress and unsolved problems. In *Earth's earliest biosphere: its origin and evolution*, J. W. Schopf (ed.), 361–84. Princeton: Princeton University Press.

Skyring, G. W. and T. H. Donnelly 1982. Precambrian sulfur isotopes and a possible rôle for sulfite in the evolution of biological sulfate reduction. *Precambrian Res.* 17, 41–61.

Sokolov, B. S. and M. A. Fedonkin 1984. The Vendian as the terminal system of the Precambrian. *Episodes* 7, 12–19.

Stackebrandt, E. and C. R. Woese 1981. The evolution of prokaryotes. *Soc. Gen. Microbiol. Symp.* 32, 1–32.

Thode, H. G. and A. M. Goodwin 1983. Further sulfur and carbon isotope studies of late Archean iron-formations of the Canadian Shield and the rise of sulfate reducing bacteria. *Precambrian Res.* 20, 337–56.

Tiffney, B. H. and E. S. Barghoorn 1974. The fossil record of the Fungi. *Farlow Herbarium Cryptogamic Bot.*, Occ. Papers 7, 1–42.

Tornabene, T. G. 1978. Non-aerated cultivation of Halobacterium cutirubrum and its effect on cellular squalenes. *J. Mol. Evol.* 11, 253–7.

Tyler, S. A. and E. S. Barghoorn 1954. Occurrence of structurally preserved plants in pre-Cambrian rocks of the Canadian Shield. *Science* 119, 606–8.

Veizer, J. 1976. Evolution of ores of sedimentary affiliation through geologic history; relations to the general tendencies in evolution of the crust, hydrosphere, atmosphere and biosphere. In *Handbook of strata-bound and stratiform ore deposits*, K. H. Wolf (ed.) 1–41. Amsterdam: Elsevier.

Veizer, J. 1983. Geologic evidence of the Archean-Early Proterozoic Earth. In *Earth's earliest biosphere: its origin and evolution*, J. W. Schopf (ed.) 240–59. Princeton: Princeton University Press.

Veizer, J. (in press). Recycling on the evolving Earth: geochemical record in sediments. *27th International Geological Congress*. Utrecht: VNU Science Press.

Vidal, G. and A. H. Knoll 1983. Proterozoic plankton. *Mem. Geol. Soc. Am.* 161, 265–77.

Walker, J. C. G., C. Klein, M. Schidlowski, J. W. Schopf, D. J. Stevenson and M. R. Walter 1983. Environmental evolution of the Archean-Early Proterozoic Earth. In *Earth's earliest biosphere: its origin and evolution*, J. W. Schopf (ed.), 260–90. Princeton: Princeton University Press.

Walter, M. R. (ed.) 1976. *Stromatolites*. Amsterdam: Elsevier.

Walter, M. R. 1977. Interpreting stromatolites. *Am. Sci.* 65, 563–71.

Walter, M. R. 1983. Archean stromatolites: evidence of the Earth's earliest benthos. In *Earth's earliest biosphere: its origin and evolution*, J. W. Schopf (ed.), 187–213. Princeton: Princeton University Press.

Walter, M. R. and G. R. Heys 1985. Links between the rise of the Metazoa and the decline of stromatolites. *Precambrian Res.* 29, 149–74.

Walter, M. R., J. Bauld and T. D. Brock 1976a. Microbiology and morphogenesis of columnar stromatolites (Conophyton, Vacerrilla) from hot springs in Yellowstone National Park. In *Stromatolites*, M. R. Walter (ed.), pp. 273–310. Amsterdam: Elsevier.

Walter, M. R., J. H. Oehler and D. Z. Oehler 1976b. Megascopic algae 1300 million years old from the Belt Supergroup, Montana: a reinterpretation of Walcott's Helminthoidichnites. *J. Paleont.* 50, 872–81.

Wang Guiziang, Zhou Benhe and Xiao Ligong 1983. Late Precambrian megascopic fossils from Huainan, Anhui and their significance. *Abstracts, Int. Symp. Late Precambrian Geol., China, Acad. Geol. Sci. China*, 109–10.

Whittaker, R. H. 1969. New concepts of kingdoms of organisms. *Science* 163, 150–60.

Woese, C. R. 1982. Archaebacteria and cellular origins: an overview. *Zbl. Bakt. Hyg. I. Abt. Orig.* C3, 1–17.

Woese, C. R. and G. E. Fox, 1977. Phylogenetic structure of the prokaryotic domain: the primary kingdoms. *Proc. Nat. Acad. Sci. USA*, 74, 5088–90.

Yan Yong-Kui and Qian Maiping 1983. The paleobiota of the Upper Proterozoic strata in northern Jiangsu and Anhui Provinces and its significance. *Abstracts, Int. Symp. Late Precambrian Geol., China, Acad. Geol. Sci. China*, 103–04.

3

Rates and modes of evolution in the Mollusca

BRUCE RUNNEGAR

ABSTRACT

Molluscs have an excellent fossil record extending some 570 million years from the earliest Cambrian to the present and therefore may be expected to provide some of the best information about evolutionary processes that occur over long periods of time. For example, some molluscan lineages have undergone little structural (morphological) change in hundreds of millions of years, whereas others have evolved so rapidly that the events may be regarded as instantaneous in geological time. Although conservative lineages are characterised by their low diversities, the production of the gross morphological differences that lead to new higher taxa does not seem to require more than limited diversity in the ancestral group. It is therefore necessary to distinguish between the production of disparity (morphological difference) and the production of diversity (the number of lower taxa) when assessing rates of evolution. Macroevolution (the origin of new higher taxa) appears to depend largely on the production of disparity, whereas microevolution (the origin of new lower taxa) results mainly in increased diversity.

Molluscs exhibit many of the features long observed in the histories of other animal groups. The highest taxa appeared early in the history of the phylum when diversity was low (seven of the eight classes had evolved by the end of the Cambrian some 500 million years ago). Secondly, the oldest representatives of higher taxa are small and there has been a continuous increase in maximum body size over the past 570 million years. Thirdly, most higher taxa appeared long before their first substantial radiations. Fourthly, most higher taxa are recognised largely by hindsight after a period of diversification has occurred. Fifthly, rates of diversification vary between different higher taxa. Sixthly, accelerated rates of evolution occur under special conditions such as geographic isolation, the development of symbiotic relationships, or the removal of competition as a result of mass extinction.

INTRODUCTION

In contrast to many other groups of organisms, molluscs have an excellent fossil record extending some 570 million years from the earliest Cambrian to the present (Fig. 3.1). (The isotopic time scale used in this chapter is that of Palmer (1983).) This long and successful history has produced the greatest variety in body form of any animal phylum and has allowed molluscs to colonise more marine and terrestrial environments than any other major group of animals. There are more different kinds of molluscs living today than ever before, and all but one of the eight classes of the Mollusca are abundantly represented in the modern biota. For all these reasons, studies of fossil molluscs may be expected to provide some of the best information about evolutionary processes that occur over long periods of time ($>10^6$ years). Such information is only directly available from the fossil record although it may be inferred by other means. The evolutionary history of the Mollusca may therefore be used to test or support phylogenetic hypotheses derived from studies of the comparative anatomy, comparative physiology, comparative embryology and comparative biochemistry of living organisms.

Molluscs are solitary animals, most of which used a calcareous exo-skeleton (shell) at some time in their evolutionary history. However, there is no obvious single diagnostic character that is found in all molluscs. For example, most molluscs have calcareous shells but many are naked. Most have a rasping organ (radula) used for obtaining food but this is lost in the Bivalvia. Most molluscs have a mantle cavity containing specialised gills called ctenidia, but scaphopods, land snails and some small marine molluscs lack gills and the mantle cavity of aplacophorans is, at best, rudimentary. Many molluscs are bilaterally symmetrical but gastropods are not. Many molluscs have free-living larvae but others hatch directly from large eggs. Many molluscs construct their skeletons from aragonite but others use calcite or a combination of calcite and aragonite. Some molluscs use haemo-cyanins as respiratory pigments, whereas others use haemoglobins.

Given this great diversity of form and organisation, the only acceptable definition of the phylum must be a phylogenetic one. Thus the Phylum Mollusca is unified by the idea that all members are descended from a single lineage of bilaterally symmetrical, unsegmented, acoelomate animals that developed a dorsal exoskeleton in the form of an organic (proteinaceous) cuticle, an array of calcareous spicules or a single continuous calcareous shell. In addition to these derived features, molluscs inherited an open-ended gut, dorsoventral muscles and a developmental programme that begins with the spiral cleavage of the fertilised egg (Anderson 1982, Run-negar & Pojeta 1985).

As with other kinds of organisms, some molluscs have evolved quickly whereas others appear to be almost unchanged after hundreds of millions of years. In order to understand these and other different rates of evolution it is necessary to estimate the evolutionary distance between related animals, for the rate of evolution is equal to the evolutionary distance divid-ed by time.

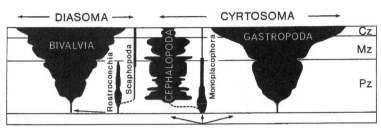

Figure 3.1 Fossil record and evolution of marine families of the molluscan Subphyla Cyrtosoma and Diasoma, slightly modified and redrawn from Sepkoski (1980). Pz, Palaeozoic; Mz, Mesozoic; Cz, Cainozoic; the beginning of the Palaeozoic occurred about 570 million years ago.

There are two main ways of measuring evolutionary distance in living and fossil organisms, although the degree of complexity may also provide a rough indication of the amount of evolution that has occurred. The first is *taxonomic diversity* (the number of kinds of phyla, classes, species, individuals, proteins, genes etc.) and the second is *morphologic disparity* (the amount of difference between related phyla, classes, species, individuals, proteins, genes etc.). Taxonomic diversity is more easily quantifiable although meaningful data may be difficult to obtain for extinct taxa; morphological disparity is often best described in qualitative terms.

Jaanusson (1981) has provided a useful explanation of the importance of morphological disparity in evolutionary progress. He argued that innovative developments – such as the origin of the mammalian jaw – occurred abruptly in single populations. Because no intermediates can be imagined between the ancestral state (reptilian jaw) and the derived state (mammalian jaw), such transformations did not take place gradually over many generations. Instead, some individuals of a critical *dithyrial* population had breached the functional threshold and had a mammalian jaw; others retained the jaw of their reptilian ancestors. As the mammalian jaw proved to be more effective, the progeny of individuals having it survived to give rise to the Mammalia. This example and the others chosen by Jaanusson (1981) deal with complex anatomical structures of extinct organisms; however, the same logic applies at other levels, such as the origin of new kinds of proteins (Runnegar 1984b).

It is possible to incorporate Jaanusson's ideas and terminology into a more general model of evolutionary processes (Fig. 3.2). Innovation – the crossing of a functional threshold – gives rise to morphological disparity and, commonly, to increased diversity. Diversity generates greater diversity (microevolution) at a rate that may be characteristic for a given group of organisms in a relatively stable environment, but the main source of new higher taxa (macroevolution) lies in the pathway that leads to morphological disparity. Such developments involve the crossing of one or more new functional thresholds and the organisms that approach these thresholds develop morphologies that make this possible. These developments are called *proximations* and the structures are 'predisposed' (preadapted) to serve

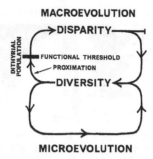

Figure 3.2 Model of evolutionary processes based on the ideas of Jaanusson (1981). Disparity (gross morphological difference) results from the crossing of functional thresholds and normally leads to greater diversity (the number of lower taxa). Diversity leads to greater diversity and to proximations that enable another functional threshold to be crossed. Dithyrial populations contain individuals that differ significantly in morphology but are capable of interbreeding. See text for further explanation.

in their new rôle once the threshold has been crossed (Jaanusson 1981). Some innovations are successful but lead nowhere; organisms possessing them tend to have long histories but low diversities and are often described as 'living fossils'.

I shall now attempt to illustrate these principles with examples from the evolutionary history of the Mollusca. In doing so I shall draw mainly on examples from the Class Bivalvia for three reasons. First, I know more about these kinds of animals and their extinct relatives – the Class Rostroconchia (Pojeta & Runnegar 1976) – than other groups of molluscs. Secondly, it is possible to infer a great deal about the soft anatomy and mode of life from the bivalve shell. Thirdly, the bivalves have, in general, evolved rather slowly in comparison to some other well-studied groups of molluscs such as the cephalopods. It is therefore likely that their evolutionary history will be better illustrated by the fossil record, it being a kind of 'time-lapse' version of the history of more rapidly evolving animal groups.

Molluscs appear abruptly with a substantial standing diversity in the oldest Cambrian deposits so far discovered (Runnegar 1983, Runnegar & Pojeta 1985). Because about 15 genera belonging to several families, four orders, three classes and two subphyla are found in these beds, it is clear that the phylum has a significant, but as yet invisible, Precambrian history. Consequently, there is little palaeontological evidence that can be used to suggest how and when the phylum may have originated, and it is therefore necessary to use information from other sources for this purpose.

ORIGIN AND PRECAMBRIAN HISTORY OF THE MOLLUSCA

Although it is possible that molluscs are descended from an unrecognised extinct phylum of animals, it is more parsimonious to hypothesise that they are directly related to the ancestors of one of the extant phyla. Embryological studies have shown that the pattern of spiral cleavage development that includes 4d mesoderm and a subsequent larva of the trochophore type, links many phyla including molluscs, annelids, pogonophorans, uniramians (arthropods), nemerteans and platyhelminths (Anderson 1982). As there is no hard evidence to indicate that the ancestors of the Mollusca were either segmented or possessed a spacious body cavity (coelom), the annelids, pogonophorans and uniramians may be excluded as potential ancestral groups. Thus there is now a general consensus that molluscs were derived from Precambrian animals having the grade of organisation and body form found in modern free-living flatworms and nemerteans (Stasek 1972, Trueman 1975, Salvini-Plawen 1980). Additional support for this hypothesis comes from comparisons of a limited number of invertebrate 5S ribosomal RNA sequences (Komiya *et al.* 1983, Kumazaki *et al.* 1983a, Ohama *et al.* 1983, Walker & Doolittle 1983); the nucleotide sequences of the 5S rRNAs from two nemerteans and three molluscs suggest that the Nermertea is closely related to the Mollusca (Kumazaki *et al.* 1983a,b).

Because almost all Early Cambrian molluscs are very small (1–5 mm, Fig. 3.3), the ancestors of the Mollusca were probably about a millimetre in size. As they were so small, it is unlikely that they were conspicuously flattened, used muscles for locomotion, or relied on anything more than

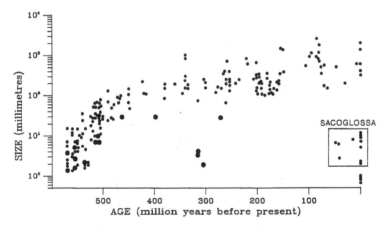

Figure 3.3 Maximum size (longest dimension) of living and fossil molluscs. Most of the points represent the largest specimens of any species in the literature or examined by the author in museum collections; they demonstrate that the maximum size of shelled molluscs has increased steadily through the Phanerozoic. The larger points represent the largest known specimens of the oldest representatives of various higher taxa (classes, subclasses and orders). Note that most higher taxa developed in animals of small body size. The boxed points represent the maximum sizes of living and fossil bivalved snails and some closely related living univalves; the points below the box represent the maximum sizes of some of the smallest living shelled molluscs. Data from numerous sources.

facilitated diffusion to obtain and circulate oxygen within their bodies (Alexander 1971). The ancestral molluscs were probably of a similar size and thus bore little resemblance to hypothetical primitive molluscs shown in most zoological textbooks. When this point is appreciated, it becomes much easier to see how a transition from one 'phylum' to another might have occurred.

The smallest living bivalves (species of *Condylonucula*) are sexually mature at a size of 500–600 μm (Moore 1977) and there are several other kinds of living molluscs that grow to less than a millimetre in size (Fig. 3.3). These tiny animals are greatly simplified in comparison with their larger counterparts (e.g. Morse 1976), and thus provide better models for the ancestral states than do the more conspicuous larger forms. Thus we might expect that the ancestral mollusc lacked gills, moved by means of cilia, had poorly differentiated body layers, few organs, and relied mainly on processes operating at the cellular level. Such organisms have no chance of being fossilised and we must therefore use other evidence to attempt to date the origin of the phylum. It is important to attempt to do so because the first major event in the history of the phylum was the period of development that led to the diversification into classes at the beginning of the Phanerozoic some 600–500 million years ago.

With the possible exception of the vertebrates, all of the extant phyla of animals appear to have existed by Early Cambrian times. As the molluscs lie about halfway up the evolutionary ladder leading from an animal-like protist to the Vertebrata, the origin of the Mollusca lies in the later part of the Precambrian (1000–600 million years ago). If it is assumed that the evolution of the phylum took place during the primary radiation of the multicelled animals (Metazoa), there are several independent lines of evidence that may be used to estimate when this radiation – and thus the origin of the Mollusca – took place.

It is first necessary to know whether the Metazoa have a common ancestry and hence represent a monophyletic group. Most students of this problem have opted for a single origin of the multicellular animal phyla but some, such as Anderson (1982), have preferred to derive the metazoans from different animal-like protists at somewhat different times. This question appears to be settled by the following evidence.

The extracellular protein collagen is the principal structural component of metazoan connective tissues and the most abundant protein in large animals. It has been found in representatives of every metazoan phylum studied and appears to be restricted to the Metazoa (Adams 1978, Towe 1981), although an enzyme required for the post-translational hydroxylation of proline seems to have been inherited from the common ancestor of animals and plants (Ashford & Neuberger 1980). Consequently, if the collagens of all animal phyla could be shown to be homologous, this would provide strong support for the idea that all metazoans share a common ancestry. As there is now good evidence that fibrillar collagens of the mesogloea of the cnidarian *Actinia equina*, the body wall of the parasitic platyhelminth *Fasciola hepatica*, and the byssus of the mollusc *Mytilus edulis* are homologous to vertebrate Type I collagens (Nordwig & Hayduk 1969, DeVore *et al.* 1984, Runnegar 1985, 1986), there can be little doubt that the Metazoa is a monophyletic taxon.

Given this fact, it is possible to use differences between other common protein and nucleic acid molecules to attempt to estimate the time of origin of the metazoan phyla (Runnegar 1982). Because of the limited amount of information so far available, such estimates are no more than 'guesstimates' but they do point to an unobserved Precambrian history for the Mollusca of the order of 100–200 million years (Runnegar 1986). This date ties in fairly well with estimates of the time of divergence of the ancestral genes for different kinds of vertebrate collagens (Bernard *et al.* 1983, Runnegar 1985) and with an observed decline in both abundance and diversity of algal stromatolites attributed to grazing by metazoan animals (Awramik 1971, Walter & Heys, 1986). It also coincides approximately with a backwards extrapolation of Sepkoski's (1978) curve of the diversity of Late Precambrian and Cambrian metazoan orders, an extrapolation based on the premise that the fossil record represents only about 1–10% of the metazoan orders existing at the time (Runnegar 1986). Thus, in summary, the first molluscs appear to have been minute non-shelled animals that evolved from nemertean-like metazoans some 100–200 million years prior to the beginning of the Cambrian. The events that led to the development of the phylum were probably little more than the formation of a toughened and perhaps spiculose surface on the dorsal part of the body and the development of a specialised organ (radula) for obtaining food. If such animals had not given rise to the kinds of animals we now call molluscs it is probable that they would be classified with their ancestors in another phylum. This is because higher taxa are recognised largely by hindsight after a protracted amount of evolution has resulted in increased diversity.

MACROEVOLUTION IN THE MOLLUSCA: EVOLUTION OF HIGHER TAXA

There are eight classes of the Mollusca, one of which (Rostroconchia) became extinct at the end of the Palaeozoic (a different view is given by Yochelson (1978, 1979)). Members of two of the classes (Aplacophora and Polyplacophora) never evolved a continuous calcareous shell and are not discussed here. Their history is reviewed in Runnegar *et al.* (1979) and Pojeta (1980). The remaining six classes are conveniently divided into two subphyla, one of which (Cyrtosoma) gave rise to the other (Diasoma) prior to the beginning of the Cambrian (Fig. 3.1; Runnegar & Pojeta 1974, Runnegar 1983).

All eight classes of the Mollusca had evolved by the end of the Ordovician and all but the Scaphopoda had appeared prior to the end of the Cambrian about 500 million years ago. There are two obvious features of this primary radiation of the phylum; almost all molluscs were very small (Fig. 3.3) and

EVOLUTION IN THE MOLLUSCA

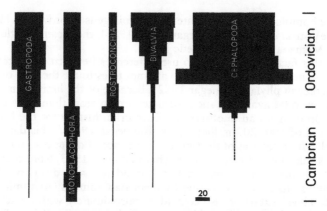

Figure 3.4 Cambrian and Ordovician histories of the molluscan subphyla Cyrtosoma and Diasoma. The approximate diversity of each class is indicated by plots of the number of species versus time (Rostroconchia) or the number of genera versus time (other classes). The single genus of scaphopods known from the Late Ordovician is not shown. The dashed line leading to the Cephalopoda represents the known stratigraphic range of tall septate monoplacophorans.

most lineages contained few genera and species (Fig. 3.4). These facts tell us two things. First, that macroevolution (the origin of new higher taxa) is not caused by the kinds of radiations that result in great diversity, and secondly, that small animals are more likely to originate new higher taxa than are large animals (Stanley 1973). The second point is well illustrated in Fig. 3.3; the oldest representatives of a number of molluscan classes, subclasses and orders have a body size of less than about 10 mm.

The origins of each of the six classes of the Subphyla Cyrtosoma and Diasoma can be tied to particular functional thresholds that were crossed by the first representatives of each class. For the Monoplacophora the functional threshold was the formation of a continuous dorsal exoskeleton (univalved shell). For various reasons explained elsewhere (Runnegar & Pojeta 1985), the first shells must have been limpet-shaped cones similar to those of the Cambrian genus *Scenella* and the living genera *Neopilina* and *Vema* (McLean 1979). In fact, the living monoplacophoran *Vema hyalina*, which grows to a maximum size of about 2 mm, displays an anatomy that has existed in an almost unchanged form since the beginning of the Cambrian (Pojeta & Runnegar 1976). Because of the limited opportunities presented by a limpet-shaped shell (such as the restriction to hard substrates that act as a second 'valve'), limpet-shaped monoplacophorans have not diversified greatly during their 600 million year history.

There are two main geometric possibilities for converting a limpet-shaped shell into a shell of another shape. Either the relative size of the shell aperture can be changed so that the shells become taller and perhaps coiled; or the shape of the aperture may be varied to produce flattened cones. The first of these modifications led via the Helcionellacea to the planispirally coiled bellerophont monoplacophorans (Runnegar & Jell 1976, Runnegar & Pojeta 1985); the second led to the Rostroconchia and hence to the Bivalvia (Pojeta & Runnegar 1976, Runnegar 1983) and Scaphopoda (Pojeta & Runnegar 1979). In contrast, asymmetric coiling led to torsion and the origin of the Gastropoda (Runnegar 1981), whereas partitioning of a tall uncoiled monoplacophoran shell ultimately produced the Cephalopoda (Yochelson et al. 1973). Thus the functional thresholds that had to be crossed in each case were: the origin of a univalved shell (Monoplacophora); asymmetric coiling and torsion (Gastropoda); development of a buoyancy control device (siphuncle, Cephalopoda); lateral compression of a univalve shell (Rostroconchia); origin of the bivalved condition (Bivalvia); and ventral fusion of an elongate laterally compressed univalved shell (Scaphopoda).

As will be explained below, the lateral compression of the shells of primitive rostroconchs may also be regarded as a proximation that led one member of this group to cross the functional threshold that gave rise to the Bivalvia. Some perspective on the nature and significance of these events may be gained from an examination of the analogous history that gave rise to bivalved snails within the last 60 million years. This relatively insignificant event in the history of the Mollusca would seem to be wholly comparable to the events that led to two classes of the Diasoma during the Cambrian.

Evolution of the bivalved snails

Opisthobranch gastropods belonging to the Order Sacoglossa have a specialised radula consisting of a single row of dagger-like teeth. This allows them to pierce plant cells, predominantly algae, to obtain sap, so they are best described as 'marine aphids'. Like aphids they are generally quite small.

A few members of the Sacoglossa have developed bivalved shells (Kawaguti & Baba 1959, Keen & Smith 1961, Kay 1968). These shells are so similar to those of true bivalves that fossil bivalved sacoglossans had been placed in the Bivalvia for many years. Their correct affinities were only realised after the first living animals had been discovered and described (Kawaguti & Baba 1959).

At best, bivalved snails are given the rank of a superfamily within the Sacoglossa. This taxonomic decision reflects their rarity and low diversity. It is therefore instructive to examine briefly the evolutionary history of this small group of organisms, for they have achieved the grade of organisation – the degree of disparity – that characterised other higher taxa of the Mollusca at the time of their first appearance. In particular, they are directly comparable to the first bivalves and they therefore well illustrate the point that higher taxa are recognised largely by hindsight.

Perhaps surprisingly in view of their small size (< 8 mm, Fig. 3.3) and delicate shells, bivalved sacoglossans have a fossil record that extends for some 50 million years to the Middle Eocene of the Paris Basin. During that time they have achieved a global distribution, having somehow reached southern Australia by the Early Pliocene (Ludbrook & Steel 1961). However, during the past 50 million years the group has spawned no more than a half a dozen genera and not many more species (Kay 1968). Again, these developments parallel the Cambrian history of the Bivalvia (Pojeta 1975, Runnegar & Bentley 1983).

Bivalved gastropods secrete a univalved larval shell like that of other gastropods. The larval shell begins to develop into left and right valves when it has grown to a size of about 200 μm (Kawaguti 1959). Although the adult valves are superficially similar in size and shape, the asymmetry of the body that is characteristic of gastropods is reflected in the position of the larval shell (apex of the left valve) and in the fact that the shell-attached muscles lie in quite different positions in each valve (Kawaguti & Yamasu 1960). But the most dramatic difference between the shells of the bivalved snails and primitive bivalves lies in the fact that the elastic ligament, which connects the valves and acts as a springy hinge, lies on the anterior side of the larval shell. The reverse is true in all primitive bivalves; the ligament is always posterior. Thus, although the shells of the bivalved snail Berthelinia and early Cambrian bivalves are remarkably similar in external form (Kay 1968, Pojeta 1975, Runnegar & Bentley 1983), the shell of Berthelinia is reversed on the body. This is because all gastropods undergo torsion, an evolutionary innovation now incorporated into the ontogeny that rotates the head foot about 180° with respect to the rest of the body. The point of interest in this case is that the orientation of the bivalved shell refects an historical event in the history of the class but the form of the shell is wholly convergent with that of a primitive bivalve.

How did the bivalved condition arise in the Sacoglossa? Fortunately, there are other living sacoglossan genera that appear to represent the transitional condition (Kay 1968). The shells of the genera Volvatella and Cylindrobulla are loosely coiled and weakly calcified (Marcus & Marcus 1956, Burn 1966). Each animal has an adductor muscle that runs from left to right sides of the younger part of the shell. Contraction of this muscle narrows the shell aperture by increasing the spiral curvature of the whole structure. Thus it is clear that the formation of an adductor (closing) muscle preceded the development of a bivalved shell.

Because a shell must be either univalved or bivalved, there can be no intermediate state between those observed in the bivalved sacoglossans and genera such as Cylindrobulla. The required evolutionary step is large but obviously not unbridgeable as it has happened at least three times in the history of the Mollusca (Runnegar 1983) and in a variety of other animal groups. Because of the prior existence of a functional adductor muscle there is no need to invoke the synchronous occurrence of improbable macroevolutionary events in the evolution of the bivalved snails and it is reasonable to suppose that at the time of transformation both phenotypes could have coexisted in a single interbreeding (dithyrial) population (Jaanusson 1981).

The lesson to be learned from this example is that an event of comparable magnitude to the one that led to the Class Bivalvia took place in the Late Mesozoic or Early Tertiary. Because the flexibility required for the development of an adductor muscle has to be distributed over the whole shell in genera such as *Cylindrobulla*, the shells of the ancestral group are almost uncalcified and consequently are unknown as fossils. In contrast, bivalves work best if the valves are rigid so that the force of the adductor muscles may be transmitted without loss to the intervening ligament. As a result, bivalved snails have relatively strongly calcified shells and a long fossil record.

It might be argued that the small size and low diversity of living bivalved snails show that the evolution of this group of animals is not comparable to the event that led to the highly successful Bivalvia. This argument is likely to be incorrect for two reasons. First, the history of the Bivalvia tells us that the primary radiation of the bivalved sacoglossans cannot be expected for another 10 million years at least (Figs 3.4 and 3.5), and secondly, it may be a matter of opportunity. It is at least possible to imagine an event that would allow the 'marine aphids' to replace their present terrestrial counterparts. I use this example, not because it is likely to happen, but because it illustrates once again how higher taxa must be viewed by hindsight; the events that produce higher taxa may be judged to be unimportant at the time. Viewed in this way, there is little to distinguish 'micro' and 'macro' evolutionary events except for the fact that some innovations provide a potential that would be beyond the capacity of organisms lacking them.

Evolution of the Bivalvia

Two genera of bivalves, *Fordilla* and *Pojetaia*, are widely distributed in Early Cambrian strata but no bivalves have yet been discovered in later Cambrian deposits (Pojeta 1975, Jell 1980, Runnegar & Bentley 1983). This apparently puzzling stratigraphic distribution is explained below.

The two Early Cambrian bivalves share many features such as small size, bilateral symmetry, simple hinge structures, two adductor muscles and similar shell microstructures (Runnegar & Bentley 1983, J. Pojeta, Jr, personal communication). They are so similar that they may be placed in the same family (Fordillidae; Pojeta 1975) and yet they may also be referred to different subclasses because each may be traced through the Phanerozoic to quite different groups of living bivalves (Runnegar & Bentley 1983). This paradoxical situation illustrates the artificiality of higher taxa which merely serve to 'render diversity manageable through grouping' (Newell & Boyd 1978). If this symposium had been held in the Early Cambrian, it is probable that *Fordilla* and *Pojetaia* would, at best, be ranked as a family or superfamily of the Monoplacophora.

The obvious ancestors of the Early Cambrian bivalves are the earliest Cambrian rostroconchs, of which *Heraultipegma* is the best known genus Pojeta & Runnegar 1976, Runnegar 1983). This laterally compressed univalve has a suitable size and shape, and a shell microstructure that consisted largely of spherulitic prisms of aragonite (Runnegar 1983). A similar microstructure was present in the shells of *Pojetaia* (Runnegar & Bentley 1983) and *Fordilla* (J. Pojeta, Jr, personal communication), and it survives in an unmodified form in the ligaments of all living bivalves, including those with otherwise wholly calcitic shells (Runnegar 1983).

By analogy with the presumed ancestors of the bivalved snails, *Heraultipegma* may have had at least one adductor muscle and a flexible hinge. It had therefore approached the bivalved condition (the required proximation, Fig. 3.2), but it was the crossing of the functional threshold to the truly bivalved condition seen in *Fordilla* and *Pojetaia* that marked the origin of the Class Bivalvia. As the only fundamental difference between the ligament and valves in these early bivalves would have been the proportion of intercrystalline matrix (glycine-rich proteins) to mineral, the crossing of this functional threshold was a relatively small step. It is also one that was proximated by at least one other kind of monoplacophoran (Runnegar 1983).

The five anatomical features required by early bivalves that are unlikely to have been present in any or all rostroconchs are: 1, left and right shell-attached pedal (foot) muscles; 2, anterior and posterior adductor muscles; 3, a flexible ligament (postero-dorsal); 4, hinge teeth; and 5, a collagenous larval attachment structure (byssus). It has already been shown that the byssus of living mussels is constructed from collagen encoded by a gene that was inherited from the ancestral metazoan and that adductor muscles may have been pesent in some rostroconchs. The division of the shell-attached muscles, the production of hinge teeth and the restriction of the junction between the valves to the area of the ligament could all have happened synchronously within a single dithyrial population in the Early Cambrian. All of these features are found in the bivalved snails but not in the related univalves that are thought to represent the ancestral condition. Thus the dithyrial population that led to the Bivalvia could be expected to yield fossils that would be identified as representatives of two different classes (Rostroconchia and Bivalvia). Unfortunately, no single sample containing both *Heraultipegma* and early bivalves has yet been discovered. If it is, it should be examined carefully with Jaanusson's hypothesis in mind.

MICROEVOLUTION IN THE MOLLUSCA: EVOLUTION OF LOWER TAXA

At least six higher taxa of the Mollusca originated in the Early Cambrian and yet did not begin to radiate in any substantial way until the Late Cambrian or Early Ordovician (Fig. 3.4). By contrast, the Class Cephalopoda appears to have begun its first explosive radiation almost as soon as it evolved (Fig. 3.4; Chen & Teichert 1983).

These patterns may be best explained by postulating different rates of diversification and different chances of success for different classes of the Mollusca. It is possible to approximate a radiation by assuming that the number of lower taxa (in this case genera) doubles every so many million years. Plots of this kind show that the primary diversification of the Bivalvia is approximated by a curve drawn on the assumption that the number of genera in the class doubled every 20 million years (Fig. 3.5). This crude estimate is unlikely to be realistic in detail but it does help to explain the long delay in the diversification of the class and the poor fossil record of Cambrian bivalves.

The value of this crude model is also supported by data from the Rostroconchia (Fig. 3.5). The oldest known rostroconch (*Heraultipegma*) pre-dates the oldest known bivalve (*Pojetaia*) by some 10 million years. If the rate of diversification of the Rostroconchia were approximately the same as the rate exhibited by the Bivalvia, the rostroconchs could be expected to begin their substantial radiation about 10 million years before the Bivalvia. This is in fact what is observed in Late Cambrian strata, and it may be concluded that the decline in rostroconch diversity after a peak in the earliest Ordovician was due to increasing competition from the more efficient bivalves (Fig. 3.5). A comparison of Figures 3.4 and 3.5 shows that this model could also explain the observed Cambrian–Ordovician diversities of the classes Monoplacophora and Gastropoda. The monoplacophorans are analogues of the rostroconchs and the gastropods are analogues of the bivalves. Thus all four classes appear to have diversified slowly and relatively consistently following their origins in the Late Precambrian or Early Cambrian.

In contrast, the diversification of the Cambrian–Ordovician cephalopods

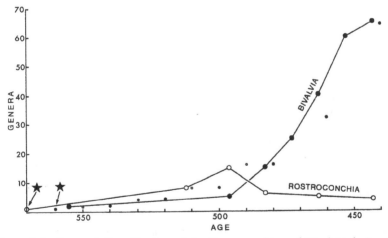

Figure 3.5 Rates of diversification of the Rostroconchia and Bivalvia during the Cambrian and Ordovician. The rates of diversification of both classes may be modelled by curves representing a doubling time of 20 million years (small filled and unfilled circles) and the drop in diversity of the rostroconchs that occurred between about 500 and 480 million years ago is thought to be due to the effect of competition from the bivalves. The approximate times of origin of the classes are indicated by stars.

is better modelled by doubling times of the order of 3–4 million years
(Fig. 3.6). These figures are likely to be even more unrealistic because the
event modelled includes one or more major periods of extinction (Crick
1981, Chen & Teichert 1983). Nevertheless, it is clear that the total
number of shelled cephalopod genera increased during this period of time
in a way that is reasonably well described by a model in which the rate of
evolution of shelled cephalopods was about an order of magnitude greater
than that of the other classes.

This suggestion may be tested by a comparison of the rates of diversification of younger shelled cephalopods with those of their Cambrian–
Ordovician relatives. For example, the ceratites – which are characteristic
of Triassic marine strata – had their origins in Middle Permian discoidal
goniatites (Spinosa *et al.* 1975). A plot of the diversity of Permian–Triassic
ceratites is best modelled by a curve in which doubling takes place every
3–5 million years (Fig. 3.6), a rate comparable to that estimated for
Cambrian-Ordovician nautiloids. Perhaps more surprisingly, the rate of
diversification of the Permian–Triassic ceratites seems to be unaffected by
the major extinction event at the end of the Permian; the measured drop in

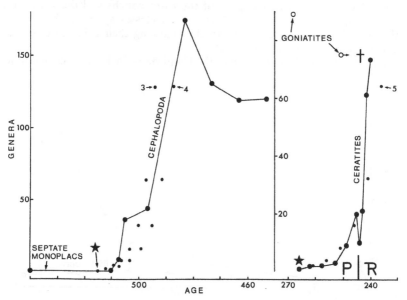

Figure 3.6 Rates of diversification of Cambrian–Ordovician cephalopods and
Permian–Triassic ceratites. The rates of diversification of these groups may be
modelled by curves representing doubling times of the order of 2–5 million years
(small filled circles). The measured decline in diversity at the Permian/Triassic
boundary may be an artifact resulting from the rarity of earliest Triassic marine
strata. However, the rate of diversification of the ceratites appears to have increased
after the extinction of the goniatites at the end of the Permian (cross). Stars indicate
the origins of the higher taxa.

diversity shown in Figure 3.6 is likely to be a preservational artifact resulting from the unusual nature and rarity of earliest Triassic marine strata. On the other hand, it is unlikely that the Triassic radiation of the ceratites would have proceeded as far as it did if the ceratites had been unable to invade the ecospace vacated by the extinction of the Palaeozoic goniatites and other organisms.

The idea that there is a fundamental difference in the rates of evolution of different higher taxa cannot be carried too far because it is also possible to show that under some circumstances rates of diversification have varied greatly within a single class. However, these differences in rates of diversification may be superimposed effects caused largely by special circumstances. It is possible to illustrate this point by using the histories of several different groups of bivalves (Fig. 3.7).

Different rates of diversification within the Bivalvia

Some genera of living bivalves – if broadly interpreted – have fossil records that extend for some hundreds of millions of years into the Palaeozoic. The two best examples of this phenomenon are the genera *Solemya* and *Pinna*, both of which have existed in an almost unmodified form since the mid-Palaeozoic and can be traced to similar forerunners (Fig. 3.7). Both of these genera have highly specialised morphologies and both are adapted for unusual and particular modes of life (Owen 1961, Reid & Bernard 1980, Felbeck *et al.* 1981, Yonge 1953). For example, in *Solemya* the gut is either greatly reduced or lacking altogether and the animals may either absorb dissolved organic molecules from the local environment or participate in a symbiotic relationship with sulphide-oxidising chemoautotrophic bacteria. *Solemya* often lives in deep sealed burrows under near-anoxic conditions and its occurrence in organic-rich shales of Early Permian age in Western Australia (Dickins 1963) suggests that it may always have done so.

By contrast, *Pinna*, (and its close relative *Atrina*) inhabit well-oxygenated sublittoral environments in tropical and subtropical climatic zones. The shells are triangular in shape, have an inflexible ligament, and are embedded vertically in the sediment in such a way that the posterior part of the shell is often subject to mechanical damage. As a result, the body is small relative to the shell and is largely confined to the part that is buried. The exposed part of the shell is weak, easily broken and easily repaired, and flexible enough to be closed by the adductor muscle without the use of a normal ligament.

The extraordinary features of both *Solemya* and *Pinna* represent substantial departures from the morphology of their presumed ancestors and average bivalves (Pojeta & Runnegar 1985). They therefore exhibit a large amount of morphological disparity, but disparity that has not led to increased diversity. Both represent successful designs that have no potential for further development and they have therefore survived in an unmodified form for hundreds of millions of years. Viewed in this way *Solemya* and *Pinna* are aptly described as 'living fossils', but the information contained in the genes of the living forms is likely to be very different from the information in the genes of their Palaeozoic ancestors. Just as the morphology of haemoglobin molecules has been maintained despite major changes in the haemoglobin-coding sequences of distantly related animals (Lesk & Chothia 1980), so the morphologies of Solemya and Pinna have survived because no change was possible. Only neutral or near-neutral mutations were fixed in the genomes of these animals and there must have been intense selection against both evolutionary progression and regression. Based on morphology alone, Pinna and Solemya could be said to have a rate of evolution close to zero. On the basis of other criteria, such as the number of base changes in their DNA sequences or the arrangement of coding and noncoding sequences in their genes, Pinna and Solemya may have evolved as quickly as an average animal. Neither genus has an unusually high amount of DNA per cell (C-value; Hinegardner 1974) so that in this respect these bivalves do not resemble some other organisms that have a long fossil record (Hinegardner 1976).

A second pattern of diversification is shown by the pectinoid bivalves, commonly termed scallops (Fig. 3.7). For reasons that are explained in Gould (1971), Stanley (1972) and Waller (1978), the scallop shell represents an optimum design. As a result, it has changed little in superficial appearance during the past 400 million years.

However, during this time some significant improvements were possible. These included the development of a highly efficient elastic polymer that could act as a responsive energy store in the ligament (Alexander 1966), the replacement of weaker and heavier aragonitic shell microstructures with lighter and stronger structures made of calcite (Waller 1978), and the development of the ctenolium (Waller 1984). When these changes are coupled with various kinds of structures used to strengthen the shell (ribs), it becomes obvious that scallops could be expected to display a greater standing diversity than specialised genera such as Solemya and Pinna. However,

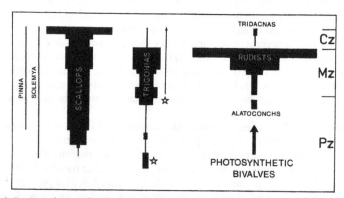

Figure 3.7   Fossil records of selected groups of bivalves that illustrate different rates of diversification. The approximate diversity of each group is indicated by plots of the number of genera versus time. The stars indicate the presence of secondary teeth; see text for further explanation. Pz, Palaeozoic; Mz, Mesozoic; Cz. Cenozoic.

because the basic design was so good, scallops achieved a substantial diversity soon after the form first evolved (Fig. 3.7). Since then the number of scallop genera has remained much the same (the larger number of scallops recorded from the Neogene may reflect better sampling). Thus, although scallops have obviously undergone more morphological evolution than either *Solemya* or *Pinna*, the rates and amounts of change have not been spectacular compared with those that were required to produce the initial form.

The Australian genus *Neotrigonia* is the only living representative of a group of bivalves that shows a third pattern of diversification. Trigoniids were abundant and diverse during the Mesozoic but most disappeared from the record at the end of that period.

The ancestors of the trigoniids were descended from Ordovician bivalves called lyrodesmatids (Pojeta 1978, Pojeta & Runnegar 1985). The lyrodesmatids gave rise to a genus called *Schizodus* that, in turn, diversified into the ancestors of the trigoniids and other sister genera at the end of the Palaeozoic (Newell & Boyd 1975). In an elegant study of the functional morphology of living and Mesozoic trigoniids, Stanley (1977) has shown that the success of the Mesozoic trigoniids can be attributed to two factors: the evolution of a muscular T-shaped foot and the concomitant development of large hinge teeth with secondary denticles. Because secondary denticles may be used to infer the presence of the functionally important trigoniid foot, the stratigraphic distribution of this character is of some importance (Fig. 3.7).

All Ordovician and Early Silurian lyrodesmatids have primary and secondary teeth but the later Palaeozoic descendants of these forms have only small primary teeth (Newell & Boyd 1975). Secondary teeth reappear in some Permian and most Triassic genera in an atavistic fashion. It is therefore tempting to assume that a gene required for the production of these structures had been retained in an unexpressed form through much of the Palaeozoic. The now-famous experiment of Kollar and Fisher (1980) which led to the production of 'reptilian' teeth in chicken embryonic tissue shows that this suggestion is at least a biological possibility.

The post-Mesozoic history of the trigoniids is little different from the Cambrian history of the class or the Middle Palaeozoic history of *Schizodus*. As the northward movement of Australia has now allowed *Neotrigonia* to invade the Indopacific zoogeographic realm, it is possible that this relict of a once-diverse family will again achieve the success of its ancestors in the Mesozoic. There has probably been too little time since the Mesozoic for more than minimal diversity to develop.

Some support for this statement is provided by the history of tridacnid bivalves. These extraordinary photosynthetic animals (Yonge 1980) appear to be ecological analogues of the bizarre rudists of the later Mesozoic. However, by contrast to the rudists, which diversified into some 90 genera in the Late Cretaceous, there are only two living genera of tridacnids (Fig. 3.7; Rosewater 1965, Yonge 1980).

In evolving the tridacnid form, ancestral cockles (Cardiids) underwent a remarkable transformation of body form (Stasek 1961, Yonge 1980). In this case, the functional threshold that resulted in great morphological disparity appears to have been the acquisition by infection of symbiotic zooxanthellae. A similar event in the ancestors of the rudists (Vogel 1975) may account for both their success and their extraordinary shapes (Kauff-man & Sohl 1974), and a similar argument may be advanced to explain the origin of the remarkable alatoconchids now known from a number of tropical Permian sites (Yancey & Boyd 1983). Thus, in each of these cases, abrupt morphological changes may have been triggered solely by the development of a symbiotic relationship with photosynthetic algae. The low diversity of living tridacnids may merely reflect their recent ancestry and they may yet have the potential to achieve the diversity found in Mesozoic rudists. It follows that the alatoconchids may also have become more significant components of the biosphere if they had not had the misfortune to evolve just prior to the faunal crisis at the end of the Palaeozoic.

Finally, unusually rapid rates of diversification have occurred in regions that have become geographically isolated. Two good examples are the Neogene bivalves of the eastern Mediterranean (Gillet 1946) and the Late Permian bivalves of the Paraná-Karroo system of eastern South America and southern Africa (Runnegar and Newell 1971, Runnegar 1984a). Both of these environments seem to have been large brackish 'lake-seas'. In each case, rates of evolution within the 'lake-sea' have greatly outstripped the rates of evolution of related genera in nearby normal marine environments. These accelerated rates of evolution in the isolated basins may be explained in several ways, but the most obvious answers are reduced population sizes, unstable conditions and an absence of competition from organisms exclud-ed because of their intolerance to lowered or variable salinities.

CONCLUSIONS

The data obtained from the fossil record of the Mollusca support the idea that macroevolution – major changes that may ultimately give rise to diverse groups of completely new kinds of animals – results from steps that lead to the crossing of significant functional thresholds. These events tend to occur in small animals because such organisms are less complex, more common and less specialised. Most of the required transformations are likely to be abrupt and therefore may be regarded as instantaneous in geological time. They do not necessarily lead to new higher taxa which are recognised by hindsight after a subsequent period of diversification.

In contrast, microevolution — the production of lower taxa – is a con-tinuous process that normally occurs at a rate that may be characteristic for a particular animal group. Exceptional circumstances, including the pro-duction of new ecospace by mass extinction or isolation, may lead tem-porarily to accelerated rates of diversification, but ultimately there is a limit which may only be breached by another innovation. Such innovations are relatively rare so most higher taxa have long histories.

REFERENCES

Adams, E. 1978. Invertebrate collagens. *Science* **202**, 591–8.

Alexander, R. McN. 1966. Rubber-like properties of the inner hinge-ligament of Pectinidae. *J. Exp. Biol.* **44**, 119–30.

Alexander, R. McN. 1971. *Size and Shape.* London: Edward Arnold.

Anderson, D. T. 1982. Origins and relationships among the animal phyla. *Proc. Linn. Soc. N.S.W.* **106**, 151–66.

Ashford, D. and A. Neuberger 1980. 4-Hydroxyl-L-proline in plant glycoproteins. *Trends Biochem. Sci.* **5**, 245–8.

Awramik, S. M. 1971. Precambrian columnar stromatolite diversity: reflection of metazoan appearance. *Science* **174**, 825–7.

Bernard, M. P., M. Chu, J. C. Myers, F. Ramirez, E. F. Eikenberry and D. J. Prockop 1983. Nucleotide sequences of complementary deoxyribonucleic acids for the Proα1 chain of human Type I collagen. Statistical evaluation of structures that are conserved during evolution. *Biochemistry* **22**, 5213–23.

Burn, R. 1966. The opisthobranchs of a caulerpan microfauna from Fiji. *Proc. Malacol. Soc. Lond.* **37**, 45–65.

Chen, J. and C. Teichert 1983. Cambrian cephalopods. *Geology* **11**, 647–50.

Crick, R. E. 1981. Diversity and evolutionary rates of Cambro-Ordovician nautiloids. *Paleobiology* **7**, 216–29.

DeVore, D. P., G. H. Engebretson, C. F. Schachtele and J. J. Sauk 1984. Identification of collagen from byssus threads produced by the sea mussel, *Mytilus edulis. Comp. Biochem. Physiol.* **77B**, 529–31.

Dickins, J. M. 1963. Permian pelecypods and gastropods from Western Australia. *Bur. Min. Resour. Geol. Geophys. Bull.* **63**, 1–203.

Felbeck, H., J. J. Childress and G. N. Somero 1981. Calvin-Benson cycle and sulphide oxidation enzymes in animals from sulphide-rich habitats. *Nature* **293**, 291–3.

Gillet, S. 1946. Lamellibranches dulcicoles les Limnocardiidés. *Rev. Sci.* **3258**, 343–53.

Gould, S. J. 1971. Muscular mechanics and the ontogeny of swimming in scallops. *Palaeontology* **14**, 61–94.

Hinegardner, R. 1974. Cellular DNA content of the Mollusca. *Comp. Biochem. Physiol.* **47A**, 447–60.

Hinegardner, R. 1976. Evolution of genome size. In *Molecular Evolution*, Ayala, F. J. (ed.), pp. 179–99. Sunderland, Massachusetts: Sinauer Associates.

Jaanusson, V. 1981. Functional thresholds in evolutionary progress. *Lethaia* **14**, 251–60.

Jell, P. A. 1980. Earliest known pelecypod on earth – a new Early Cambrian genus from South Australia. *Alcheringa* **4**, 233–9.

Kauffman, E. G. and N. F. Sohl 1974. Structure and evolution of Antillean Cretaceous rudist frameworks. *Verhandl. Naturf. Ges. Basel.* **84**, 399–467.

Kawaguti, S. 1959. Formation of the bivalve shell in a gastropod, *Tamanovalva limax. Proc. Japan Acad.* **35**, 607–11.

Kawaguti, S. and K. Baba 1959. A preliminary note on a two-valved Sacoglossan gastropod, *Tamanovalva limax*, n. gen., n. sp., from Tamano, Japan. *Biol. J. Okayama Univ.* **5**, 177–84.

Kawaguti, S. and T. Yamasu 1960. Formation of the adductor muscle in a bivalved gastropod, *Tamanovalva limax*. *Biol. J. Okayama Univ.* 6, 150–9.

Kay, E. A. 1968. A review of the bivalved gastropods and a discussion of evolution within the Sacoglossa. *Symp. Zool. Soc. Lond.* 22, 109–34.

Keen, A. M. and A. G. Smith 1961. West American species of the bivalved gastropod genus *Berthelinia*. *Proc. California Acad. Sci.* 30, 47–66.

Kollar, E. J. and C. Fisher 1980. Tooth induction in chick epithelium: expression of quiescent genes for enamel synthesis. *Science* 207, 993–5.

Komiya, H., M. Hasegawa and S. Takemura 1983. Nucleotide sequences of 5S rRNAs from sponge *Halichondria japonica* and tunicate *Halocynthia roretzi* and their phylogenetic positions. *Nucleic Acids Res.* 11, 1969–74.

Kumazaki, T., H. Hori and S. Osawa, 1983a. The nucleotide sequences of 5S rRNAs from two Annelida species, *Perinereis brevicirrus* and *Sabellastarte japonica* and an Echiura species, *Urechis unicinctus*. *Nucleic Acids Res.* 11, 3347–50.

Kumazaki, T., H. Hori and S. Osawa 1983b. The nucleotide sequences of 5S rRNAs from two ribbon worms: *Emplectonema gracile* contains two 5S rRNA species differing considerably in their sequences. *Nucleic Acids Res.* 11, 7141–4.

Lesk, A. M. and C. Chothia 1980. How different amino acid sequences determine similar protein structures: the structure and evolutionary dynamics of the globins. *J. Mol. Biol.* 136, 225–70.

Ludbrook, N. H. and T. M. Steel 1961. A Late Tertiary bivalve gastropod from South Australia. *Proc. Malacol. Soc. Lond.* 34, 228–30.

McLean, J. H. 1979. A new monoplacophoran limpet from the continental shelf off southern California. *Contrib. Sci. Nat. Hist. Mus. Los Angeles County* 307, 1–19.

Marcus, E. and E. Marcus 1956. On the tectibranch gastropod *Cylindrobulla*. *Anais Acad. Brasileira Cienc.* 28, 119–28.

Moore, D. R. 1977. Small species of Nuculidae (Bivalvia) from the tropical western Atlantic. *Nautilus* 91, 119–28.

Morse, M. P. 1976. *Hedylopsis riseri* sp. n., a new interstitial mollusc from the New England coast (Opisthobranchia, Acochlidiacea). *Zool. Scripta* 5, 221–9.

Newell, N. D. and D. W. Boyd 1975. Parallel evolution in early trigoniacean bivalves. *Bull. Am. Mus. Nat. Hist.* 154, 53–162.

Newell, N. D. and D. W. Boyd 1978. A palaeontologist's view of bivalve phylogeny. *Phil. Trans. R. Soc. Lond.* B284, 203–15.

Nordwig, A. and U. Hayduk 1969. Invertebrate collagens: isolation, characterization and phylogenetic aspects. *J. Mol. Biol.* 44, 161–72.

Ohama, T., H. Hori and S. Osawa 1983. The nucleotide sequences of 5S rRNAs from a sea-cucumber, a starfish and a sea-urchin. *Nucleic Acids Res.* 11, 5181–8.

Owen, G. 1961. A note on the habits and nutrition of *Solemya parkinsoni* (Protobranchia: Bivalvia). *J. Microsc. Sci.* 102, 15–21.

Palmer, A. R. 1983. The decade of North American geology 1983 geologic time scale. *Geology* 11, 503–4.

Pojeta, J. 1975. *Fordilla troyensis* Barrande and early pelecypod phylogeny: Studies in Paleontology and Stratigraphy. *Bull. Am. Paleont.* 67, 363–85.

Pojeta, J. 1978: The origin and early taxonomic diversification of pelecypods. *Phil. Trans. R. Soc. Lond.* B 284, 225–46.

Pojeta, J. 1980. Molluscan phylogeny. *Tulane Studies Geol. Paleont.* 16, 55–80.

Pojeta, J. and B. Runnegar 1976. The paleontology of rostroconch mollusks and the early history of the phylum Mollusca. *USGS Prof. Pap. 968*, 1–88.

Pojeta, J. and B. Runnegar 1979. *Rhytiodentalium kentuckyensis*, a new genus and new species of Ordovician scaphopod, and the early history of scaphopod mollusks. *J. Paleont.* 53, 530–41.

Pojeta, J. and B. Runnegar 1985. The early evolution of diasome mollusks. In *The Mollusca*, vol. 7, *Evolution of the Mollusca*, E. R. Trueman and M. Clarke (eds). New York: Academic Press.

Reid, R. G. B. and F. R. Bernard 1980. Gutless bivalves. *Science* 208, 609–10.

Rosewater, J. 1965. The family Tridacnidae in the Indo-Pacific. *Indo-Pacific Mollusca* 1, 347–96.

Runnegar, B. 1981. Muscle scars, shell form and torsion in Cambrian and Ordovician univalved molluscs. *Lethaia* 14, 311–22.

Runnegar, B. 1982. A molecular-clock date for the origin of the animal phyla. *Lethaia* 15, 199–205.

Runnegar, B. 1983. Molluscan phylogeny revisited. *Mem. Ass. Australas. Palaeontols* 1, 121–44.

Runnegar, B. 1984a. The Permian of Gondwanaland. *Proc. 27th Int. Geol. Congr.* 1, 305–39.

Runnegar, B. 1984b. Derivation of the globins from type *b* cytochromes. *J. Mol. Evol.* 21, 33–41.

Runnegar, B. 1985. Collagen gene construction and evolution. *J. Mol. Evol.* 22, 141–9.

Runnegar, B. 1986. Molecular palaeontology. *Palaeontology* 29, 1–24.

Runnegar, B. and C. Bentley 1983. Anatomy, ecology and affinities of the Australian Early Cambrian bivalve *Pojetaia runnegari* Jell. *J. Paleont.* 57, 73–92.

Runnegar, B. and P. A. Jell 1976. Australian Middle Cambrian molluscs and their bearing on early molluscan evolution. *Alcheringa* 1, 109–38.

Runnegar, B. and N. D. Newell 1971. Caspian-like relict molluscan fauna in the South American Permian. *Bull. Am. Mus. Nat. Hist.* 146, 1–66.

Runnegar, B. and J. Pojeta 1974. Molluscan phylogeny: the paleontological viewpoint. *Science* 186, 311–17.

Runnegar, B. and J. Pojeta 1985. Origin and diversification of the Mollusca. In *The Mollusca*, vol. 10, *Evolution of the Mollusca*, E. R. Trueman and M. Clarke (eds), pp. 1–57. New York: Academic Press.

Runnegar, B., J. Pojeta, M. E. Taylor and D. Collins 1979. New species of the Cambrian and Ordovician chitons *Matthevia* and *Chelodes* from Wisconsin and Queensland: Evidence for the early history of polyplacophoran mollusks. *J. Paleont.* 53, 1374–94.

Salvini-Plawen, L. V. 1980. A reconsideration of systematics in the Mollusca (phylogeny and higher classification). *Malacologia* 19, 249–78.

Sepkoski, J. J. 1978. A kinetic model of Phanerozoic taxonomic diversity. I. Analysis of marine orders. *Paleobiology* 4, 223–51.

Sepkoski, J. J. 1980. *The marine fossil record* (poster). Chicago: University of Chicago.

Spinosa, C., W. M. Furnish and B. F. Glenister 1975. The Xenodiscidae, Permian ceratitoid ammonoids. *J. Paleont.* 49, 239–83.

Stanley, S. M. 1972. Functional morphology and evolution by byssally attached bivalve mollusks. *J. Paleont.* 46, 165–212.

Stanley, S. M. 1973. An explanation for Cope's Rule. *Evolution* 27, 1–26.

Stanley, S. M. 1977. Coadaptation in the Trigoniidae, a remarkable family of burrowing bivalves. *Palaeontology* 20, 869–99.

Stasek, C. R. 1961. The form, growth and evolution of the Tridacnidae (giant clams). *Arch. Zool. Exp. Gen.* 101, 1–40.

Stasek, C. R. 1972. The molluscan framework. In *Chemical Zoology*, vol. 7, *Mollusca*, M. Florkin and B. T. Scheer (eds), pp. 1–43. New York: Academic Press.

Towe, K. M. 1981. Biochemical keys to the emergence of complex life. In *Life in the Universe*, J. Billingham (ed.) pp. 297–306. Cambridge and London: MIT Press.

Trueman, E. R. 1975. *The locomotion of soft-bodied animals*. London: Edward Arnold.

Vogel, K. 1975. Endosymbiotic algae in rudists? *Palaeogeog. Palaeoclimat. Palaeoecol.* 17, 327–32.

Waller, T. R. 1978. Morphology, morphoclines and a new classification of the Pteriomorphia (Mollusca: Bivalvia). *Phil. Trans. R. Soc. Lond.*, B. 284, 345–65.

Waller, T. R. 1984. The ctenolium of scallop shells: functional morphology and evolution of a key family-level character in the Pectinacea (Mollusca: Bivalvia). *Malacologia* 25, 203–19.

Walker, W. F. and W. F. Doolittle 1983. 5S rRNA sequences from four marine invertebrates and implications for base pairing models of metazoan sequences. *Nucleic Acids Res.* 11, 5159–64.

Walter, M. R. and G. Heys 1986. Links between the rise of the metazoa and the decline of stromatolites. *Precambrian Res.* 29, 149–74.

Yancy, T. E. and D. W. Boyd 1983. Revision of the Alatoconchidae: a remarkable family of Permian bivalves. *Palaeontology* 26, 497–520.

Yochelson, E. L. 1978. An alternative approach to the interpretation of the phylogeny of ancient mollusks. *Malacologia* 17, 165–91.

Yochelson, E. L. 1979. Early radiation of Mollusca and mollusc-like groups; In *The origin of major invertebrate groups*, M. R. House (ed.), pp. 23–58. London and New York: Academic Press.

Yochelson, E. L., R. H. Flower and G. F. Webers 1973. The bearing of the new Late Cambrian monoplacophoran genus *Knightoconus* upon the origin of the Cephalopoda. *Lethaia* 6, 275–310.

Yonge, C. M. 1953. Form and habit in *Pinna carnea*. *Phil. Trans. R. Soc. Lond.*, B 237, 335–74.

Yonge, C. M. 1980. Functional morphology and evolution in the Tridacnidae (Mollusca: Bivalvia: Cardiacea). *Rec. Aust. Mus.* 33, 735–77.

4

Rates of evolution among Palaeozoic echinoderms

K. S. W. CAMPBELL and C. R. MARSHALL

ABSTRACT

Echinoderms provide some of the best palaeontological material for the study of evolution on a large scale. Their skeletons closely reflect soft anatomy, they are easily recognised because of their complexity, they are well dated and they are now very well studied by modern techniques. Using taxonomic and morphological approaches it can be shown that the major Palaeozoic evolutionary changes in this phylum took place in the Cambrian and Ordovician. During this interval new Baupläne appeared several times. The group shows no correlation between rates of cladogenesis and rates of morphological change. The availability of physical space did not trigger morphological change. The crossing of adaptive thresholds did not produce diversification, at least until long periods of time had elapsed, and many higher taxa probably became extinct by chance before diversification could take place. A model that involves rapid phenotypic expression of accumulated genetic change seems to fit the data best. Such change seems to have been possible only within a limited period after a major new genome was established.

INTRODUCTION

In general, the major subdivisions of a higher taxon such as the classes in a phylum seem to originate early in its history, whereas the minor subdivisions seem to originate at any time (Nicol *et al.* 1959, Sepkoski 1978, 1979, Raup 1983). It is sometimes claimed that this is a necessary consequence of classification in terms of nested sets. However, such a claim is based on two preconceptions. The first is that evolution is a gradualistic process that provides no data other than divergence points on which to base classificatory schemes; the second is that classification by nested sets is the only way in which classifications have been drawn up, or indeed should be drawn up. The first of these views is commonly held, but is quite unempirical. The only way to discover what has happened is to examine the record, and it may well be that discrete 'morphological blocks' rather than interminable transitions are the outstanding results of the evolutionary process. If this were true, it would not imply the absence of transitions, nor would it make a classification that drew attention to such transitions useless. It would, however, place emphasis on the distinctness of the morphological blocks and to a lesser extent on their internal patterns of diversification. Explicit or not, this is the way palaeontologists have interpreted their data, and taxa given equal high rank have been represented as appearing at any time. Problems related to the grouping of taxa into superclasses, subphyla, etc., and the provision of names for these units, have not been a matter of much consequence because they have been regarded as artifacts of the classification scheme rather than real features of the data themselves. Difficulties arise when issues of representation and naming determine the interpretation of the observations.

If the record is correctly interpreted as showing that discrete functional morphological blocks appeared at any time but the majority of them appeared early in the history of their phyla, and if these blocks were established over periods of time that were short relative to those during which they were subsequently elaborated, these would be phenomena of significance for any study of rates of evolution (Whyte 1965). Further, if such large discrete blocks, which are recognised by the coexistence of a complex of morphological characters, could be shown to be similarly discrete genetically and to have originated during periods of low taxonomic diversity, the factors controlling evolutionary rates and the modes of their evolution could be justifiably considered to have been different during their establishment and elaboration phases. No adequate genetic data are available to test this hypothesis; even if existing organisms could be examined from this point of view they could never produce data for a critical test because they represent such a small fraction of the diversity that has existed through geological time. Despite this limitation, work on molecular genetics will almost certainly be able to provide a check on many aspects of the interpretation of morphological data. On the other hand, it may be possible to check the other aspects of the hypothesis by examining rates of diversification at various taxonomic levels, the degrees of morphological difference between selected higher taxa such as classes shortly after their establishment, and the relationship between rates of morphological change and physical and biological environments as reflected in the geological record. We have chosen the Palaeozoic echinoderms as a suitable group for such an examination.

THE QUALITY OF THE FOSSIL RECORD

Much criticism of the use of fossils for the study of evolutionary rates in the early Palaeozoic stems from doubts about the adequacy of the fossil record. The most commonly raised issue concerns the absence of metazoan body fossils throughout most of the Proterozoic despite estimates that metazoans began about 1000 million years (Ma) ago (Runnegar 1982). Most invertebrate phyla appear abruptly during the Cambrian, and their level of complexity suggests a long Proterozoic history during which no body fossils were preserved. Only one genus from the Late Proterozoic Ediacara fauna has been thought to have echinoderm affinities, viz. *Tribrachidium*, but this interpretation now has little support (Glaessner 1984). Derstler (1981) has examined this problem and reached the conclusion that echinoderms had only a short Precambrian history, if any. We accept his arguments (see also Paul & Smith 1984). In particular, we are impressed with the fact that many cryptogenic classes appear during the Ordovician as well as the Early Cambrian, and a cryptogenic origin of classes cannot be regarded as a strong argument for an inadequate Precambrian record. Moreover, the work of Smith (1984b) supports the view that forms ancestral to both the Pelmatozoa and the Eleutherozoa occur in the Early Cambrian, thus increasing the probability that these two major groups arose after the Precambrian. Finally, the fact that no echinoderms are yet known from the Tommotian or most of the Atdabanian (Paul & Smith 1984), an interval of about 30 Ma that has been poorly defined and insufficiently explored, suggests that there is still room for the discovery in the earliest Cambrian of morphological types ancestral to echinoderms proper and the 'carpoids'.

Several methods are available for checking the extent to which the fossil record of Palaeozoic rocks has been sampled by palaeontologists. For example, new material of only two of the currently recognised 21 classes has been discovered *in the field* since 1918 (see addendum at end of chapter) – the Helicoplacoidea and the Ctenocystoidea. All other classes erected since that time have been based on previously collected material. Since 1918 the effort expended on palaeontological field work has been perhaps a hundred times greater than that expended previously, and vast new areas have been systematically explored. Moreover, much effort has been concentrated on the search for intermediate morphologies. Further field exploration is unlikely to produce a significant number of new classes.

At another level, work by Paul (1982) on 'cystoids' has shown that the number of families, genera and species in that group have increased by 10%, 50% and 100% respectively since 1900. These figures include taxa generated by analysis of existing material as well as new field discoveries. Although the number of genera and species will no doubt continue to rise, the prospect of finding new *morphological designs*, or Baupläne, which would have to represent new higher taxa, becomes progressively more remote.

Success in finding new types of Cambrian molluscs (Runnegar, this volume) has suggested that new techniques for examining silicified or phosphatised specimens from early Phanerozoic limestones may produce primitive echinoderm morphologies represented by small relatively undifferentiated organisms. Such a possibility cannot be discounted, but we should point out that the small echinoderms discovered to date from the

Australian Middle and Late Cambrian have been assigned without difficulty to well-known edrioasteroid, eocrinoid or ctenocystoid groups (e.g., see Jell *et al.* 1985). As a result of work of this kind the ranges of known major groups have been extended further back in time, but clear intermediates have not been found. The effect of this has been to make the time interval in which new intermediate morphologies can be discovered progressively more restricted.

These points deal with the problem of how adequately we have sampled the preserved record, but they do not address the question of how adequately the record itself has sampled the life forms that have existed in the past. Many publications refer to the small percentage of species that have become fossilised and to the small area of the earth's surface on which ancient rocks are preserved. For example, it is said that early Phanerozoic rocks are more likely to be eroded or metamorphosed than later representatives. We know of no study that has shown conclusively that Cambrian and Ordovician sediments are seriously under-represented. In fact, all continents have considerable areas of such rocks preserved in a relatively undeformed state. Also, although the percentage of species preserved is no doubt small, we can choose to study a group that inhabited shallow seas and had a robust skeleton, thus improving the chances of preservation and recovery. Even if the probability of preserving any given species is small, the probability of preserving and finding an identifiable representative of a family is high even if the family is only moderately diverse. An estimate of this probability may be obtained from the study by Smith and Paul (1982) of the Cyclocystidae. This family consists of small multiplated organisms that would be very easily broken up and made unrecognisable. However, the generic ranges (Fig. 4.1) show gaps that made up about 35% of their total spans. This figure is a minimum because the ranges may be incomplete at each end, and occurrences do not occur throughout the stages in which they are recorded. Note that the family is known to be represented in 13 of the 16 stages covered by its total known range. Thus even a family that would be thought *a priori* to have a low probability of preservation has a good record at the generic level, making the chances of finding a representative of the family high. If our analysis uses organisms from appropriate environments and uses taxa at the family level and above, the record itself is not likely to skew the results.

Figure 4.1 Possible phylogeny and time distribution of genera of the Family ▷ Cyclocystidae. The ranges do not imply the representatives are known at every level within the black range bars, most of which are plotted to correspond with the whole extent of the Ages shown on the left, even though specimens may be known from only one or two levels. Shorter range bars are used where more precise dates are available. Dotted lines indicate stratigraphic intervals from which no representatives are known either because they have not been preserved or they have not been found (after Smith & Paul 1982).

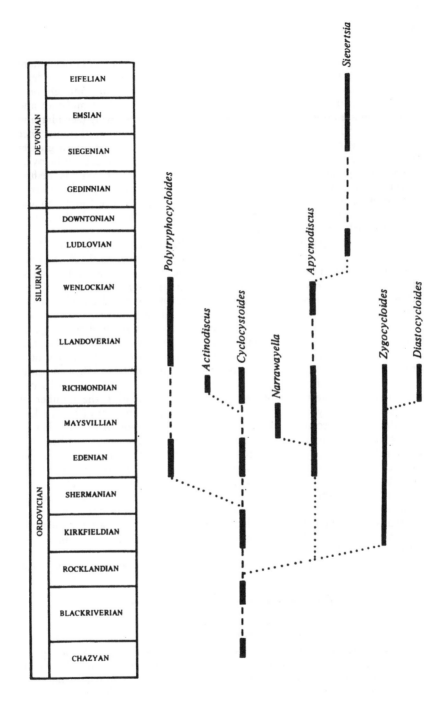

DATING FOSSIL OCCURRENCES

The use of palaeontological data for the study of rates is highly dependent on accurate dating of fossil occurrences. Most Palaeozoic echinoderm occurrences can now be dated at the Epoch level (i.e. approximately one-third of a Period). The shortest and longest Epochs in the Cambrian and Ordovician are approximately 17 Ma and 27 Ma respectively (see Harland *et al.* 1982). Silurian values are much shorter but they are not of significance for our present argument. Most occurrences can be assigned with greater precision than an Epoch. Of course, the problem of dating cannot be overcome by using higher taxa. The first appearance of a higher taxon is the first appearance of a species. Almost all first and last occurrences can be referred to as an Age, as can be seen by reference to Sepkoski (1982). Cambrian and Ordovician Ages range in length between 8 and 17 Ma. This is the lowest level at which our analysis takes place.

It is true that worldwide correlation with this degree of precision has been a matter for debate until recently. However, international efforts have resulted in widely accepted correlations. These have been effected by the rise of a refined taxonomy of the trilobites, particularly agnostids, for the Cambrian, and by the moderation of the older graptolite-based correlations of the Ordovician by new data provided by conodont and ostracode studies.

One obvious difficulty for rate studies is that our method of assigning fossil occurrences to discrete time blocks may impose apparent discontinuity on what is essentially a continuous process. For example, if new taxa or new structures are produced more or less continuously and the time units used are not of uniform length, they may appear to be produced discontinuously. This effect cannot be overcome by calculating the number of events per million years, as is obvious from an examination of Figure 4.2. Nor does it help to use progressively smaller time units if the number of events to be counted is small; when one or more of the time units becomes shorter than the time interval between events, the distribution would appear to be discontinuous. Nor is it possible to overcome the problem by the use of radiometric dates to pinpoint each relevant occurrence because dates assigned to Cambrian and Ordovician rocks usually have an order of accuracy of ±12 Ma (Harland *et al.* 1982), and in any case the number of reliable stratigraphically controlled dates is totally inadequate for the task. Even the limits placed on the Ages and Epochs in the early Palaeozoic are subject to significant error.

Does this place a veto on any analysis of rates? It certainly does if we wish

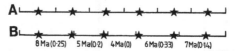

Figure 4.2 Hypothetical diagram showing the possible effects of varying the lengths of geological time units on measures of rates of innovation of taxa. A, uniform time intervals (say 5 Ma) with innovations occurring at regular intervals. B, the same innovations with intervals as shown and rates per Ma in parentheses.

Table 4.1 Table of innovations, extinctions and total number of taxa represented during the various Palaeozoic time intervals shown. Only those classes discussed in the text have been included. Figures on the right side show the number of taxa crossing into the Mesozoic.

	Cambrian			Ordovician			Silurian			Devonian			Carboniferous		Permian		
	L	M	U	L	M	U	L	M	U	L	M	U	L	U	L	U	cross
Classes																	
in	4	3	1	5	3	2	—	—	—	—	—	—	—	—	—	—	
out	1	2	—	—	—	—	1	1	1	1	3	2	—	1	—	1	
total	4	6	5	10	13	15	15	14	13	12	11	8	6	6	5	5	4
Orders																	
in	4	5	2	13	6	8	1	2	1	2	1	—	—	—	—	—	
out	1	5	—	—	1	4	2	—	2	3	4	4	2	2	2	5	
total	4	8	5	18	24	31	28	28	29	29	27	23	19	17	15	13	8
Families																	
in	7	13	2	29	48	46	7	27	18	22	18	3	43	21	4	8	
out	3	12	3	8	18	48	4	18	15	17	34	15	31	19	18	49	
total	7	17	7	33	73	101	60	83	83	90	91	60	88	78	63	53	4
Duration (Ma)	20	17	18	27	20	20	10	7	13	21	13	14	27	47	28	10	

At "total" rows the rightmost value is labelled "cross".

Table 4.2 The same data as that used for the Ordovician in Table 4.1 replotted with the Early and Late Ordovician subdivided into their standard Epochs.

	Tremadoc	Arenig	M. Ordovician	Caradoc	Ashgill
Classes					
in	3	3	3	2	—
out	—	—	1	—	—
total	8	11	14	15	15
Orders					
in	3	10	6	7	1
out	—	—	1	2	2
total	8	18	24	30	29
Families					
in	9	20	48	32	14
out	5	2	18	21	27
total	13	28	74	88	81
Duration (Ma)	17	10	20	10	10

to examine rates of establishment of lower level taxa in the Early Palaeozoic, and perhaps even for any part of the Palaeozoic. It would be impossible to examine rates of species production in any significant way because the available time units are too long. However, for some purposes the dating problems considered above are not significant. This is illustrated by reference to the data given in Table 4.1. For example, no new classes appear during the Palaeozoic after the end of the Ordovician and no new orders after the end of the Middle Devonian. These data are obviously significant. It is also possible to show that within the broad intervals when new higher taxa appeared there are barren periods comparable in duration to the adjacent productive periods. For example, Table 4.1 shows that there was an interval of about 25 Ma in the late-Middle and Late Cambrian when relatively few echinoderm taxa of family rank and above were produced, and this interval was approximately equal in duration to the preceding Early–Middle Cambrian productive period. (For further comment on the reality of this gap, see below.) To this extent at least, rates were not uniform. Finally, by making the assumption that classes, orders and families appeared at a uniform rate through the Cambrian–Ordovician, Cambrian–Middle Devonian and Cambrian–Permian respectively, it is possible to examine the probability that the distribution of originations in Tables 4.1 and 4.2 are a consequence of partitioning the time sequence into units of the lengths indicated. In each case, the probability of such an hypothesis being correct is <0.05.

THE MEANING OF 'RATES OF EVOLUTION'

Theoretically, rates of evolution may be measured in terms of changes in various attributes of the phenotype, the production of new taxa or clades, or the properties of the genome. Not only are these three types of rates conceptually distinct, they are not necessarily correlated at any level of analysis. For example, though the proliferation of species over an interval of time may produce a large amount of morphological change, much of this may be diffuse and repetitive, and hence of less consequence for morphological advancement than a smaller proliferation in which the morphological change is more nearly unidirectional. Cladogenetic rates may not match morphological rates. On the other hand, it is well known in modern organisms that small genomic changes can produce large morphological changes, and large genomic changes may produce almost undetectable morphological changes (Baverstock & Adams, Miklos & John, this volume).

It is clear that no one measure of evolutionary rate is more 'correct' than another – they are just different. Great care must be exercised when variation in one type of rate is related causally to variation in another. For example, the production of a large number of classes of echinoderms in the early Palaeozoic is considered by some workers (Paul 1979) to be the result of a high rate of diversification made possible by radiation into largely unoccupied physical space. This hypothesis seems to imply that the large number of classes results from a high rate of speciation. This is not true, as will be shown below. So far as we can determine, high rates of cladogenesis in this instance are assumed but not demonstrated. Nor would it be possible to *demonstrate* that high morphological rates are the result of high genomic rates, simply because the fossil record cannot supply appropriate data. It may be possible to show, however, that a number of morphological changes take place consecutively over a relatively short interval of time to produce a new higher taxon, and that these changes are not correlated with the production of a large number of lower level taxa. In these circumstances we have to assume a high rate of genomic change, or a number of small-scale genomic changes with large-scale morphological effects, or a single large- or small-scale genomic change with large pleiotropic effects. It may never be possible to decide between these options with certainty, but they should be explored in terms of probabilities.

THE STOCHASTIC CAVEAT

The above discussion assumes that rates of cladogenesis *may* be related to such phenomena as biologically unused physical space, and rates of morphological change *may* be related to rates of genomic change, though identifying such relationships will be difficult. Approaches of this type are deterministic. Another approach indicates that variations in environmental factors are not necessary to explain variations in rates of cladogenesis or morphological change; they may result from purely stochastic processes. These arguments are essentially of two types.

The first claims to show that, given empirically determined probabilities of cladogenesis (speciation and extinction), the assumption that change takes place in a continuing series of small steps, and a program for defining clades, it is possible to produce by purely stochastic processes sets of clade diversities that will match many patterns observed in the fossil record. (See Gould *et al.* (1977) for a succinct statement of the method which was first developed by Raup *et al.* (1973)). A similar type of approach has been used by Raup and Gould (1974, p. 320) to show that 'the basic order of morphology on an evolutionary tree need not reflect any special biological process, but arises inevitably (given such assumption as gradual change) from topological properties of the abstract form of the tree itself'. The features of morphological order referred to include the regular unfolding of morphology, marked evolutionary trends *and large variation in the rates of evolution.*

These pieces of work were intended to show that even if diversity and morphological patterns vary within a group in an apparently systematic way, there is no need to seek correlations between changes in the observed patterns and such environmental phenomena as flooding of the continents by the seas, withdrawal of the seas from the continental shelves and major ice ages. The effects of such phenomena are not discounted. The point is that their significance can never be estimated because of the complexity of the variables involved; but even if their predicted effects could be estimated, they could be produced equally well by random processes (Gould & Eldredge 1977, Schopf 1979, Hoffman 1981).

The work on diversity has come under attack by Stanley *et al.* (1981) who claimed that the parameters used in the simulations were not appropriately scaled and that the variations in real clade diversities are almost certainly the result of changes in the probabilities of speciation and extinction. These changes could be genomic or environmental or both. The most general criticism of the use of stochastic models for the study of palaeontological diversities is given by Thomas and Foin (1982). If these authors are correct, it remains worthwhile to seek correlations between patterns and changes in the physical environment despite the fact that causal connections can be established only with various degrees of probability. In fact, it may be possible to show from studies of both the physical evidences recorded in the rocks and the evolutionary behaviour of other groups of organisms preserved in the same stratigraphic sequences, that no major changes in environment were taking place at precisely the time when the group under study was undergoing maximum change. Such an instance would suggest that the controlling factor producing the change was genetic.

The second approach starts from the assumption that the fossil record demonstrates that classes originate early in the history of a phylum and lower taxonomic categories arise at *any* subsequent time. Assuming that evolution proceeds by speciation, that splitting and extinction rates for species can be established empirically and extrapolated indefinitely into past time, and that standing diversity of marine invertebrate species at the present time can be reasonably estimated, Raup (1983) has shown that the early divergence of major groups is an inevitable outcome of tree topology. He is careful to stress two points about this conclusion. The first is that different values for splitting, extinction and standing diversity would produce trees very different in shape from those observed; the second is that the analysis says nothing about the point that the higher taxa are morphologically distant from one another from the time of their appearance. His main conclusion was that there is no *need* to seek ecological or genetic causes for the early establishment of major groups. The fact that some higher groups appear later than predicted by the model is explained by sudden massive extinctions, which reduce standing diversity and hence reset the branching systems from a later date.

Raup (1985, and this volume) has more recently moved away from time-homogeneous models of the above types, largely because of work on the punctuational nature of extinction, but also because difficulties arising from the application of the molecular clock to palaeontological data have encouraged him to consider the possibility of high genomic rates of change at certain stages of evolution. Both environmental and genetic fluctuations are seen as sufficiently probable for him to consider time-homogeneous models as only time-averaged descriptions of the record. A further comment is that the models are largely posited on the assumptions that change takes place only in small steps and that cladogenetic and morphological rates of evolution are correlated. These assumptions are both in need of further investigation.

THE CHOICE OF ECHINODERMS

The echinoderms are one of the best groups to use for rate studies as has been pointed out by Paul (1977). The reasons may be summarised as follows.

(a) All extant and fossil representatives are marine. A high proportion of the representatives of all extant classes occur in shallow seas, although all classes also have many deep-water representatives. Palaeozoic echinoderms are rarely found in rocks that are considered to be of deep-water origin on the basis of independent criteria, whereas they are often extremely abundant in sediments of shallow-water origin. It would therefore be very difficult to argue that the record of the group is significantly incomplete because many major evolutionary events took place in deep water.

(b) Their tests have a characteristic 'stereom' that makes recognition easy in comparison with other groups that also have a carbonate skeleton. The presence of echinoderms in sediments can be detected from fragments and from petrological sections. This pinpoints targets for further study.

(c) Most of them are morphologically complex, making the discrimination of taxa relatively easy and providing a larger number of characters to use for the erection of a range of higher taxa. This is an important factor in rate studies as Schopf *et al.*(1975) have shown. The availability of a large number of characters should be viewed positively for within-phylum studies of echinoderms. Large numbers of characters

permit us to make assessments of fossil taxa similar to those that would
be used by students of extant taxa.

(d) Their basic functions are represented in skeletal features. For example,
parts of the skeletons in most groups can be recognised as structures
related to reproduction (gonopores), respiration (rhombs, diplopores),
locomotion (spines, pores for tube feet), food gathering (ambulacra,
lanterns), excretion (anal pores and tubes) and attachment (stems,
cementing surfaces). This gives the group a considerable advantage over
most molluscs and brachiopods. It would not be possible to ascribe
abrupt changes in the structure of an echinoderm skeleton to slow
accumulation of changes in the soft anatomy of the organism and the
subsequent accommodation of the skeleton when some threshold was
crossed.

(e) A vast amount of work has been done recently on unravelling the morphology of Palaeozoic, and particularly early Palaeozoic, representatives of the group. This work has resulted in new analyses of taxonomy and functional morphology. References to the main studies can
be obtained from Ubaghs (1975), Bell (1976), Sprinkle (1983), Smith
(1984a, b), Paul and Smith (1984), Paul (1972, 1977, 1984) and Jell
(1983). The relationships of the various groups are now moderately
well understood. Although some marked differences of opinion remain,
there is increasing agreement on the close relationship of a number of
higher taxa—echinoids and ophiocistioids, blastoids and coronoids,
edrioasteroids and cyclocystoids, for example.

THE NEED FOR MONOPHYLETIC COMMENSURATE GROUPS

If the evolution of higher taxa (we take the class as an example) is to be used
to imply the establishment of new morphological designs that differ in a
number of characters from all other similar designs, and the rate at which
these are produced is to be used as a measure of rate of morphological
evolution, two requirements must first be met – each class has to be
monophyletic, and the classes have to be approximately commensurate.
The first of these points is obvious; the 'amount' of evolution required to
produce a polyphyletic class may be several times the 'amount' required to
produce a monophyletic one, because the defining characteristics will have
to be produced independently at least twice. On the other hand, rates
estimated using the assumption of monophyly will be minimal, and for present purposes that would not be undesirable. In any case there is little
dispute over the monophyly of most of the groups regarded herein as
classes, except for the rhombiferans and the diploporitans (Sprinkle 1983,
Smith 1984b).

The issue of the commensurate size of the various taxa is more difficult
to resolve. Is the class Blastoidea really the equivalent of the class
Echinoidea? For that matter, what is meant by 'equivalent' in this context?
After all, if a class is only a segregated part of a complex systematic struc-

ture, isolated from all other parts at some unique branching point, their dimensions, like those of other higher taxa, are quite arbitrary. This view stems from the generally held belief, which of course has never been demonstrated, that evolution has proceeded by a large number of essentially comparable small steps. If there have been periods in the history of many, or all, groups when the rate of change was accelerated, and the number of lower taxa involved on the production of such change was small, the great majority of species and genera would fall into higher groups that were *naturally and not arbitrarily* delineated. Of course, we are conditioned to think of evolutionary change in a uniformitarian way, but we should acknowledge that this is a philosophical, and not a scientific, point of view. We need to attempt to establish our models from the records of evolution, and the fossil record provides the only accessible data base at present, though molecular clocks may provide another in the not too distant future.

Several recent publications deal with the major subdivisions of the Echinodermata (see Sprinkle 1983, Smith 1984b for references). Until the work of Smith, most authors agreed that there were about 20 higher taxa commonly referred to as classes. Sprinkle (1983) has discussed the principles used by various authors to rank taxa, and concluded that morphological uniqueness rather than taxonomic diversity should be taken as the dominant factor. But all organisms are unique to some extent, and judgement is still required to determine the magnitudes of the various unique differences. Perhaps the presence of some unique organ or group of organs is what he meant to imply, so that blastoids, for example, would be regarded as a class because of their hydrospire/lancet plate/side plate system. In the view of some workers this leads to other problems. For example, Smith (1984b, p. 441) objected to the way Sprinkle (1983) and Paul (1979) used 'prominent autapomorphies' rather than shared derived characters to recognise higher taxa, because this practice not only obscures relationships, it also leads to the purely subjective judgement that such taxa are of equal rank. It also results in a large number of low diversity classes, which is contrary to expectations if larger morphological changes are the consequence of a large number of small steps; and some classes such as the Eocrinoidea that are formed from relicts after the forms with distinctive organ systems have been removed. For these reasons, Smith (1984b) attempted to use some aspects of cladistic methods to clarify relationships. His work, though salutary in respect of relationships, leaves untouched the problem of subjective judgements that have to be made to produce a taxonomic ranking, an exercise that he would regard as futile (Paul & Smith 1984, p. 477). He used the plesion concept, which may be acceptable if all we wish to do is to establish relationships and avoid the problem of making the Linnaean hierarchy excessively unwieldy. For example, the helicoplacoids were acknowleged by Smith to be a monophyletic group, sister to all other echinoderms except 'carpoids', but he gave them nominal family rank as a plesion, without further discussion. Such an approach is of significance only for the study of relationship, because the ranking ultimately depends on a judgement, the grounds of which are unspecified. In our view, the conclusion reached by cladistic analysis, which is the same as that reached by other analyses, is that helicoplacoids have so many distinctive features (autapomorphies) that they cannot be assigned rationally to any known higher taxon, fossil or extant, apart from the Echinodermata. Hence there is no systematic argument against erecting a higher taxon to contain them if we judge this to be appropriate on other grounds.

Smith's general view (1984b, p. 454) is that 'nominal categorial rank is *unimportant* [our italics] and is best based on diversity or historical precedence'. Presumably the helicoplacoids were considered to have too low a diversity to be given more than family status. It is true that the within-group differences are such that the known genera could be encompassed by a single family, but this sidesteps what for us is the main issue, viz. how different is this family from its nearest relative, and how should this difference be expressed in Linnaean terms? It also unduly emphasises the 'success' of the group. Suppose that it persisted until the end of the Ordovician and produced another ten genera by increasing or decreasing the size of the interambulacral plates overall or on discrete parts of the test, the number and dimensions of the strengthening ridges, the organisation of the cover plates, the position of the mouth relative to the poles, the division of one or more ambulacra at their tips, the tightness of the coiling of the plates and the relative elongation of the whole skeleton. Would it then qualify for a higher taxonomic status? If diversity is the criterion, then the answer is positive. A pattern of genera would be recognised and the elements of the pattern would be given suprageneric status and this would have to be accommodated. But all that would have happened is an alteration of the proportions and relationships of the features of the group that were in existence at its time of formation. The essential character of the group would remain the same – its basic design would remain unaltered. This is defined by the collapsible nature of the test, the spiral form of the plating, the absence of dorsoventral differentiation, the lateral position of the mouth, the triradiate ambulacra, the lack of a defined anus, and the lack of attachment structures at either pole. The point that some of these characters may be inferred to be primitive for the phylum is not relevant to this issue. The fact is that this group has a number of characters not known to occur in any other group and has built them into a distinctive morphological design.

Having said this, we acknowledge that there is still no objective way to assign a rank to the Helicoplacoidea. All that can be said is that its sister group relationship, together with the magnitude of its morphological differences from the Cystoidea and Crinoidea, which are the two groups generally given at least class status within the nearest major sister taxon, the Pelmatozoa, suggest that it should be given a high rank. Class status would not be inappropriate, but even if it were given ordinal rank our point would be made.

It is not possible to treat all the groups that have in the past been given class rank in this way. However, it is clear from the fossil record that a number of monophyletic echinoderm groups, five of which are still extant, can be traced over considerable periods of geological time. These groups are distinguished from one another by the possession of one or more non-homologous characters related to some vital function such as respiration, attachment, food gathering etc., or some highly transformed character with homologues in other groups, in association with a distinctive cluster of characters that individually are found in other groups. However, as has been recognised for centuries, not all seemingly possible character clusters occur in nature. The discrete morphological blocks defined in this way have conventionally been given high taxonomic ranks, usually orders or classes. Not that this reasoning has been consciously followed by taxonomists; nor has it necessarily resulted in the 'equivalence' of taxa given the same rank. A case may be made that some 'designs' have been overvalued, particularly those with attached calyces that carry brachioles. Lower taxa have been identified by some distinctive modification of the number or arrangement of the characters that define the higher group to which they belong. Diversity has not been a factor in these definitions (cf. Strathmann & Slatkin 1983). Despite the lack of precision in the definition of these units, an examination of the distribution of higher taxa through geological time should give a guide to the pattern of morphological diversification in a phylum.

Note that in making this statement we are not implying that higher taxa are absolutely discrete entities – clearly they will have been produced by the transformation of some pre-existing group. We see the significance of the work of Paul and Smith (1984) with regard to the present exercise not in its valuable contribution to the understanding of these transformations, but rather in the way they try to minimise the extent of such transformations and imply that their production is of a piece with intra-group transformation. This is a response to workers in the American 'School' (see Sprinkle 1983, for references) and to Paul's own previous work, which has tended to maximise these features and emphasise their difference from intra-group changes.

THE MAJOR ECHINODERM TAXA

We accept the widely held view that 'carpoids' form a monophyletic group of echinoderms, and that from the point of view of evolutionary rates they may be regarded as a major taxon that includes four major groups – Stylophora, Ctenocystoidea, Homostelea and Homoiostelea. The relationships of these groups have never been elucidated in detail, but Dr R. P. S. Jefferies has a cladistic analysis in press, and Philip (1979) has presented a tentative phylogenetic chart in which they are all independently derived. We propose to regard them as monophyletic taxa that are sufficiently discrete and distinctive morphologically to be regarded as classes. They began their short histories in the Early or Middle Cambrian.

At the other extreme are the extant classes, Crinoidea, Echinoidea, Asteroidea, Ophiuroidea, Somasteroidea and Holothuroidea. Holothuroidea have left a very indefinite fossil record, and hence we do not consider them further. The Somasteroidea are interpreted differently by a number of authors (Paul & Smith 1984), but we accept them as a group within the Asteroidea (Nichols 1962); they are counted as one order within that class in the numerical analysis. Although the Crinoidea is well defined from the Ordovician onwards, there are difficulties with its recognition in the Cambrian. Derstler (1981, p. 73) recorded a hybocrinid-like inadunate from the Late Cambrian. We do not accept *Echmatocrinus* from the Middle Cambrian Burgess Shale as a crinoid – it has too many primary arms, no organisation of the calical plates and an undifferentiated holdfast. In addition, the structures referred to as tube feet may have a different interpretation. Like many of the arthropods from the Burgess Shale, this genus is best regarded as an abortive development that is not part of the sequence that led to the Crinoidea. The first occurrences of the Echinoidea, Asteroidea and Ophiuroidea are well documented.

This leaves a large number of extinct Palaeozoic groups commonly referred to as classes. Several of these are distinctive and their status is generally agreed. We deal with them first.

The Edrioasteroidea is a well defined group whose first members are *Stromatocystites* from the Early Cambrian and *Cambraster* from the Middle Cambrian (Paul & Smith 1984). These two genera were included in one family by Jell et al. (1985). Though the cyclocystoids have been assigned to a single-family class for a number of years by several authors, Smith (1984b) has suggested that they should be regarded as an order within the Edrioasteroidea because they share uniserial ambulacra and a single-layered marginal-plate ring with the isorophids, an acknowledged edrioasteroid group. The large number of distinctive features exhibited by cyclocystoids ensure that as a group they are well separated from their nearest relatives, even if these are the isorophids. From the present point of view it matters little if they are regarded as an order or a class; they represent a significant new design that appeared abruptly in the record.

Among the remaining pelmatozoans the Blastoidea stands out as a well-defined unit characterised by the form of the calyx, the organisation of its plates and the structure of its ambulacra. All modern authors concur in granting it class status. There is no unanimity, however, about the handling of the parablastoids, edrioblastoids and the coronates, all of which have at some time or other been included in the Blastoidea. The parablastoids and the edrioblastoids together include less than five known genera, none of which have the defining characters of the blastoids listed above. Their main similarities with the Blastoidea are the bud-shaped calyx and the disposition (but not the structure) of the ambulacra. They could be equally well regarded as allies of the rhombiferans as the blastoids. Paul and Smith (1984) doubtfully place them with some diploporites. In any case, they are small distinctive groups that should be recognised as such. The same comment applies to the Coronata, recently given class status by Brett et al. (1983). This group is discussed further below.

The Paracrinoidea is a small monophyletic group according to Paul and Smith (1984, p. 468). Its members have distinctive peristomial grooves, a reduced number of uniserial arms or recumbent ambulacra that are simple or branched and carry brachioles on one side only, and a respiratory pore system. They are 'cystoids', but their closer affinities are unknown. Once again it is a small coherent group separated by a group of distinctive innovations involving respiratory and feeding systems. It should be given high tax-onomic rank despite its low diversity.

The remaining pelmatozoans make up a bewildering morphological array and have been assigned over the years to the Class Cystoidea. Other workers have split them into three classes – Rhombifera, Diploporita and Eocrinoidea – largely on the basis of their respiratory structures. However, Paul has argued that rhombs and diplopores have evolved several times and that the eocrinoids are a catch-all group made of remnants after the rhomb and diplopore bearers have been removed. Details of this analysis are not yet available (Paul & Smith 1984), but no matter what the best taxonomic perspective turns out to be, the morphological range of these groups is im-mense. For present purposes it would be possible to group the hemicosmitid and the glyptocystitid rhombiferans together as the 'Rhombifera'. If we ac-cept that fistulipores, which involve a mechanism of pumping body fluids out through the test wall, are more similar to diplopores than pore rhombs, but have evolved independently of either, they could be ignored from the point of view of numerical analysis. We regard the Diploporita as a group including both stemmed and stemless types. This leaves only the Eocrinoidea which certainly is a heterogeneous group. We are unable to use it at present, a point that causes some difficulty because of the position of *Gogia*, a key Middle Cambrian genus often taken as a reference point for the eocrinoids.

Given the above, we accept the following as classes organised into higher categories as shown. We intend disregarding the Eocrinoidea, and tent-atively regard the Parablastoidea and Eridoblastoidea as unassigned orders.

Helicoplacoidea	Homostelea	Rhombiferida	Edrioasteroidea
	Homoiostelea	Diploporita	Cyclocystoidea
	Ctenocystoidea	Blastoidea	Ophiuroidea
	Stylophora	Coronata	Asteroidea
		Paracrinoidea	Ophiocistioidea
		Crinoidea	Echinoidea

GROSS FEATURES OF THE ECHINODERM RECORD

The main features of the distribution of classes, orders and families are set out on Tables 4.1 and 4.2. Note that our correlations place some Scandina-vian and North American occurrences in the Caradocian, which is regarded as Late Ordovician for purposes of our analysis though some workers refer it to the Middle Ordovician. The following points are considered significant.

All classes originated in the Cambrian or the Ordovician. The decline in originations in the Late Cambrian and the latter part of the Middle Cam-brian, an interval of about 25 Ma, is real in our opinion. This interval also produced significantly smaller numbers of new orders and families as Table 4.1 shows. This reduction cannot be explained as the result of the absence of suitable sediments or some global phenomenon inimical to invertebrate life. Families of trilobites and molluscs were more diverse than ever before, brachiopod families showed no diminution, and even the echinoderm families showed only slight reduction (Sepkoski 1979, Fig. 3). These other phyla produced many *new* families, and the nautiloid cephalopods appeared. However, it could be argued that no 'echinoderm beds' similar to those found in later rocks have yet been found in the Upper Cambrian, or that many Upper Cambrian limestones that are rich in echinoderm plates have not yet yielded whole specimens. Jell *et al.* (1985), for example, record an Upper Cambrian isorophid edrioasteroid from Tasmania, and the limestones from western USA are thought to have more new high-level taxa. However, for the present, the existence in rocks of this age of sufficient new groups to smooth out the origination rates has not been demonstrated.

At least eight new classes originated between the Arenig and the Caradoc, and over that period of about 55 Ma the origination rate does not vary significantly. But there are obvious differences between the Cambrian and Ordovician episodes. Of the Cambrian classes, four are 'carpoids' (or homalozoans as they are sometimes known) and one is the Helicoplacoidea, all of which are only remotely similar to living echinoderms and had only short existences. Of the remaining classes, the crinoids and the edrioasteroids are represented by only a few specimens that are extremely primitive and are not fully representative of the developed morphologies of their groups. The remainder, which we can group as 'cystoids', were far more successful in producing a variety of types, but most of these were short lived. The new Ordovician classes include the ophiuroids, asteroids and echinoids, which give the group a more modern aspect, and the rhombiferans, diploritans and blastoids, which diversify to some extent shortly after their establishment. This information is summarised in graphical form by Paul and Smith (1984, Fig. 14). The average longevities of the classes originating in the two episodes are approximately the same, whether the classes that persist to the present day are included or excluded. On the other hand, the orders and families that originate in the Ordovician have a higher average longevity than those that originate in the Cambrian. This is unlikely to be the result of any systematic bias on the part of taxonomists.

This analysis demonstrates that though the Cambrian classes differed widely in morphology they were unable to take advantage of environments that were open for exploitation. Most Ordovician classes were comparable in this respect, but more of their orders and families were able to diversify and persist despite the fact that the physical environment was less open.

Classes became extinct at a more or less uniform rate throughout the Palaeozoic, with an increase of doubtful significance in the Middle and Late Devonian. It is not possible to ascribe the relatively short histories of some of the classes to catastrophic extinction events. Nor is it possible to support the view that some classes became extinct as the result of rapid diversification of other echinoderm classes or other phyla. Non-diverse groups are always at the mercy of randomly occurring events, and the extinction pattern observed is consistent with such an explanation.

Orders continued to originate until the Middle Devonian, but like the classes they reached their peak of origination in the Ordovician. They also continued a steady decline by extinction till the end of the Permian.

The origination rates of families, though low in the Cambrian, were high from the Ordovician until the Late Carboniferous. The high figures for the Devonian and Carboniferous come from only four classes – Crinoidea, Asteroidea, Ophiuroidea and Blastoidea – though the 'cystoids' influence the Early Devonian figures. The crinoids account for more of the families from the Ordovician–Permian than all the other classes together. No class that originated in the Cambrian or the Tremadoc could have resulted from the diversification of a pre-existing class; no diverse classes were available. Not only were there few families, many families contained less than five genera. For the classes that originated after the Tremadoc it can be shown that no possible ancestor was diverse in any preceding epoch. New classes – new structural designs – do not appear out of a matrix of diversification.

Nor did new designs diversify during the epoch in which they originated as is shown on Table 4.3. Eight classes produced only one order and less than ten families throughout their history (Table 4.4). The three classes that are diverse today – asteroids, ophiuroids and echinoids – did not diversify during the Palaeozoic. Their diversities remained low until the Mesozoic or Cenozoic. This is particularly well shown for echinoids at the species level by Smith (1984a, Fig. 8.1). Such data indicate that the classes did *not* cross some threshold that enabled a new structural design to exploit new parts of the environment and thus undergo a radiation. Large morphological change and high rates of cladogenesis are not coupled in any sense.

Nor were the early representatives of most of the classes, particularly those from the Cambrian, abundant or widespread. For example, the work of Smith and Paul (1982) on cyclocystoids concluded that they really were minor elements in the fauna; this probably accounts for the point that the genera have a limited known geographical distribution. Some northern hemisphere Cambrian taxa have recently turned up in Australia (Jell *et al.*

Table 4.3 Number of orders and families produced per class during the Epoch in which the class originated. If each class originated towards the end of an epoch the numbers would inevitably be low. However, correlations are sufficiently good to demonstrate that this is not a significant factor in producing the skewed distribution observed.

Classes		Orders
15	with	1
1	with	2
1	with	3
Classes		Families
9	with	1
4	with	2
2	with	3
1	with	4
1	with	5

Table 4.4 Total number of orders and families in the recognised classes throughout the Palaeozoic.

	Orders	Families		Orders	Families
Crinoidea	7	167	Diploporita	1	9
Asteroidea	6	26	Homoiostelea	1	6
Ophiuroidea	4	18	Ophiocistioidea	1	5
Rhombifera	3	16	Homostelea	1	2
Edrioasteroidea	3	12	Helicoplacoidea	1	1
Stylophora	3	9	Ctenocystoidea	1	1
Echinoidea	3	7	Coronata	1	1
Blastoidea	2	13	Cyclocystoidea	1	1
Paracrinoidea	2	5			

1985). These include the genera *Cambraster* and *Ctenocystis* which have their previously recorded geographical range considerably extended. These discoveries well illustrate two important points – contrary to what might be expected they can be assigned to existing low level taxa and do not expand the morphological range of any higher taxon and, where found, they are still minor elements of the fauna.

We can summarise the above points as follows in so far as classes represent major morphological designs, and lower taxa represent modifications of such designs:

(a) The basic designs were established in two bursts of about 35 and 55 Ma during the early history of the phylum.
(b) Within these two intervals, originations were more or less uniformly dispersed. There is no evidence of some periodically operating mechanism controlling their origin.
(c) They are not generated during periods of high cladogenetic rates, and they cannot be said to be formed during periods of evolutionary radiation.
(d) Nor did their establishment result in cladogenesis. They do not mark morphological thresholds which, when crossed, permitted evolutionary radiations to take place immediately or even after intervals of 100–200 Ma.

DO CLASSES CONVERGE TOWARDS THEIR TIMES OF ORIGIN?

The above analysis implies that classes were established quickly. The time involved cannot be determined with precision, but it may have taken significantly less than 5 Ma for their designs to be assembled. Subsequent evolution during the Palaeozoic just rings the changes within the limits imposed by these designs. We are not suggesting that classes sprang fully formed from the brow of Zeus, but it does seem possible that class designs were established so quickly, and with so few steps in groups of such low diversity, that they will appear to have been obviously distinct from one another from the time of their inception.

This raises the difficult problem of measuring the morphological distances between classes at various times during their histories and demonstrating that these do not decrease towards the early Palaeozoic. Estimates of the morphological distance between ranked extant vertebrate taxa have been attempted by several workers (see Cherry et al. 1982), but their approach is not applicable to the current problem. The vertebrate skeleton is essentially composed of elements that are represented in all groups, and hence measures of similarity can be estimated without too much difficulty. Echinoderm classes have many elements that are incommensurate. For example, the coelomic respiratory structures – hydrospires, diplopores and coronal canals – are unique. The problem extends even to character clusters that are thought to be homologous but are grossly different in detail, for example lanterns in echinoids and ophiuroids which appeared at about the same time. No attempts have yet been made to make such estimates of morphological distance, and this remains a most significant task.

Meanwhile, qualitative approaches to this problem suggest that there are few grounds for believing that related classes show significant morphological convergence towards their times of separation. Obviously it is not possible to canvass the whole spectrum, but an example of two groups that have been regarded as close relatives is instructive. We refer to the classes Blastoidea and Coronoidea. Sprinkle (1980), Brett et al. (1983) and Paul and Smith (1984) argued that the blastoids were derived from coronoids, and Paul (1985) has recently suggested that the Coronata should be regarded as a subclass of the class Blastoidea. It is worth quoting Paul and Smith (p. 647) on the basis for this relationship.

Coronates and blastoids have identical cup plating, pore structures across radial: radial sutures (although of different types) and similar ambulacral structures in which homologous structures can be recognised. Coronate ambulacra are erect, pinnate, biserial arms, but arise from a single large primary ambulacral plate which we believe is the homologue of the blastoid lancet. To derive a blastoid ambulacrum merely requires elongating the primary ambulacral plate and developing a recumbent ambulacrum upon it.

The calyces of the two groups are remarkably similar, but the essential point is that there is no evidence from the record that the groups are any less distinctive when they first appear than later in their history. The recent description of Macurdablastus from the Caradoc (Broadhead 1984), which is apparently a true blastoid, has considerably extended the time range of the group, but it only emphasises the point that the blastoids and the coronates were distinct from early in their histories. Macurdablastus has typical hydrospires (though they have a reduced number of folds), the ambulacra are totally recumbent, the lancet is present and is probably complex. Unfortunately, nothing is known of its brachioles or its side plates, and so it is not known if the genus is spiraculate or fissiculate. However, it is clearly a member of the Blastoidea and shows no convergence on the earliest Coronoidea which date from the Caradoc also.

In addition to this lack of convergence, we must also take into account two further points. It would be impossible to transform adult coronal canals into hydrospires or vice versa, and the ambulacra of the coronates are multibranched arms which apparently bear neither pinnules or brachioles, whereas blastoid ambulacra consist of unbranched side plates with brachioles. Hence the suggestion of Paul and Smith that coronate arms could be easily transformed into blastoid ambulacra faces considerable problems. It is far more likely that the two types of respiratory structures and ambulacra developed separately from some unknown ancestral group. This is supported by the fact that they appear at almost the same time in the geological record.

We conclude that these two classes, which we acknowledge are probably as closely related as any, do not converge towards their times of origin; they were morphologically widely separated in features related to major functions, namely feeding and respiration. In our opinion, similar cases could be made for other pairs of echinoderm classes.

CHARACTER ANALYSIS

The taxonomic analysis given above may be considered objectionable because it involves many judgements about character combinations and their relative value in determining relationships. Some of these objections may be overcome by an analysis of the time taken for the Palaeozoic character transformations of homologous structures to occur. This would be independent of taxonomic considerations, and hence would give an independent estimate of evolutionary rates. For such an examination to be valid, it would obviously have to consider characters that arose at any time during the Palaeozoic. All characters must have appeared either *ab initio* from previously undifferentiated tissue, or as stages in a transformation series from one of these characters. Take, for example, the covers over the ambulacra in echinoderms. According to Paul and Smith (1984, Figs 5, 13) these were initiated as sheets which subsequently broke down into irregular plates; these plates became organised into pairs and subsequently some of the pairs formed the basis of biserial brachioles; so far as we are aware, they tranformed no further. Cover sheets appeared in the Early Cambrian and biserial brachioles were fully developed by the end of the Early Cambrian. Biserial brachioles may have arisen more than once, but there is no strong evidence to support such a view. Brachiole-bearing echinoderms persisted till the end of the Palaeozoic, and so far as is known they retained the same basic structure throughout this long period, apart from the development of branches in *Bockia* and its allies (Bockelie 1981), and the development of uniserial brachioles, e.g. in the Rhipidocystidae. The cover plates that did not become modified to form brachioles underwent a variety of changes of pattern, becoming adapted to the different styles of ambulacra evolved. For example, blastoids have paired rows of plates corresponding in position with the food grooves on the side and lancet plates (Beaver *et al.* 1967); the diploporite *Protocrinites* has four plates attached to each brachiolar plate (Bockelie 1984, p. 18); and many groups develop a simple biserial pattern of plates. Cover plate patterns are not well known for many groups, but so far as we are aware no new patterns were introduced after the evolution of the blastoids, assuming that all blastoids had similar patterns. All these changes therefore took place by the Middle Ordovician, after which there was a long period of stability.

A second example involves coelomic respiratory structures. Most extant echinoderms are able to respire across the body wall where it is thin enough, and probably this was the primitive echinoderm condition. Modification of the body wall to produce more effective exchange surfaces is a feature of many groups. These modifications are essentially of two types, those that involve pumping body fluids through thin-walled structures outside the main thecal surface, and those that involve pumping sea water through thin-walled structures inside the body cavity. Each of these has secondary level structures that are modifications of the basic pattern, e.g. open or spiraculate hydrospires, conjunct or disjunct pore rhombs, diplopores etc. Some of these secondary level structures may have been derived from another one; for example, spiraculate hydrospires may have evolved from open hydrospires. Within each of these types, modifications involve only minor changes such as an increase or decrease in the number of hydrospire folds, but the changes between the two types – that is from open to closed hydrospire chambers – require major modifications to allow circulation of the water.

We have attempted to examine a number of skeletal features in this way. The choice of the starting points for the transformation series may be a matter for some concern. They ought to be new structures in the sense that they have arisen from previously undifferentiated structures, but it is sometimes difficult to apply this criterion. For example, if we accept that lanterns evolved from mouth marginal plates with spines (Smith 1984a) we have still to discover the origin of the spines. Were these spines part of an even more primitive skeletal wall or did they appear *de novo*? We cannot answer such questions in the absence of suitable material. This is a matter of some moment for the present study because all such problems relate to features in the early Palaeozoic.

Our first attempts at an analysis of the temporal distribution of structural initiations were in terms of functional groups, following the method used by Paul (1977, 1979). However, many structures have two or three functions which were assumed at different times. For example, tube feet are used for feeding, respiration and locomotion, and there is good reason to believe that these functions were adopted in that sequence; muscles in crinoid brachia were first used to produce better controlled mobility for feeding and respiration, but subsequently were used for swimming. Such functional changes obviously would have required some structural changes, but in most instances these need not have been great. For example, muscular crinoid arms used for swimming are structurally little different from those

Table 4.5 Table showing the first known occurrence of the primitive and progressively more evolved characters in 14 sequences which cover the main aspects of Palaeozoic echinoderm morphology. The name of the taxon in which the first occurrence is known is shown in parentheses. This is generally a higher taxon, but the genus is given for several Early Cambrian occurrences. Arrows indicate the sequence of character origination, *not* proposed taxon relationships.

Character	Sequence of character origination

1 *Gross skeletal organisation*

irregularly multiplated flexible theca without ambulacra (Hypothetical) ⟶ irregularly multiplated flexible theca with ambulacra (Helicoplacoidea E. Camb.) ⟶ irregularly multiplated flexible theca with oral and aboral surfaces differentiated (Stromatocystitids E. Camb.)

theca free, oral side under, 'anus' on oral side (Ophiuroidea E. Ord.) ⟶ suppression of dorsal plating, test globular, oral side under, anus aboral (Echinoidea Late Ord.)

marginal plates and skirt of attachment, anus on oral side (*Cambraster* M. Camb.) ⟶ dorsoventral differentiation maintained, oral side under, anus aboral (Ophiocistioidea E. Ord.)

irregularly multiplated calyx, stem on aboral side, anus on oral side (Lepidocystoids E. Camb.) ⟶ same, but with more regularly plated rigid calyces as in blastoids, crinoids and some rhombiferans (Late Camb. —E. Ord.)

2 *Plated ambulacra*

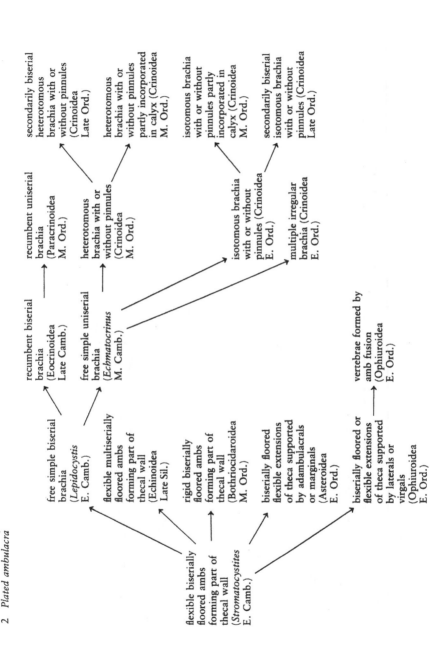

(*Continued*)

Table 4.5 (Continued).

Character	Sequence of character origination

3 *Ambulacral symmetry of integrated ambulacra*

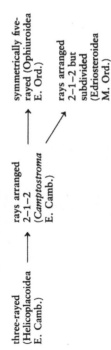

three-rayed
(Helicoplacoidea
E. Camb.) → rays arranged
2–1–2
(*Camptostroma*
E. Camb.) → symmetrically five-
rayed (Ophiuroidea
E. Ord.)

rays arranged
2–1–2 but
subdivided
(Edriosteroidea
M. Ord.)

4 *Ambulacral movement systems*

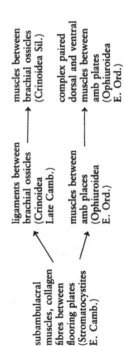

subambulacral
muscles, collagen
fibres between
flooring plates
(Stromatocystites
E. Camb.) → ligaments between
brachial ossicles
(Crinoidea
Late Camb.) → muscles between
brachial ossicles
(Crinoidea Sil.)

muscles between
amb places
(Ophiuroidea
E. Ord.) → complex paired
dorsal and ventral
muscles between
amb plates
(Ophiuroidea
E. Ord.)

5 *Ambulacral cover plates*

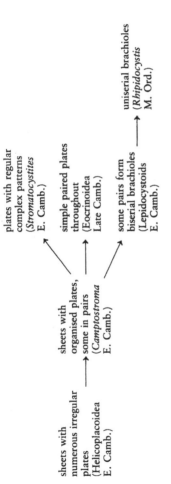

sheets with numerous irregular plates (Helicoplacoidea E. Camb.) → sheets with organised plates, some in pairs (*Camptostroma* E. Camb.)

→ plates with regular complex patterns (*Stromatocystites* E. Camb.)

→ simple paired plates throughout (Eocrinoidea Late Camb.)

→ some pairs form biserial brachioles (Lepidocystoids E. Camb.) → uniserial brachioles (*Rhipidocystis* M. Ord.)

6 *Radial water vessels and tube feet*

external vessel with ampullae through amb plates (*Camptostroma* E. Camb.)

→ totally external vessel with no amb perforations (Eocrinoidea M. Camb.)

→ vessel buried in thecal wall, but with openings for ampullae and tube feet (Echinoidea Late Ord.) → vessel internal but with simple openings for tube feet (Echinoidea E. Sil.)

→ vessel buried in thecal wall but with openings for tube feet only (Ophiuroidea E. Ord.)

(*Continued*)

Table 4.5 (Continued).

Character	Sequence of character origination

7 Brachioles

biserial brachioles on each side of ambulacrum (*Kinzercystis* E. Camb.) → uniserial brachioles on each side of ambulacrum (*Rhipidocystis* M. Ord.) → uniserial brachioles on one side of ambulacrum (Paracrinoidea M. Ord.)

8 Hydropores

simple opening through flexible surface (Helicoplacoidea E. Camb.) → opening through sutures (Lepidocystoids E. Camb.) →
- hydropore plate in C–D interray (Stromatocystitoids E. Camb.)
- madreporite plate (Ophiuroidea E. Ord.)
- opening into spiracle (Blastoidea M. Ord.)

9 Lanterns

mouth margin plates unmodified but bearing spines (Hypothetical) → modified first amb mouth angle plates and spines (Ophiuroidea E. Ord.) →
- ?odontophores (Ophiuroidea Ord?)
- hemipyramids, epiphyses, rotulae and teeth (?Echinoidea E. Sil.) → same plus compasses (Echinoidea Sil.?) → upright pyramids with modified tooth structure (Echinoidea Carb.?) → same but with perignathic girdle (Echinoidea Perm.)

10 *Coelomic respiratory structures*

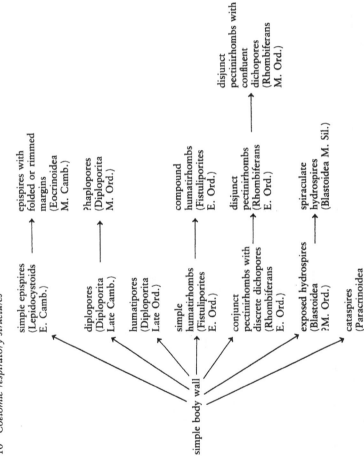

(Continued)

Table 4.5 (Continued).

Character	Sequence of character origination

11 *Genital structures*

multiple internal
gonads opening
through multiple
apical genital pores
(Echinoidea
E. Dev.)

multiple internal
gonads opening
through bursal
around mouth
(Ophiuroidea
E. Ord.)

multiple external
gonads on brachia
(Crinoidea
Late Camb.?)

single internal
gonad opening
through pore in
C–D interray
(*Stromatocystites*
E. Camb.)

(Continued)

12 *Excretory structures*

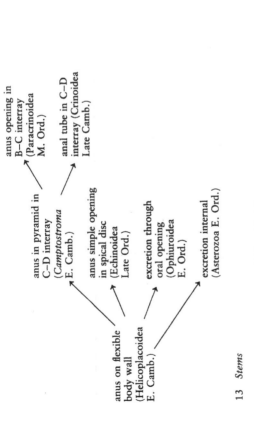

anus on flexible body wall (Helicoplacoidea E. Camb.)

anus in pyramid in C–D interray (*Camptostroma* E. Camb.)

anus opening in B–C interray (Paracrinoidea M. Ord.)

anus simple opening in spical disc (Echinoidea Late Ord.)

anal tube in C–D interray (Crinoidea Late Camb.)

excretion through oral opening (Ophiuroidea E. Ord.)

excretion internal (Asterozoa E. Ord.)

13 *Stems*

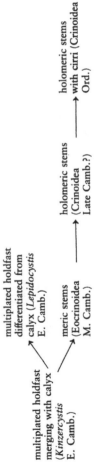

multiplated holdfast merging with calyx (*Kinzercystis* E. Camb.)

multiplated holdfast differentiated from calyx (*Lepidocystis* E. Camb.)

meric stems (Eocrinoidea M. Camb.)

holomeric stems (Crinoidea Late Camb.?)

holomeric stems with cirri (Crinoidea Ord.)

Table 4.5 (Continued).

Character	Sequence of character origination
14 *Spines*	

differentiated cover plates (Hypothetical) ——→ simple marginal adambulacral spines (Echinoidea Late Ord.) ——→ larger ambulacral spines on bosses (Echinoidea Late Sil.)

simple spines on interambulacral plates (Bothriocidaroids M. Ord.) ——→ larger interambulacral spines on bosses (Echinoidea Late Sil.)

? ——→ simple pedicellariae (Echinoidea Late Sil?) ——→ complex pedicellariae (Echinoidea Carb.)

? ——→ paxillae (Ophiuroidea Ord.?)

in related attached forms. Consequently, we have chosen to examine the characters independently of their function.

The results are given in Table 4.5. It is obvious that almost all of the structures were represented before the end of the Ordovician, though one evolved as late as the Permian. This outcome is not a consequence of terminating the series arbitrarily. Take the example of calyx shape, which is terminated at the end of the Ordovician with 'regularly plated calyces as in blastoids, crinoids and rhombiferans'. The basic number of blastoid plates was present in *Macurdablastus* and subsequent changes involved a few minor variations in the deltoids, though some details of the ambulacral plates may also have become modified; camerate, inadunate and flexible crinoids with all the variety of monocyclic and dicyclic cups, tegmens, and tubes and incorporated brachia, were already present, and subsequent changes involved only a reworking of these basic elements. Similar comments apply to other end points of our series. Nor have we excluded any distinctive structures that appeared for the first time in the later Palaeozoic. We are confident therefore that by far the greatest morphological changes in the phylum took place before the end of the Ordovician.

A second point is that many of the series were complete within an even shorter period of time. For example, the range of ambulacral symmetries, ambulacral cover plates, stems, steles, etc. were developed over much shorter periods of time.

One issue that is not brought out by this method, however, is the speed with which new characters in the series are transformed from their predecessors. An ancestral character may have persisted for a long period and then changed rapidly. In fact, such seems to be the norm even for the most persistent series. For example, take the coelomic respiratory structures which are shown as developing from an undifferentiated wall mainly in the Early and Middle Ordovician. There is no evidence that these structures evolved gradually through the Cambrian. They apparently differentiated rapidly from the primitive condition which persisted in separate stocks through most of the Cambrian. What is clear from Table 4.5 is that, once initiated, the new structure was elaborated rapidly.

Hence we conclude that, so far as the Palaeozoic history of the phylum is concerned, major structural changes are concentrated in the Cambrian and Ordovician, that within this interval these changes appeared abruptly, and that once initiated many of them were quickly elaborated. This is what would have been expected if the taxonomic analysis given in the preceding section reflected the morphological history of the group rather than some arbitrary arrangement favoured by taxonomists.

OTHER GROUPS

We draw attention to the fact that the Echinodermata is not anomalous in the large number of structural associations and distinctive structures that appeared early in the history of the group. Striking examples are to be found among the early jawed vertebrates as well as various invertebrates. For example, a variety of non-placoderm bony fishes – the 'rhipidistians', lungfishes and actinopterygians – appeared unheralded in the Early Devonian. Quite remarkably the earliest representatives are well ossified, the lungfishes in particular being more heavily ossified then than at any time in their subsequent history. It is difficult to argue that they had a long sequence of unpreserved antecedents, especially as other 'fishes' are well known in older rocks which are of a type that would record any bony organisms present at the time.

Perhaps the best comparison, however, is with the flood of organisms from the Burgess Shale. These include many taxa that would normally be referred to the Arthropoda but which cannot be included in any of the classes defined on the basis of subsequent fossil or extant organisms (see Whittington 1980, for a review). These Burgess organisms are not only characterised by unusual character combinations, they have a number of 'bizarre' characters which have been grossly misinterpreted by workers who use extant taxa as the basis for interpretation (Whittington & Briggs 1985). The arthropods clearly show a remarkable short-lived burst of structural and taxonomic diversification in the Cambrian on a scale similar to that shown by the echinoderms in the Cambrian and Ordovician.

POSSIBLE MECHANISMS

Assuming that these results are real, there are four fundamentally different ways of interpreting them.

(1) The first is the model of Paul (1979) which suggests that colonisation and radiation took place under conditions of low competition in the early Palaeozoic when even relatively inefficient designs were successful: then followed a period of competition and retrenchment of the least efficient ones. The model was tested against functional data related to protection, respiration and feeding. Radiation was discussed in terms of diversification following the evolution of structures to permit high- and low-level filter feeding and deposit feeding.

In our view, there is little evidence in the fossil record to support the concept of radiations in the sense of a large number of taxa exploiting new environments. For example, the suggestion that the evolution of lanterns enabled the echinoids, asteroids and ophiuroids to radiate because they had adopted a deposit feeding habit, is not supported by the distribution of genera or species in the fossil record (Smith 1984a, Fig. 8.1). The crossing of a threshold seems to have resulted in no diversity explosion. Nor is there evidence to support the hypothesis that open space is *the* critical factor in producing rapid large-scale change. And it remains to be demonstrated that any of the small groups that disappeared in the early and middle Palaeozoic were less efficient than the ones that remained, despite attempts to prove this (Paul 1968, 1972, 1975, 1979). Although it may be possible to show that endothecal rhomb-type respiratory structures became more efficient with time, or that endothecal mechanisms were inherently more efficient than exothecal ones, it is difficult to show that any given species with endothecal rhombs was respiring with greater or less efficiency than one with diplopores or hydrospires. This is partly because many echinoderms are known to use more than one method of respiration, but also because efficiency measures that permit precise comparisons between types have not been developed. The first point is emphasised by the fact that some of the latest rhombiferans, instead of producing more efficient rhombs lost them completely (Jell 1983); they presumably respired by some quite different means, such as a pumping mechanism using the flexible multi-plated wall.

On the other hand, small groups are prone to random extinction (Valentine 1980) and of course this could happen at any time. As indicated above, the pattern of class extinctions suggests a random rather than a systematic process. The interpretation of Paul, therefore, which emphasises the primary rôle of the environment in producing fluctuations in evolutionary rates, has little to commend it. It involves the reading of a Neo-Darwinian interpretation into the data.

(2) The second interpretation is also Neo-Darwinian. Its protagonists are impressed by the point that apparently large morphological changes may occur with minimum genetic change; it is known that experimental animals can produce large morphological changes over a few generations with little genetic change. Therefore the time limits proposed by palaeontologists for the development of new classes are sufficient for any required genetic and morphological changes to accumulate. In their view, rates may be high because the sequence of such changes is unusually continuous over the period of time concerned, but the mode remains the same (Carson & Templeton 1984). Some workers are also impressed with the view that these changes may occur in small representatives of a phylum, before size imposes physiological constraints on morphology (Runnegar, this volume); because such small organisms are difficult to find as fossils, there is a sequence of connecting morphologies that remains to be discovered. Still others are concerned that developmental constraints are so strong that the size of any possible viable genetic or morphological change is highly restricted (Anderson, this volume).

It is interesting to note that these approaches represent three quite different philosophical positions. The first is that mechanisms known at present are adequate and others are therefore unnecessary to explain the morphological rates required to produce classes; the second endorses this view but adds that the record, or our knowledge of it, is faulty, and hence the time involved and the number of steps available have been seriously underestimated; the third is that known mechanisms will not permit any changes other than small ones. The first stresses that there is *no need* for a new hypothesis, the second that a faulty *data base* impels people to seek a new hypothesis, and the third that present understanding makes certain types of postulated changes *impossible*.

We see none of these objections as vetoes. Historical science is always subject to such criticisms, and rightly so. But current understanding of mechanisms should never be allowed to veto the search for new hypotheses, and under no circumstances should they cause us to abandon tested observations. These are lessons that the controversies over the age of the earth and continental drift should have taught us. Some of our observations, particularly the absence of a large number of small steps, are opposed to the main tenet of proposed Neo-Darwinian hypotheses. It is no defence to find fault with the fossil record; it is always possible to fall back on this expedient without clear justification. Finally, we still have to find a justification for the observation that 5–10 Ma in the early Palaeozoic produced morphological changes of a greater magnitude than 5–10 Ma at any subsequent time in the Palaeozoic. There is ample room for new hypotheses.

(3) We have dealt with the stochastic interpretation above. In summary, the proponents of this view are moving away from time-homogeneous models towards descriptions that allow variation in rates of extinction and splitting (Raup, this volume). The former models are being regarded as time-averaged descriptions. Some workers have difficulty with both homogeneous and non-homogeneous models on two counts. They *assume* that large-scale morphological change takes place in a series of small steps, and they extrapolate from speciation models to patterns derived from studies of higher taxa so that there may be a scaling problem (Stanley *et al.* 1981).

(4) A fourth type of explanation approaches the problem with the flexible point of view that the factors controlling both rate and mode of evolution may themselves vary during different evolutionary phases (Valentine & Campbell 1975). This approach has a great deal to commend it, because it opens the greatest possible number of areas to further investigation. Valentine (1980), for example, accepted the proposition that the record shows a rapid diversification of higher taxa and subsequent fill-in by lower taxa, and he attempted to show by means of modelling that given (a) the possibility of change by a series of morphological step sizes the frequency of which is described by a logistic curve, the small steps being the most frequent, and (b) an equilibrium number of taxa (including higher taxa) for any given area, then only a few large changes will take place before the environment is filled. Given these assumptions, the patterns of taxonomic diversification are to be explained by a process that is deterministic and involves regulation by massive variations in the interaction between environmental (ecological space) and genetic (step size) factors through the Phanerozoic. Of these two factors, Valentine considered that the latter was probably 'the most significant single factor regulating diversity at higher taxonomic levels'. As well as demonstrating how diversification may take place, the model also predicts that small classes or phyla will occasionally be lost, and that this will occur shortly after appearance. The probability of small higher taxa persisting is low, though the record has many examples of persistent low-diversity classes. The model is heavily dependent on the possibility of large changes and on the concept of a taxonomic diversity equilibrium.

Such an approach also has its difficulties. It provides no genetic mechanism for large-step changes, though these are said to be probably the most significant single factor regulating diversity at higher taxonomic levels. (We do not regard this as a serious criticism; rather it is a spur to explore the possible genetic mechanisms.) There is no strong theoretical or observational evidence that we know of for the view that only a few large steps will take place before the environment is filled; on the contrary, intuitively it seems probable that if classes and orders did not become internally diverse, the number of such higher taxa could become quite large. Nor is there any evidence that the environment in the Cambrian was rapidly filled before some of the classes became extinct. The issues have been discussed recently by Hoffman (1985).

CONCLUSIONS

In summary, therefore, we reach the conclusion that none of the above explanations is satisfactory. We are not in a position to offer a better one, but we believe that it is now possible to state more clearly the points to be encompassed by any explanation. These are as follows. It must:

(a) provide a means for rapidly differentiating new structures from previously undifferentiated ones; that is, it must show how new organ systems such as respiratory trees or hydrospires appear rapidly from previously undifferentiated tissue;
(b) indicate why it is that only a few of the mathematically possible combinations of previously evolved characters are found in association with each of these new structures;
(c) enable these character combinations to become established during the early phases of the evolution of the phylum, but become much less frequent during later phases;
(d) decouple rates of cladogenesis from rates of morphological change;
(e) indicate how high morphological rates can be related to, but not initiated by, low standing diversities at all taxonomic levels;
(f) indicate how functional thresholds can be crossed without any evidence of rapid diversification.

There seems to be no alternative but to seek some unusual feature of the primitive genome that would allow it to change in such a way that large coordinated viable morphological changes could take place over short periods of geological time. Such changes would not be initiated by environmental factors; nor would selection, in the normal population biology sense, take place to establish the resultant groups against competing groups. The environment would act in a relatively passive way to accept or reject a group rather than as a filter to remove marginally less efficient competing groups. Such changes would apparently be possible only during the early stages of the development of the genome. Subsequent evolution would limit the possibility of such changes while increasing the possibility of diverse small-scale changes.

ACKNOWLEDGEMENTS

We wish to acknowledge assistance from Dr Peter Jell, National Museum, Melbourne, who discussed several matters with us and provided manuscripts in advance of publication; and Dr Andrew Smith, British Museum (Natural History), who provided a great deal of information following a detailed criticism of an earlier draft. We remain entirely responsible for the data and their interpretation. We have benefited from discussions with Dr John Tipper of our own department.

REFERENCES

Beaver, H. H., R. O. Fay, D. B. Macurda, Jr., R. C. Moore and J. Wanner 1967. Blastoids. In *Treatise on Invertebrate Paleontology*, Part S Echinodermata 1(2), R. C. Moore (ed.) pp. 297–455. The University of Kansas and Geological Society of America.

Bell, B. M. 1976. A study of North American Edrioasteroidea. *NY State Museum Sci. Service, Mem.* 21, 1–447.

Bockelie, J. F. 1981. The Middle Ordovician of the Oslo region, Norway, 30. The eocrinoid genera *Cryptocrinites, Rhipidocystis* and *Bockia*. *Norsk. Geol. Tidsskrift* 2, 123–47.

Bockelie, J. F. 1984. The diploporita of the Oslo region, Norway. *Palaeontology* 27, 1–68.

Brett, C. E., T. J. Frest, J. Sprinkle and C. R. Clement 1983. Coronoidea: A new class of blastozoan echinoderms based on taxonomic re-evaluation of *Stephanocrinus*. *J. Paleont.* 57, 627–51.

Broadhead, T. W. 1984. *Macurdablastus*, a Middle Devonian blastoid from the Southern Appalachians. *Univ. Kansas Paleontol. Contrib.* 110, 1–9.

Carson, H. L. and A. R. Templeton 1984. Genetic revolutions in relation to speciation phenomena: the founding of new populations. *Ann. Rev. Ecol. Syst.* 15, 97–131.

Cherry, L. M., S. M. Case, J. G. Kunkel, J. S. Wyles and A. C. Wilson 1982. Body shape metrics and organismal evolution. *Evolution* 36, 914–33.

Derstler, K. L. 1981. Morphological diversity of Early Cambrian echinoderms. In *Short papers for the second international symposium on the Cambrian system, 1981*, M. E. Taylor (ed.), US Geol. Surv., Open File Report, No. 81–743, 71–5.

Glaessner, M. F. 1984. *The dawn of animal life: a biohistorical study*. Cambridge: Cambridge University Press.

Gould, S. J. and N. Eldredge 1977. Punctuated equilibria: the tempo and mode of evolution reconsidered. *Paleobiology* 3, 115–51.

Gould, S. J., D. M. Raup, J. J. Sepkoski, Jr., T. J. M. Schopf and D. S. Simberloff 1977. The shape of evolution: a comparison of real and random clades. *Paleobiology* 3, 23–40.

Harland, W. B., A. V. Cox, P. G. Llewellyn, C. A. G. Pickton, A. G. Smith and R. Walters 1982. *A geologic time scale*. Cambridge: Cambridge University Press.

Hoffman, A. 1981. Stochastic versus deterministic approach to paleontology: The question of scaling or metaphysics? *N. Jb. Geol. Palaont. Abh.* 162, 80–96.

Hoffman, A. 1985. Biotic diversification in the Phanerozoic: diversity independence. *Palaeontology* 28, 387–92.

Jell, P. A. 1983. Early Devonian echinoderms from Victoria (Rhombifera, Blastoidea and Ophiocistioidea). *Mem. Assoc. Australas. Palaeontols* 1, 209–35.
Jell, P. A., C. F. Burrett and M. R. Banks 1985. Some Cambrian and Ordovician echinoderms from eastern Australia. *Alcheringa* 9, 183–208.

Nichols, D. 1962. *Echinoderms.* London: Hutchinson University Library.
Nicol, D., G. A. Desborough and J. R. Solliday 1959. Paleontologic record of the primary differentiation in some major invertebrate groups. *J. Wash. Acad. Sci.* 49, 351–66.

Paul, C. R. C. 1968. *Macrocystella* Callaway, the earliest glyptocystitid cystoid. *Palaeontology* 11, 580–600.
Paul, C. R. C. 1972. Morphology and function of exothecal pore-structures in cystoids. *Palaeontology* 15, 1–28.
Paul, C. R. C. 1975. A reappraisal of the paradigm method of functional analysis in fossils. *Lethaia* 8, 15–21.
Paul, C. R. C. 1977. Evolution of primitive echinoderms. In *Patterns of evolution,* A. Hallam (ed.), pp. 123–58. Amsterdam: Elsevier.
Paul, C. R. C. 1979. Early echinoderm radiation. *Syst. Assoc. Spec. Vol.* 12, 415–34.
Paul, C. R. C. 1982. The adequacy of the fossil record. *Syst. Assoc. Spec. Vol.* 21, 75–117.
Paul, C. R. C. 1984. British Ordovician cystoids, Part 2. *Palaeont. Soc., Monogr.,* 65–153.
Paul, C. R. C. 1985. Ordovician and Silurian coronates (Echinodermata) from Czechoslovakia. *Geol. J.* 20, 21–9.
Paul, C. R. C. and A. B. Smith 1984. The early radiation and phylogeny of echinoderms. *Biol. Rev.* 59(4), 443–81.
Philip, G. M. 1979. Carpoids – echinoderms or chordates. *Biol. Rev.* 54, 439–71.

Raup, D. M. 1983. On the origins of major biologic groups. *Paleobiology* 9, 107–15.
Raup, D. M. 1985. Mathematical models of cladogenesis. *Paleobiology* 11, 42–52.
Raup, D. M. and S. J. Gould 1974. Stochastic simulation and evolution of morphology. *Syst. Zool.* 23, 305–22.
Raup, D. M., S. J. Gould, T. J. M. Schopf and D. S. Simberloff 1973. Stochastic models of phylogeny and the evolution of diversity. *J. Geol.* 81, 525–42.
Runnegar, B. 1982. A molecular-clock date for the origin of the animal phyla. *Lethaia* 15, 199–205.

Schopf, T. J. M. 1979. Evolving paleontological views on deterministic and stochastic approaches. *Paleobiology* 5, 337–52.
Schopf, T. J. M., D. M. Raup, S. J. Gould and D S. Simberloff 1975. Genomic versus morphologic rates of evolution: influence of diversity I. Analysis of marine orders. *Paleobiology* 4, 223–52.
Sepkoski, J. J., Jr 1978. A kinetic model of Phanerozoic taxonomic diversity I. Analysis of marine orders. *Paleobiology* 4, 223–51.
Sepkoski, J. J., Jr 1979. A kinetic model of Phanerozoic taxonomic diversity II. Early Phanerozoic families and multiple equilibria. *Paleobiology* 5, 222–51.
Sepkoski, J. J., Jr 1982. A compendium of fossil marine families. *Milwaukee Public Museum Contrib.* 51, 1–125.
Smith, A. B. 1984a. *Echinoid palaeobiology.* London: Allen & Unwin.

Smith, A. B. 1984b. Classification of the Echinodermata. *Palaeontology* **27**, 431–59.

Smith, A. B. and C. R. C. Paul 1982. Revision of the class Cyclocystoidea (Echinodermata). *Phil. Trans. R. Soc. Lond.* (B) **296**, 577–684.

Sprinkle, J. 1980. Origin of blastoids: new look at an old problem. *Geol. Soc. Am. Abstr. Prog.* **12**(7), 528.

Sprinkle, J. 1983. Patterns and problems in echinoderm evolution. *Echinoderm Studies* **1**, 1–18.

Stanley, S. M., P. W. Signor III, S. Lidgard and A. F. Karr 1981. Natural clades differ from 'random' clades: simulations and analyses. *Paleobiology* **7**, 115–27.

Strathmann, R. R. and M. Slatkin 1983. The improbability of animal phyla with few species. *Paleobiology* **9**(2), 97–106.

Thomas, W. R. and T. C. Foin 1982. Neutral hypotheses and patterns of species diversity: fact or artifact? *Paleobiology* **8**, 45–55.

Ubaghs, G. 1975. Early Paleozoic echinoderms. *Ann. Rev. Earth Planet. Sci.* **3**, 79–98.

Valentine, J. W. 1980. Determinants of diversity in higher taxonomic categories. *Paleobiology* **6**, 444–50.

Valentine, J. W. and C. A. Campbell 1975. Genetic regulation and the fossil record. *Am. Sci.* **63**, 673–80.

Whittington, H. B. 1980. The significance of the fauna of the Burgess Shale, Middle Cambrian, British Columbia. *Proc. Geol. Assoc.* **91**, 127–48.

Whittington, H. B. and D. E. G. Briggs 1985. The largest Cambrian animal, *Anomalocaris*, Burgess Shale, British Columbia. *Phil. Trans. R. Soc. Lond.*, B **1141**, 569–609.

Whyte, L. L. 1965. *Internal factors in evolution*. London: Tavistock.

Addendum

* In June 1986, the discovery of a new class of living echinoderms was announced. It has been compared with the Cyclocystoidea. A full discussion is awaited. This is the first new class of extant echinoderms discovered since 1821.

Baker, A. N., F. W. E. Rowe and H. E. S. Clark 1986. A new class of Echinodermata from New Zealand. *Nature 321*, 862–64.

5

The initial radiation and rise to dominance of the angiosperms

ELIZABETH M. TRUSWELL

ABSTRACT

Evidence accumulated during the past two decades has shown that the appearance of the angiosperms in the fossil record was, in contradiction to previously held views, neither more sudden nor more mysterious than other groups of organisms. The principal radiation appears to have begun during the Barremian stage of the Early Cretaceous, or slightly earlier. Evolutionary patterns are best shown in sequences of pollen morphotypes from the eastern United States, where there are clear increases in diversity and morphological complexity upwards through the Barremian to Cenomanian stages. Application of advancement indices to the sequence shows the pollen flora, even at the top of the sequence in the Cenomanian, to be more limited in its morphological range than pollen of extant taxa. Leaf floras show a similar progression from early, poorly diversified floras with 'first rank' leaf architecture, to more complex forms. These data accord with the idea of an initial radiation of the angiosperms having occurred entirely within the Early Cretaceous, rather than with the postulate that the group had a long pre-Cretaceous history.

In Australia, which formed part of the Southern Gondwana Province in the mid-Cretaceous, monosulcate morphotypes appear in the early Albian, some 10 million years after their appearance elsewhere, testifying to an immigrant origin for the flowering plants on this continent. Statistical data from Queensland basins suggest the early producers of angiospermous pollen were concentrated in the coastal zones.

The acquisition of the complex of characters that collectively define angiospermy seems likely to have been initially independent of large-scale environmental change, and to have occurred through single gene mutations, evolving under selective pressures of variable moisture availability in streamside or coastal habitats. Rates of evolution were high in the early

Aptian–Albian phase, consistent with the occupation of a new adaptive zone by a group that had physical, evolutionary and ecological access to it. A slowing of rates after the first phase is compatible with adjustment of the group to new adaptive subzones.

Plots of the overall diversity of fossil plants through the geological record show the appearance of the angiosperms to be linked with a major increase in total plant species diversity beginning in the mid-Cretaceous. Their rise to dominance coincides with a decline in gymnosperm species numbers. Comparisons of all plant groups shows the angiosperms to have the highest origination rates and the shortest species durations.

Diversity plots for the Australian record are possible only in a limited number of areas because of paucity of data and uncertainties about the age of sedimentary rocks. Late Cretaceous and Tertiary pollen data from the Gippsland Basin, including graphs of total diversity, originations and extinctions, reveal little perturbation over the Cretaceous/Tertiary boundary, but show a clear response to climatic events in the Early Tertiary. The record is biased toward moister environments where pollen and other plant remains are most likely to be preserved; semi-arid habitats, where speciation rates are highest in the extant flora, are poorly represented in the fossil record.

INTRODUCTION

A survey of current literature suggests that no discussion of the origins and rise to dominance of the flowering plants is complete without reference to Charles Darwin's comments on the 'abominable mystery' of the apparently sudden appearance of this major plant group. Darwin's concern, expressed in a letter to J. D. Hooker in 1879 (Darwin & Seward 1903) was with the apparently rapid development of the angiosperms, a rapidity which he felt to be in conflict with the notion of evolution as an essentially gradual process. The problem as perceived by Darwin arose because of the apparent modernity of the earliest identifiable angiosperm fossils. Leaves known last century from the mid-Cretaceous were assigned to a diversity of extant genera, a practice that reinforced the impression that the flowering plants had sprung fully formed into the biological world at a narrowly defined time.

This view persisted well into the 20th century because of the failure of palaeontologists to identify either the ancestral angiosperm stock or convincing links between the modern-looking fossil leaves and earlier forms. The perception of the fossil record of the flowering plants as of little value in understanding their origin and evolution created a vacuum in which speculation flourished. The lack of fossil data stimulated the development of theories such as that which claimed a long pre-Cretaceous evolution of the angiosperms in highland areas far removed from depositional sites wherein fossils might be preserved. It meant, further, that angiosperm phylogenies, and concepts of primitive and specialised characters, were constructed entirely on the basis of comparative morphological studies of living taxa, a situation that still persists (see Davis & Heywood 1963, p. 35; Kanis 1981, p. 81).

Intensified study of the fossil record during the past 25 years appears to have brought about at least a minor revolution in this pattern of thought. The revolution owes its success, in large measure, to the application of palynology – in this case the study of the fossilised walls or exines of pollen grains – to the field of early angiosperm evolution. Pollen grains, because of their production in large numbers, the ease with which they are transported and incorporated in sediments, and their resistance to decay, provide a more comprehensive picture of the regional vegetation for any particular interval of time than do plant macrofossils, the leaves, stems and seeds. They can thus more readily provide clues to the composition of vegetation far removed from the sites of deposition. The absence of angiosperm pollen in pre-Cretaceous rocks, confirmed by diligent searching by palynologists, combined with the simple morphology of the oldest angiospermous pollen grains in the Cretaceous and the subsequent elaboration of these, has provided what seems now to be widely accepted evidence against the idea of a long pre-Cretaceous history of the group in areas where they have left no fossil record. In concert with the palynological analyses of Cretaceous and pre-Cretaceous rocks, there has been rigorous re-examination of macrofossils previously claimed to be pre-Cretaceous flowering plants (Hughes 1961, 1976, Scott et al. 1961). The failure of these to stand up against such scrutiny, each being found wanting on either morphological or stratigraphical grounds, has reinforced the alternative concept that the initial radiation of the angiosperms occurred entirely within the Cretaceous.

This comparatively recent intensified study and re-evaluation of the fossil record has provided insights into the timing of angiosperm origins, into the patterns and tempo of their early radiation and diversification and, perhaps more speculatively, into the geography of their early development (Doyle & Hickey 1976, Doyle 1977). As a generalisation, it is apparent that evaluation of plant fossils has lagged behind that of animals in appraising evolutionary processes through geological time. Recent compilations of the diversity of fossil plants through the Phanerozoic (Niklas et al. 1980, 1983, Tiffney 1981) have helped to redress this imbalance. These compilations have highlighted those events, both geological and biological, which have influenced the plant kingdom as a whole; they thus provide a perspective against which the development of the flowering plants can be set.

This review is concerned primarily with the origin and radiation of the group as revealed by scrutiny of the fossil record in appropriate sedimentary sequences, and the broader issues of angiosperm diversity as a function of large-scale changes in the total plant kingdom through time. Supporting data have been derived almost entirely from northern hemisphere sources; in order to present a review which is more appropriate to the location of the present forum, and to provide a preliminary test of some of the conclusions expressed in published studies, I have attempted to analyse the Australian angiosperm record.

ANGIOSPERMY: CONCEPTS AND ORIGINS

Recognition of the oldest angiosperms in the fossil record is made difficult by the nature of angiospermy itself, which can be vizualised as the end product of the accumulation of a number of adaptations. Few of these characters are confined to the group, and none can be isolated as the sole cause of its success. The survival value of individual characters is uncertain; the success of the group in the Cretaceous and post-Cretaceous world can be attributed to the integration of these traits into a functional and behavioural whole.

The characteristic that most clearly distinguishes the angiosperms is that of double fertilisation, with the consequent development of polyploid, nutritive endosperm tissue. Such a feature is unfortunately unlikely to be preserved in the fossil record. The only other feature that appears at present to be exclusively angiospermous is in the fine structure of the pollen-grain wall; most angiosperms are characterised by a sexine, or outer layer, that is columellate, and a nexine, or inner layer, that is homogeneous and lacking laminations. Gymnosperm sexine structure, by contrast, is alveolate, the nexine lamellate (Doyle *et al.* 1975). Although an increasing amount of information is becoming available on the microstructure of the grain walls in fossil pollen, the possibility that wall laminations may be destroyed during fossilisation makes it difficult to use fine detail of the inner wall to distinguish fossil angiosperm pollen from that of other types.

Most angiosperms have ovules enclosed within an ovary, the carpel. This enclosure necessitates the germination of pollen on a stigmatic surface, with access to the female egg cell provided by means of a pollen tube. The gametophytes in angiosperms are simple, with the microgametophyte consisting of a tube cell, sperm nuclei and the remains of a pollen grain; the megagametophyte, which is retained on the sporophyte, is correspondingly simple, and lacks archegonia. Stamens and the presence of accessory reproductive structures in the flower, viz. sepals and petals, are also characteristic of flowering plants. A number of vegetative features are also typical, including leaf laminae with reticulate venation (in dicotyledons),vascular tissue with vessels, and sieve tubes with companion cells. Many features characteristic of the angiosperms occur in other plant groups. The enclosure of the ovules, the presence of pollen tubes, of vessels in the conducting tissue, and the organisation of reproductive parts into a flower-like structure are all characters that occur in certain gymnosperms. Other more generalised characters of the angiosperms, such as particular chemical defences against invasion, behavioural strategies, such as a short reproductive cycle, and the herbaceous habit are characteristic of angiosperms but are not confined to them.

It is unlikely that all of these characters arose simultaneously in the history of the group. Probably, there was a continuum of morphological developments throughout an interval of geological time. This view of angiospermy as the result of the progressive accumulation of characters highlights the problem of identifying the earliest members of the group. The acquisition of individual features, and their integration into a distinctive whole apparently reached some kind of adaptive threshold that initiated radiation of the group in the Early Cretaceous. As stressed by Niklas *et al.* (1980), it is unlikely that all characters had evolved to achieve maximum reproductive and competitive efficiency by then – some may have appeared, or been 'perfected', later. Hughes's (1976) comment is salutory in considering the problem of identifying early angiosperms in the fossil record. He stressed that current concepts of angiospermy are based solely on observations made on living morphotypes. A perspective from the mid-Cretaceous would give a different, probably much simplified, impression of a group of plants less clearly distinguishable from contemporary gymnosperms than they are now, with fewer characters defining the limits of the group. Indeed, Hughes (1981) goes so far as to suggest that even the term 'angiosperm', based as strongly as it is on a modern concept, obstructs understanding when applied to Cretaceous fossils.

The problem of angiosperm ancestry is one for which the accumulation of a wealth of data from the fossil record has so far failed to find an answer. As Retallack and Dilcher (1981a) observed, almost every group of gymnosperms, living and fossil, has at one time or another been proposed as possibly ancestral. Most current literature concerns the claims of disparate groups of Mesozoic gymnosperms to be angiosperm ancestors, the problems of monophyletic versus polyphyletic origins, and the homology of a range of gymnosperm organs with angiosperm reproductive structures. I have chosen to omit discussion of angiosperm origins from the present review, except to draw attention to a number of recent papers in which it is discussed. Stewart (1983) provides a useful summary of characters among gymnosperm groups that could be considered 'preangiospermous'. Krassilov (1973, 1975) and Hughes (1976) list the claims of the Mesozoic seed plants Czekanowskiales, Caytoniales, Ginkgoales and Bennettitales as possible ancestors. In all cases, however, there are major problems in identifying fossil organs homologous to angiosperm ovulate structures. Recently, Retallack and Dilcher (1981a) revived an earlier claim (Plumstead 1956) that the Permian glossopterids may be part of the ancestral angiosperm stock, suggesting the *Glossopteris* leaf to be homologous with the angiosperm carpel. The time gap involved between the last glossopterids in the Triassic and the earliest identifiable angiosperms of the Cretaceous is a barrier to acceptance of this pteridosperm group as ancestral to the flowering plants. On the polyphyletic or monophyletic origin debate, it is clear that the same fossil data are still being interpreted in different ways. Doyle (1978) has argued that the pattern of diversification of Early Cretaceous pollen is consistent with a monophyletic origin; Hughes (1976) contends that a polyphyletic origin from different lines of Mesozoic gymnosperms remains possible.

PALYNOLOGICAL EVIDENCE FOR EARLY RADIATION

The Atlantic coastal plain sequence of the USA

The Atlantic coastal plain of Virginia, Maryland, Delaware and New Jersey has come to be recognised, during the past 15 years, as the classic area for the demonstration of early angiosperm evolution. Seaward-dipping sediments of the fluvio-deltaic Potomac Group, (the Patuxent, Arundel, Patapsco and Raritan Formations) have yielded sequences of pollen and leaves which together provide strong positive evidence that the initial radiation of the angiosperms occurred within the Cretaceous (Doyle 1969,1973, 1977; Doyle & Hickey 1976). This non-marine sequence has been dated by correlating spore and pollen zones erected there with zones defined in regions of marine sedimentation, where their relationship to standard international time scales has been established. Comparison with sequences in England (Kemp 1970), western Canada (Jarzen & Norris 1975) and the Gulf coast of the USA (Pierce 1961, Hedlund & Norris 1968) allowed the Patuxent Formation, at the base of the Potomac, to be dated as Barremian, and the Raritan Formation to be identified as of Cenomanian, or early Late Cretaceous age.

Within the Potomac Group there is a clear increase in abundance of angiosperm pollen upwards, through the five zonal units established by Doyle (1973). In the basal Zone 1, angiosperm pollen is rare, and much subordinate to gymnosperm (especially conifer) pollen and fern spores. All pollen grains believed to be angiospermous are of one morphological type, although minor diversity suggests some prior evolution. They have a single elongate germinal aperture – the monosulcate condition – a feature shared by gymnosperms such as the Cycadales, Ginkgoales and Bennettitales, and by some members of the angiosperm Subclass Magnoliidae. The wall structure of this Zone 1 pollen type, however, distinguishes it from gymnosperm pollen. It is clearly stratified, with an outer, perforate layer – the tectum – connected to a smooth inner nexine by a series of rods or columellae. The most common of these angiospermous types has been assigned to the fossil genus *Clavatipollenites*. Near the top of Zone 1, this bilaterally symmetrical monosulcate pollen type is joined by tricolpate pollen, in which three germinal furrows are arranged longitudinally to give a radial symmetry. Such a pollen morphotype is exclusively angiospermous.

A steady elaboration of pollen form, expressed chiefly in the arrangement of germinal apertures, is discernible up-sequence in the Potomac. Tricolpate pollen is succeeded by tricolporoidate forms, in which there is a hint of the presence of pores at the grain equator. Development of this into a distinct tricolporate condition follows, approximately at the level of the late Albian Stage. It is succeeded in the Cenomanian by forms in which the colpi are reduced and the pores pre-eminent, giving triangular triporate grains.

This successive appearance of new morphological groups of pollen, in a stratigraphically consistent sequence, has been interpreted as reflecting the evolutionary diversification of a natural monophyletic group. There is, however, no inference that this region represents a centre of origin for the angiosperms; it is highly likely that the monosulcate pollen types of Zone 1 reflect angiospermous parent plants which have migrated into the region. Doyle (1977) attempted to quantify changes in the frequency of pollen morphotypes through the Potomac Group, to determine the extent to which that sequence represents a pattern of evolutionary diversification. His plot of pollen 'species' or morphotypes is reproduced here as Figure 5.1. To provide continuity of section the data have been drawn from well sections rather than from scattered outcrops. Uncertainties in the calibration of the stage boundaries in the Cretaceous and in correlation of the Potomac rocks with those boundaries, meant that rates of origination could not be confidently expressed in terms of millions of years. A rough estimate for the time involved in Figure 5.1 would be of the order of 20 million years – a duration of 15.5 Ma for the Albian has been published (Harland *et al.* 1982), but the accuracy is uncertain.

The origination rates for new morphotypes, and the rate of change in total frequency, are highest early in the sequence, with peaks near the base of Subzone 11B, in about the middle Albian. Origination rates fall thereafter. This, coupled with rising extinctions, causes a levelling off in total diversity in Subzone 11C, and a diversity drop in Zone 111. This pattern, of an early 'explosive' phase, with high origination rates soon after the appearance of the group, followed by a levelling off, was claimed

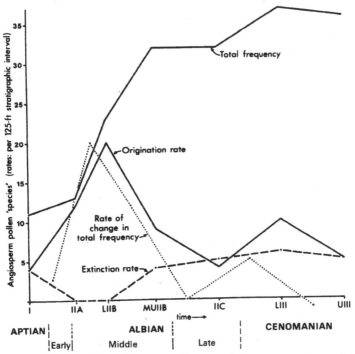

Figure 5.1 Frequency of angiosperm pollen morphotypes per unit stratigraphic interval through Potomac Group sediments (from Doyle 1977). Palynozones defined by Doyle are shown here correlated with Cretaceous Stages. Origination rate shows first appearances, extinction rate last appearances: changes in total frequency were approximated as mid points of successive intervals. (L = Lower; M = Middle; U = Upper.)

by Simpson (1953) to characterise an initial phase of rapid adaptive radiation as the group enters a new adaptive zone, followed by less rapid intrazonal evolution. The peak in origination rates in Subzone 11B may be the climax of an early phase of rapid adaptive radiation, but Doyle cautioned that the pattern may include elements that have migrated into the region from elsewhere. The peak corresponds to the influx of tricolpate pollen, and a growing body of data shows that these originated earlier in the Cretaceous tropics than in contemporary temperate zones. Hence it is probable that migration effects confuse the diagram at this level. The zero extinction rates in the 11A to 11B parts of the graph may support this; according to Simpson's model, extinction rates should be high during early phases of adaptive radiation.

To test whether the increase in the number of angiosperm species reflected *in situ* diversification or immigration, Doyle (1977) undertook further analysis of morphological change. To do this, he assigned advancement scores to grain characters. The ancestral Zone 1 condition, typified by *Clavatipollenites,* was scored zero for all five characters considered. Monosulcate apertures, bilateral symmetry, small size, a finely reticulate exine and the tendency for grains to be shed singly rather than in tetrads were all features regarded as representative of this zero level of advancement: progressively higher scores were added for trends such as elaboration of the sulcus through tricolpate to triporate conditions, for a flattening in shape, for the elaboration of the wall into a coarsely reticulate pattern, and for the shedding of the grains in adherent groups. The advancement index for each morphotype was summed from values assigned to each of these characters. In Figure 5.2, the number of morphotypes showing different levels of advancement are shown as histograms for each of the zones in the Potomac Group. The mean and maximum advancement indices move steadily to the right; there are, in general, no sudden appearances of new morphotypes. An exception is the peak in advancement index 3 in Subzone 11A, which may reflect the somewhat abrupt appearance of tricolpate pollen grains probably by migration into the area. There, as in other regions where Cretaceous pollen sequences are known, there are no morphological intermediates between the monosulcate and tricolpate apertural condition. Therefore, it seems that the data expressed as Figure 5.2 reflect processes both of *in situ* diversification of early angiosperm groups, and some mobility among certain taxa.

What Figure 5.2 does show clearly is that the flora as a whole was much less complex than modern floras. As noted by Doyle, a few non-specialised extant families in the Subclasses Rosidae and Dilleniidae have pollen with advancement indices in the 4 to 7 range on Figure 5.2, but for most living families their index would be off the scale shown. At least one living family, however, has pollen morphologically close to the zero advancement index: within the Chloranthaceae, in the allegedly primitive Magnoliidae, pollen of the genus *Ascarina* is virtually indistinguishable from *Clavatipollenites*, an occurrence considered by Muller (1981) to reflect the relictual preservation of early-developed features.

The regularity of the advancement patterns shown in Figure 5.2 is claim-

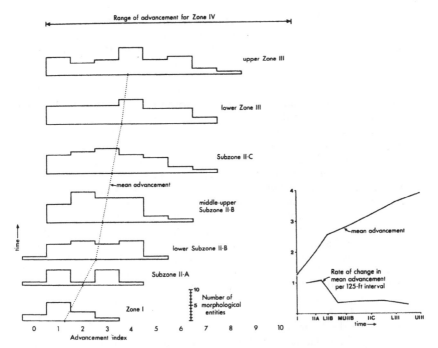

Figure 5.2 Histograms of angiosperm pollen advancement indices for Potomac Group. Shown are numbers of morphotypes with particular advancement indices for each of the stratigraphic intervals sampled (from Doyle 1977, Hickey & Doyle 1977). Advancement scores are based on mean size, aperture type, grain shape, exine sculpture and the nature of the pollen unit when shed. These five features as evident in the *Clavatipollenites* grain type are each scored zero.

ed (Doyle 1977) to preclude much earlier diversification of the angiosperms. Some prior development and diversification of the *Clavatipollenites* types cannot be ruled out; indeed this has been established in other regions. Walker and Walker (1982) consider the type to be relatively advanced on the basis of its ultrastructure, and Brenner (1982, 1984) has described forms from the Hauterivian of Israel which may have been produced by angiosperms closer to their initial diversification. Such grains may eventually provide a more appropriate datum for the measurement of advancement levels. In terms of evolutionary mode, the limited number of characters offered by pollen grains, and the difficulties in deciding on species limits, make it impossible to distinguish if patterns are gradualistic or punctuational. Doyle's (1977) analysis showed that some aspects of the data suggest phyletic gradualism; others suggest more sudden innovations.

Environmental effects and geographic variation

The sequence of pollen morphotypes in the Early Cretaceous Potomac

Group sediments has also been identified in other areas. Many of these are in marine sediments, and dating by invertebrate fossils is firmer than in the non-marine Potomac. This has allowed us to trace patterns that suggest migration of some early angiosperm taxa. In a global survey of Cretaceous palynofloras Brenner (1976) recognised four floristic provinces, defined on the basis of the non-angiosperm component of the spore and pollen floras – that is, on the pteridophyte and gymnosperm elements. The Potomac sequence forms part of the Southern Laurasian Province, encompassing the middle latitudes of the Northern Hemisphere, with abundant data from Europe and North America. Floras are characterised by diverse fern spores, notably those of the Schizaeaceae, and by common *Classopollis*, the complex pollen of the conifer family Cheirolepidaceae. Within the province, monosulcate pollen of *Clavatipollenites* morphology appears in Barremian sediments of the English Wealden, where its diverse morphological expression is evidence of prior diversification in the group (Hughes *et al.* 1979). There, as in eastern North America, monosulcate forms are succeeded by tricolpates in the early to middle Albian, and there are parallels too in the ensuing sequence of tricolporate and triporate types (Laing 1975, 1976).

The Northern Laurasian Province encompasses geographic areas now largely above 60°N latitude. Palynofloras of the Arctic are dominated by bisaccate conifer pollen, *Classopollis* is absent, and fern spores poorly diversified. In these high northern latitudes angiosperm pollen does not appear until the Cenomanian.

The Northern Gondwana Province probably holds the most critical evidence for early angiosperm radiation. This area includes the Middle East, equatorial West Africa and South America, except for the extreme south. Doyle (1984) recently suggested that this region is the most likely one for the origin of the angiosperms as a whole, a suggestion based on the greater diversity in the Barremian of angiospermous monosulcate pollen, including some more primitive than *Clavatipollenites*. In the West Africa–Brazil region, tricolpate pollen appears in the Aptian, a stage earlier than in temperate regions. The pollen is present in sequences overlain by evaporites which mark the first stages of rifting of the Atlantic Ocean, prior to late Aptian marine incursions. As well as the early appearance of tricolpate pollen, there is evidence in this region of a greater range of morphotypes in the Albian and Cenomanian than occurs elsewhere. Palynomorphs associated with the angiosperm pollen in this region are poorly diversified and dominated by pollen of the conifer family Cheirolepidaceae; members of this group, while perhaps not obligate xerophytes, certainly were well adapted to dry conditions. This, and the association with salt deposits, led initially to suggestions that conditions were at least seasonally dry, perhaps even semi-arid, during the early phases of angiosperm development (Hickey & Doyle 1977, Doyle, 1978). A recent, more thorough, review of the Northern Gondwana data (Doyle *et al.* 1982) seems to weaken this claim, as it can be demonstrated that a mosaic of arid and very wet conditions occurred through the region during those early phases.

In the Early Cretaceous, Australia, with Patagonia and southern Africa, falls within Brenner's Southern Gondwana Province. Palynomorph assemblages contain diverse pteridophyte spores and conifer pollen, and are similar in a general way to those of Southern Laurasia. For the angiosperm component, it has been maintained for some time that the earliest identifiable forms in Australia reflect an immigrant origin for the group on this continent. This view conflicts with the widely quoted postulate of Takhtajan (1969) that the angiosperms originated in a tropical region which includes eastern Australia, a postulate based on the large number of allegedly primitive angiosperms in the Australian tropical flora.

The earliest descriptive papers (Burger 1970, 1973, Dettmann, 1973) indicated that the first recognisable angiosperm records in Australia were tricolpate pollen types appearing in the mid-Albian in the Great Artesian Basin. Monosulcates of *Clavatipollenites* morphology were thought to appear contemporaneously. More recent studies (Burger 1981, and reported in Dettmann, 1981 and herein) have shown that in fact monosulcate pollen precedes tricolpate forms in Australia as it does in the northern hemisphere. *Clavatipollenites* is present in early Albian formations of the Eromanga Basin in Queensland (see Fig. 5.3, which was contributed by D. Burger). Work is still required to determine just how diverse the monosulcate forms are in this early phase of evolution in Australia; the presence of one form, *Asteropollis*, with a modified star-shaped aperture has been confirmed, and is thought to reflect the immigration of more than one angiosperm type during this initial interval. An early Albian date for the

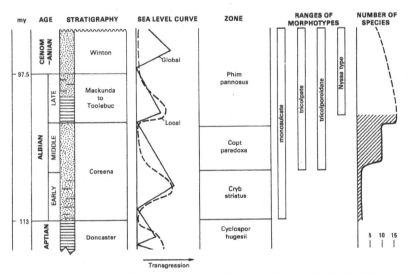

Figure 5.3 Sequence of appearance of pollen morphotypes in Early Cretaceous sediments of the Eromanga Basin, Queensland. Global sea-level curve after Cooper (1977). Centre column shows palynozones of Dettmann (1963), Dettmann and Playford (1969) and Burger (1980). Right-hand column, showing overall diversity of angiosperm pollen morphotypes, is approximate only.

first monosulcate angiosperm pollen in Australia indicates a lag time of up to 10 million years between the Barremian or pre-Barremian appearance in Africa and Europe and arrival in Australia.

The tricolpate pollen which succeeds *Clavatipollenites* in Queensland sedimentary basins perhaps represents a subsequent phase of migration. Its appearance in the mid-Albian (see Fig. 5.3) is approximately coincident with – or perhaps slightly later than – the time of its first appearance in North America and Europe. The morphological trends observed there, from tricolpate to tricolporoidate to triporate, appear to show a dislocation in the Queensland sections examined by Burger. In Figure 5.3 he shows the con-temporaneous appearance of tricolpate and tricolporoidate types, apparent-ly confirming that these types differentiated elsewhere and migrated to the Australian region. Burger (1981) suggested a migration along the northern shore of Tethys and into Australia via a south-east Asian route. Although there are no fossil data from south-east Asia to confirm such a route, it appears to be geologically possible, provided that recent data which suggest the presence of microcontinents to the north of Australia from the latest Jurassic (Pigram & Panggabean 1984) are realistically interpreted. Disper-sal via Antarctica was considered by Dettmann (1981), but early angiosperm history on that continent is also undocumented.

Although current evidence suggests migration into Australia of the very early angiosperms after their differentiation elsewhere, it would be unrealistic to deny local evolution in the phase immediately following their establishment here. Although at the level of pollen 'species' a distinct similarity has been noted between Australian species and near-contemporary ones from North America (Dettmann 1973), there is a com-ponent which is sufficiently distinct to have warranted description as new species. Also, for many species, comparisons with northern hemisphere taxa have yet to be made at the level of microstructure, a necessary step when dealing with morphological entities with a limited number of characters.

The available data from Australia are at present too sparse to construct a curve of diversity for the Early to mid-Cretaceous in a manner comparable to that of Doyle (1977). The broadly drawn stepwise diversity curve shown on the right of Burger's diagram in Figure 5.3 is the nearest approach at pre-sent. Although rough, this curve shows a clear correlation of the early angiosperms with habitat. Such an association was first reported by Dett-mann (1973) who observed that sediments deposited close to shore in marine sequences were richest in angiosperm pollen. Burger (in a talk presented to the 5th International Palynological Conference in Cambridge, England, in 1980) elaborated this theme. There, he produced statistical data on pollen densities in the Surat and Eromanga Basins of Queensland which showed that the parent plants of *Clavatipollenites* and certain tricolpate pollen occupied a restricted coastal environment; other tricolpate pollen producers may have been present in more inland areas, but in the main these areas were dominated by gymnosperms and other non-angiospermous groups. From that compilation, and from Figure 5.3, it is apparent that there were increases in the diversity of morphotypes during regressive phases immediately following sea-level peaks. The wide coastal zones exposed after withdrawal of the sea appear to be linked with bursts of adaptive radiation; the earliest angiosperms in Australia seem then to have colonised tracts characterised by habitats that were unstable in water supply and salinity.

Such a scenario is in close agreement with postulates formulated on the basis of northern hemisphere data. An early radiation in unstable habitats with a fluctuating water supply – one which was either erratically or seasonally dry – was suggested by Doyle and Hickey (1976), partly because there is an association between early angiosperm leaves and coarse-grained stream-margin facies, and partly because of the apparently early appearance of tricolpate angiospermid pollen in the semi-arid environments of West Africa. A link with coastal conditions was also claimed by Retallack and Dilcher (1981b). Their survey of both reproductive and vegetative traits showed early angiosperms, as understood from leaf and fruit remains, to have been xeromorphically adapted shrubs, small in stature, small-leaved, and abundant seed producers; such characters would have fitted them to survive as pioneer plants in disturbed coastal and streamside habitats. The fossil record suggests that the early forms migrated rapidly and extensively; this rapid dispersal may well have been facilitated by the transgressions and regressions of epicontinental seas which are now well documented for the Albian and Cenomanian.

MACROFOSSIL EVIDENCE FOR EARLY RADIATION

It was the apparent diversity and 'modernity' of mid-Cretaceous leaves that underlay earlier concepts of a mysteriously sudden rise of the angiosperms, a mystery whose aura had been deepened by the assignment of a number of those leaves to modern genera. Although this practice was undoubtedly common in monographic works of the late 19th and early 20th centuries, it is of interest to note that in some of the early descriptions of the abundant leaf flora from the Potomac Group there are suggestions that the leaves were not entirely modern in aspect. Thus, L. F. Ward, in reviewing the Mesozoic floras of the USA, described 'archaic' and 'primitive' dicotyledons among the Patapsco forms, and noted that specimens collected from Virginia by Professor William Fontaine 'certainly possess all the essential elements of dicotyledonous leaves although at the same time bearing a certain recognizable stamp of the cryptogamic and gymnospermous vegetation that characterize that early age' (Ward 1905, p. 358). In spite of this early hint that leaves with a morphology distinct from that of living plants might be present in these Cretaceous floras, it was not until Doyle's (1969) palynological studies had been accomplished that a detailed re-examination of the leaves from the Potomac Group was undertaken, with the hope that these too might disclose patterns of evolutionary interest.

The studies of L. J. Hickey (Doyle & Hickey 1976, Hickey & Doyle 1977, Hickey 1978) focused on details of venation, leaf shape and the nature of petiole attachment. These features revealed a clear pattern of morphological change through the sequence. At the level of palynological Zone 1, the rare angiosperm leaves present are of limited diversity; they are simple rather than compound, generally elliptical or ovate, and have a distinctively low level of organisation in their venation, with the tertiary and higher vein orders obscurely differentiated. These characters are those of Hickey's (1971) 'first rank' category of leaf architecture, a syndrome that he considered to be primitive among extant taxa, and which characterises modern Magnoliales, although not all features are restricted to that order. For the next youngest macrofossil-bearing horizon, Subzone IIB, new morphological complexes are discernible: venation patterns show a more rigorous organisation, leaf shapes appear that are almost compound, and are succeeded shortly thereafter by clearly compound forms. By the upper levels of Zone IIB, there are venation patterns of Hickey's 'third rank' level. At the highest stratigraphic horizon from which leaves were recovered, Subzone IIC, lobed platanoid leaves show a yet more rigorously delineated separation of secondary and tertiary veinlets. In these patterns of increasing diversity and morphological complexity, the leaf record appears to parallel that of the pollen, and is consistent with an initial radiation of the angiosperms within the Early Cretaceous. The leaves in the upper part of the sequence appear, however, to cover a greater part of the spectrum of modern morphologies than does the pollen, although most of the types fall below the organisational level of the majority of living dicotyledonous groups.

It has not yet been possible to trace evolutionary patterns satisfactorily in other plant organs. The fossil record of flowers, fruits and seeds has not been explored in a systematic manner, no doubt because of the comparative poverty of the record of these organs. In particular, the preservation of ephemeral floral parts is much less likely than the more abundant and robust leaves and pollen. The available data on reproductive characters of early angiosperms was reviewed by Dilcher (1979), who concluded that, by the end of the Early Cretaceous, the diversity of reproductive modes paralleled that seen in leaf and pollen morphology. From that review and later studies (e.g. Basinger & Dilcher 1984) it is apparent that bisexual flowers (probably insect pollinated) and catkin-like unisexual inflorescences (probably wind pollinated) had developed at least by the Early Cenomanian. This was some 20–25 million years after the initial appearance of angiospermy in the Barremian. The levels of diversity revealed by the fossil record caution against some of the concepts of primitive and advanced conditions based on comparative morphology of living taxa. For example, the unisexual, anemophilous condition is usually considered to result from reduction of bisexual forms: the fossil record available points to this being an early developed, perhaps initially simple, condition rather than a derived one.

THE RISE OF THE ANGIOSPERMS TO DOMINANCE

The Cretaceous record

The foregoing discussion outlines the currently available fossil data bearing on the early evolution of the angiosperms. Doyle (1977, 1978) suggested that these data are consistent with an evolutionary model in which most of the features of angiospermy had appeared by the Barremian–Aptian interval. The acquisition of the characteristic key biological innovations probably reflects genetic changes that were independent of large-scale environmental perturbations. Such changes could have occurred through single gene mutations, perhaps abruptly, in the manner which Hilu (1983) has suggested might have contributed to initiation of higher categories of flowering plants. Selection pressures may have been involved in this early phase of radiation, but they would have been linked with small-scale environmental changes, perhaps those of locally unstable water supplies or seasonally dry climates. Selection in these would have favoured the essentially weedy habit of small shrubs with short life cycles and opportunistic pollination mechanisms. The large-scale environmental changes associated with the breakup of Gondwana, the Albian to Cenomanian inundations on many continents, with their increased area of coastal habitat, may have stimulated the apparently rapid spread of the angiosperms after their initial radiation, and allowed diversification into new geographic areas. In Simpson's (1953) terminology, the initial radiation into the hydrologically unstable habitat, was occasioned by the early angiosperms having 'physical, evolutionary and ecological access' to it. Physical access was possible as the habitat probably developed where the group originated; access in an evolutionary sense was possible because the early angiosperms were apparently adapted to the conditions of the zone; ecological access was feasible because the existing inhabitants of the zone, the ferns and gymnosperms, were competitively inferior. In this context, the survey conducted by Retallack and Dilcher (1981b) is instructive. These authors noted that the unstable coastal or levee-bank habitats were occupied by such taxa as the ferns *Weichselia* and *Tempskya*, the conifers *Frenelopsis* and *Pseudofrenelopsis*, and certain cycads, all of which waned markedly in abundance with the increase in number of the angiosperms.

Subsequently, perhaps with access provided via stream margins, the angiosperms came to occupy open spaces in the forests, perhaps as early successional trees, or as shrubs of the forest understory. It was probably not until the Late Cretaceous that they were able to compete successfully with conifers as members of the forest canopy layer.

The magnitude of angiosperm radiations has been shown graphically by Niklas *et al.* (1980, 1983) and Tiffney (1981), who plotted the total diversity of plant species, based on data from macrofossils only, for each epoch of the Phanerozoic. After allowing for sampling biases associated with areas of outcrop for rocks of different ages, these authors noted that some intervals of high plant diversity remained. One such diversity peak is that associated with the evolution of angiospermy: the high diversities clearly associated with this event commence in the Early Cretaceous, and persist through the Cenozoic (Niklas *et al.* 1980, Fig. 8, p. 33). The relationship of angiosperm evolution to that of other plant groups (Niklas *et al.* 1983) expressed in broad divisions according to the reproductive grade of vascular plants, is reproduced here as Figure 5.4, and shows the appearance and rise

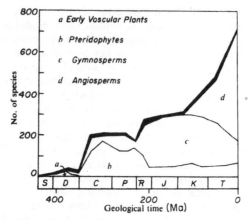

Figure 5.4 Changes in diversity of four major reproductive groups of vascular plants from the Silurian to the Tertiary. Solid black line represents non-vascular plants. Based on 18 000 fossil plant species from Northern Hemisphere citations (from Niklas *et al.* 1983).

of the angiosperms in the Cretaceous to be associated with a significant rise in total plant diversity. Further, the angiosperm curve is associated with a levelling off and then a decline in the number of species of gymnosperms, which were dominant through much of the Mesozoic. At a more detailed level, the relationship between the angiosperms and gymnosperms in the Late Cretaceous, immediately following the major angiosperm radiations, remains poorly understood. The gymnosperms, particularly the conifers, remained an important element of world vegetation for much of the Late Cretaceous and Tertiary, as they do today, in terms of biomass alone. Their continued diversification in some areas after the major bursts of angiosperm evolution is typified by the Australasian region, where a number of major genera – *Dacrydium, Phyllocladus* and *Lagarostrobus* – of the Podocarpaceae, all appear in the Late Cretaceous (Dettmann 1981, Mildenhall 1980).

One other feature of the data compiled by Niklas *et al.* (1983, Fig. 2) shows the angiosperms to have higher rates of appearance of species than the other major reproductive groups, and shorter species durations. Such high speciation rates may well be related to the small effective population sizes which are favoured by specialised mechanisms of pollen dispersal and effective dispersal of propagules: insect pollination and dispersal of propagules by a variety of animal vectors may be distinctive angiosperm characters in this regard.

The diversity curve based solely on angiosperm data – albeit that of macrofossils only, and those mostly from North America – is reproduced here as Figure 5.5 (Fig. 7 from Niklas *et al.* 1980, Fig. 6.5 from Tiffney 1981). The curve for total diversity shows a steady, but not spectacular, climb through the Cretaceous from about the Aptian onwards. The curve depicting species turnover, that is, the number of taxa appearing for the first

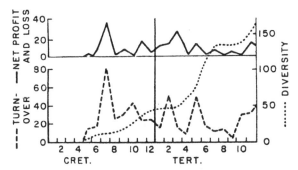

Figure 5.5 Diversity data for angiosperm macrofossils, based largely on species citations from North America (from Niklas *et al.* 1980). Turnover, net profit and loss, and diversity plotted for Cretaceous and Tertiary. The plots reflect changes in the number of species for each stratigraphic interval. Time scales are based on North American usage.

and last time, as a function of the overall diversity, is more explicit in terms of evolutionary activity. The first peak on this curve occurs in the Albian, and may reflect high evolutionary rates during the early phase of adaptive radiation. A second Cretaceous peak, at about the middle of the Late Cretaceous, is less easily explained. Niklas *et al.* (1980) suggest that it may represent the movement of the angiosperms into a new adaptive zone, that of the forest canopy.

The general increase in angiosperm diversity through the Late Cretaceous may reflect the development of a variety of plant and animal interactions. Although insect pollination mechanisms appear to have developed early in angiosperm history, it is possible that these became more effective during the Late Cretaceous. The relationship to insect evolution through the evolution of pollination mechanisms may be reflected in the patterns of species turnover, but the relationship is obscure at present. Peaks in turnover and diversity of insects (Niklas *et al.* 1980, Fig. 9) correspond approximately with the same peaks in the angiosperm curves; cause and effect are thus difficult to separate, and it may be that the insects were in fact following the plants, reflecting an increase in those insects dependent on the flowering plants rather than the insects stimulating the radiation of that group.

As well as increased plant–animal interactions with respect to pollination, fruit- and seed-dispersal mechanisms may have showed progressive diversification during the Cretaceous. Dinosaurs, birds and mammals may have furnished successive vectors for dispersal. Regal (1977) argued that the opportunity for plant–insect relationships existed long before the Cretaceous; hence the links between angiospermy and insects could not have been a dominant factor associated with the radiation of the angiosperms. Rather, he suggested that the potential for outcrossing provided by insect pollination reached its maximum advantage when animal seed vectors became available, allowing dispersal into a variety of habitats, and exploitation of new resources. The animal seed vectors would thus enable the creation of populations of widely distributed individuals. This, in a feedback mechanism, would mean that specialised, efficient pollination mechanisms were necessary to maintain genetic viability; the more haphazard wind-pollination mechanisms would have been too unreliable.

Evolution in the Tertiary

The steady rise of the angiosperm diversity curve of Figure 5.5 continues smoothly across the Cretaceous/Tertiary boundary, as does the curve for net profit and loss; the curve for turnover rate shows a minor decline. There is certainly nothing in this graph to suggest a parallel with the extinctions documented from the animal kingdom.

In the early part of the Tertiary, the diversity curve shows a slow increase. The turnover rate, however, is more erratic. The first sharp peak, that of the Palaeocene–Eocene, suggests a burst of evolutionary activity. This, it has been suggested, may reflect a 'modernisation' of floras, with the extinction of archaic forms and their replacement by taxa of more modern aspect. Another suggestion is that this period of rapid turnover may be a time lag from events affecting the Cretaceous/Tertiary boundary interval. The precise timing of the Early Tertiary event is not clear from the curve. It may, however, be linked with climatic perturbations affecting the Eocene, in the manner described below for Australian data. These may have been global in their effect, linked to factors such as the build-up of sea ice or initial ice caps on Antarctica, and thus exerted major effects on the biotas over wide areas.

The second perturbation in the turnover rate curve of Figure 5.5 occurs in the Late Oligocene–Miocene interval. This probably reflects development of the herbs, which were a new adaptive form coinciding with the advent of seasonal and unstable mid-latitude climatic conditions evident from about that time. The development then of small growth forms, coupled in many cases with an annual reproductive cycle, contrasts with the essentially arborescent forms that dominated the Early Tertiary. These developments occurred in parallel through a wide range of unrelated taxa, and ensured a second phase of success for the angiosperms, comparable to the success of their first radiation in the Cretaceous. Both of these phases of successful competition appear to have been linked with unstable habitats. The unstable habitats of the Early Cretaceous were probably local; they were microenvironments of shorelines or streamside, present at a time when global climates were comparably stable. The mid-Tertiary events, by contrast, show the success of the group at a time when oscillating global climates meant that unstable habitats were ubiquitous.

The diversity curve of Figure 5.5 excludes pollen records. For the angiosperms, this means the neglect of an abundance of potentially valuable data. There are at present no comprehensive compilations using pollen diversities; Doyle's (1977) curves are limited to the Early Cretaceous from a specific area. The pollen-based compilations of Muller (1970, 1981) are included here for completeness, although their specialised nature means they are not a true reflection of evolutionary activity, documenting, as they do,

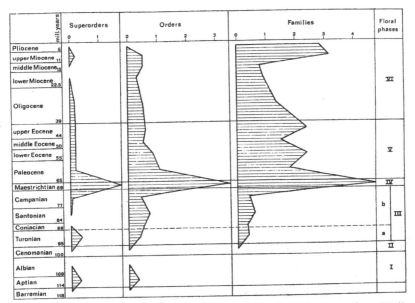

Figure 5.6 Rate of appearance of extant major taxa of angiosperms (from Muller 1981). Column on right shows major phases of angiosperm development as these are interpreted by Muller.

only the record of those fossil types that can be identified with living angiosperm taxa. The graphs from Muller's (1981) data are reproduced here as Figure 5.6. They are much biased towards concepts based on extant taxonomic groupings, and the boundaries of the character complexes representing families or orders in the Late Cretaceous, for instance, are unlikely to have been the same as they are today. Extinct taxonomic groupings are not taken into account; in fact, few such groups have been recognised above the generic level. This lack is probably most distorting in the Late Cretaceous, where abundant and diverse pollen of the Normapolles and *Aquilapollenites* groups probably represent extinct families and orders (see Batten 1984 for a review).

None the less, Figure 5.6 does give some indication of the rate of 'modernisation' of the angiosperms. The appearance of extant families shows a steady climb through most of the Late Cretaceous, followed by a climax in diversification in the Maastrichtian. There is a decline in the rate of appearance across the Cretaceous/Tertiary boundary, but fairly high levels are maintained in the Early Tertiary. The appearance of herbaceous taxa is believed to have influenced the Miocene/Pliocene peak in the rate of appearance.

AUSTRALIAN ANGIOSPERM RECORDS

The record of Early Cretaceous angiosperms in Australia was outlined earlier. For the Late Cretaceous, the palaeobotanical record is notably sparse, and based almost entirely on palynological assemblages from the Otway and Gippsland Basins (Dettmann & Playford 1968, 1969, Stover & Evans 1973, Stover & Partridge 1973). These show a steady increase in angiosperm diversity through the Late Cretaceous, but diversities appear to be generally lower than they are in coeval deposits in Europe and North America. Dettmann (1981) notes a substantial modification of the angiosperms at about the beginning of the Turonian, when triporate and polyporate pollen first appear. The Late Cretaceous saw the introduction of a number of groups which subsequently achieved prominence in the modern flora. Forms morphologically akin to, although not identical with, extant Proteaceae, and pollen allied to the *brassi* group of extant *Nothofagus* are examples.

Angiosperm history through the Tertiary rests both on palynological and on macrofossil data (Lange 1981). Although there is much information to be gained from the latter, only a few floras can be precisely dated, and many older monographs are much in need of revision. The palynological data are more abundant, in general better dated, and give a more representative picture of the floras at any point in time. In terms of plant diversity, documentation of the pollen record avoids some of the problems associated with plant macrofossils, particularly that of establishing relationships between disassociated plant parts. Neither does pollen suffer from the phenotypic plasticity of other organs, especially leaves. The disadvantages in producing a record based solely on pollen data include the fact that speciation events will not show up clearly in stenopalynous families – those such as the grasses, in which distinction between species is difficult on the basis of pollen alone. There are also some preservational problems; pollen is destroyed by oxidation, so that the record is biased towards plant communities of the wetter areas. The sheer weight of numbers may be a further disadvantage, in that whole assemblages of pollen grains are only rarely described.

This suggested recourse to the pollen record as a source of evolutionary history for the flowering plants assumes, of course, that changes in pollen morphology reflect speciation events. This is open to challenge, as the degree of correlation between pollen morphological change and morphological change in other plant organs is unresolved. However, as Muller (1981) has noted, although macromorphological change without pollen change is fairly common – in the grasses for instance – the reverse situation, changes in pollen morphology without changes in other plant parts, is reasonably rare.

With these reservations in mind, and being aware of the inadequacy of the data set, I have compiled graphs showing the patterns of change in angiosperm pollen in one Australian basin, to show the potential of this type of compilation. Figures 5.7A–C are based on pollen from the Gippsland Basin of southeastern Australia, where there is a reasonably complete Tertiary sequence for which considerable palynological data are available. Stover and Evans (1973) and Stover and Partridge (1973) originally compiled these data for stratigraphic purposes only, so that the descriptive

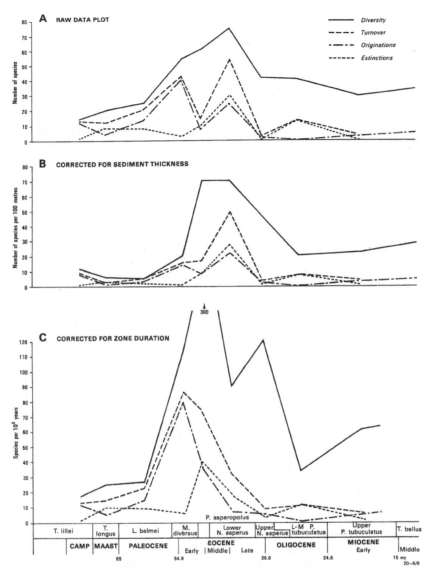

Figure 5.7 Changes in diversity, turnover, originations and extinctions for angiosperm species in the Late Cretaceous to early Miocene of the Gippsland Basin, southeastern Australia. Based on 115 species described by Stover and Evans (1973) and Stover and Partridge (1973). The palynological zonal scheme, and its correlation with the international timescale, is taken from these authors and from Partridge (1976). Originations and extinctions are based on number of first and last appearances in a given time interval; turnover is the sum of these events. A, raw data plot, uncorrected for zone thickness and duration; B, corrected for sediment thickness (zone thicknesses given in papers cited and in Partridge 1971; C, corrected for zone duration, in terms of timescale according to Harland *et al.* (1982).

coverage is unlikely to be comprehensive. For estimating the thicknesses of the palynological zones I have used the average thicknesses quoted in the aforementioned papers and in Partridge (1971); for time control I have drawn on correlations with standard Tertiary time scales quoted in the source papers and in Partridge (1976). The curves are based on plots of 115 form species of pollen.

While I do not claim accuracy of detail for these graphs, some of the observable changes coincide with known climatic fluctuations, and with evolutionary perturbations shown in the Northern Hemisphere data, suggesting that at least some features of the Australian curves may be real. There are no discernible perturbations across the Cretaceous/Tertiary boundary. Peaks in diversity occur in the Early (Fig. 5.7C) to Middle (Fig. 5.7B) Eocene, which is an interval characterised by several large temperature drops and a general perturbation of climate, as evidenced from oxygen isotope data from the south-west Pacific (Shackleton & Kennett 1975, McGowran, 1978). Peaks in extinction coincide with Middle to Late Eocene temperature drops. The lower, more stable temperatures of the Oligocene may be reflected in the flattening of the turnover, origination and extinction curves. The Eocene peak in activity appears to coincide in a general way with the broadly defined peak shown by the macrofossil data from the northern hemisphere (Niklas et al. 1980). Unfortunately, the Gippsland Basin data stop short of the mid-Miocene, so it is not possible to determine a Miocene peak reflecting the radiation of herbaceous taxa. Given that the Gippsland Basin assemblages largely reflect a swamp forest community, that element would be poorly represented in any case.

In any general survey of the angiosperm record in Australia it will be taxa from the arid and semi-arid environments that will be much under represented (Truswell & Harris 1982). Although opportunities for preservation are normally much rarer in that type of environment, in Australia the situation is exacerbated by deep weathering of much of the continent, rendering much of the onshore Tertiary record barren of pollen. This bias in the preservational record would no doubt significantly distort any comprehensive survey of angiosperm history, particularly as studies of species richness in the extant flora indicate that semi-arid regions are exceptionally favourable for speciation. Hopper's (1979) review of speciation patterns in the southwestern Australian flora demonstrates this clearly – the transitional zone between the high rainfall and the arid zones is most species-rich. Recurrent climatic stresses and erosional processes operative in the relatively recent geological past – the Neogene – have produced a mosaic of soils and habitats, resulting in fragmentation of older populations and accelerated rates of speciation. Such a stressed dynamic environment is unlikely to be adequately reflected in the fossil record.

CONCLUSIONS

Since the 1960s, the fossil record has produced a great deal of evidence to indicate that the appearance and rise to dominance of the angiosperms was no sudden phenomenon. If the time of first appearance is taken to be Barremian or slightly older, the time from then to the first flush of adaptive success in the Albian is about 25 million years; it is about another 10 million years from the Albian to the Turonian, when some cautious claim for dominance can be made. In those areas that have contributed most to understanding the patterns of angiosperm evolution, that is, the analyses of morphological lineages of pollen and leaves in the Cretaceous, and the compilations of overall angiosperm diversity through geological time, there may be discerned a persistent thread, namely that the environment has exerted a dominant control on evolutionary change. The acquisition of the characters that collectively define angiospermy may be attributable to internal genetic change, but the theme of environmental control becomes apparent in analyses of the early history of the group; the early angiosperms were believed to have competitive advantages in local unstable habitats. Subsequently, the global expansion of epicontinental seas is believed to have influenced the spread of the group in the mid-Cretaceous; there is evidence too that the environments of the Tertiary have controlled diversity in that interval – diversity increases that reflect the herbaceous habit are believed to be linked with the climatic instability that characterised the Neogene.

Another persistent line of enquiry concerns the place of migration in the developmental history of the angiosperms. Divergences in times of appearance of certain pollen morphotypes and discontinuities in morphological trends in local sequences have been interpreted as evidence for migration of early developed forms, presumably along paths such as coastal zones and streamsides. The clearest evidence for disparate times of appearance lies with the producers of tricolpate pollen for which there is evidence of a progressively later appearance with higher latitudes: predecessors of these – the producers of the monosulcate, *Clavatipollenites* type of pollen grains – apparently dispersed rapidly at an earlier date. The precise timing and pathways of this early phase of migration are unexplained. Vicariance biogeographers might well regard it as representative of a widely distributed ancestral angiosperm taxon. However, the fact that the form is succeeded in widely separated areas by a consistent sequence of pollen morphotypes seems more in accord with successive waves of migration than with evolution in isolation within a number of widely separated populations.

In 1977, James Doyle remarked that insufficient critically evaluated basic taxonomic data were available to perform the kind of quantitative evolutionary analysis that is possible for many animal groups. I judge that this state of affairs still persists, even though there has been an increase in the number of quality papers documenting fossil pollen, leaf and, notably, flower morphology. A number of obstacles prevent palaeobotanists from constructing evolutionary analyses for the angiosperms. The first is common to all plant fossils, that of interpreting whole organisms from disassociated parts and of evaluating how changes observed in one organ can be related to changes in others. The nature of the record thus imposes considerable uncertainties about the limits of taxa. The second problem is more peculiarly angiospermous, and involves the fact that extant taxa are classified almost exclusively on the basis of their floral parts, but floral parts are ephemeral and only exceptionally preserved. Modern orders and families have been identified in the fossil record by a variety of isolated organs – by leaves, wood, fruit and pollen – rarely by whole flowers; such identifications involve assumptions that parts were always correlated in the same manner as they are today.

In spite of the general lack of the essential floral parts in most fossils, there has been, throughout the history of palaeobotany, a strong tendency to force fossil organs, usually leaves, but more recently pollen, into modern taxonomic groupings. Establishing the earliest occurrence of extant families remains a favourite pastime among palaeobotanists. This has led to a situation where virtually no extinct angiosperm taxa are recognised above generic level. The unreality of this situation is apparent from comments made by those palaeobotanists who have described fossil flowers which comprise associations of floral parts that are not known to occur in any living group (see e.g. Tiffney 1977, Basinger & Dilcher 1984). The fossil record at present offers no hint as to how many extinctions at the family level there may have been since the Early Cretaceous.

There appears to be no immediate remedy for this situation. Progress towards understanding angiosperm evolution can only be made by adoption of the approaches recommended by Hughes (1976) and followed independently by such workers as Doyle and Hickey. That approach is to concentrate on the documentation of the fossils themselves, with attention to morphological detail, variation and fine resolution stratigraphy, allowing clear comparisons with material from older and younger horizons. Such methodology will enable trends to be recognised which would be obscured by forcing fossils into extant groupings that have little reality in the past.

ACKNOWLEDGEMENTS

I am grateful to Dennis Burger for provision of Figure 5.3, and to James Doyle and Karl Niklas for permission to reproduce figures from their published papers. Mary Dettmann and Gavin Young offered constructive comments on a first draft of this chapter. Publication is by permission of the Director, Bureau of Mineral Resources, Canberra.

REFERENCES

Basinger, J. F. and D. L. Dilcher 1984. Ancient bisexual flowers. *Science* 224, 511–13.

Batten, D. J. 1984. Palynology, climate and the development of Late Cretaceous floral provinces in the Northern Hemisphere; a review. In *Fossils and climate*, P. Brenchley (ed.), pp. 127–64. New York : Wiley.

Brenner, G. J. 1976. Middle Cretaceous floral provinces and early migrations of angiosperms. In *Origin and early evolution of angiosperms*, C. B. Beck (ed.), pp. 23–47. New York: Columbia University Press.

Brenner, G. J. 1982. Pre-sulcate angiosperm pollen from the late Neocomian of Israel. *Abstracts, Bot. Soc. Am., Misc. Publ.* 162, 56.

Brenner, G. J. 1984. Late Hauterivian angiosperm pollen from the Helez Formation, Israel. *Abstracts, 6th Int. Palynological Conf., Calgary, 1984,* 15.

Burger, D. 1970. Early Cretaceous angiospermous pollen grains from Queensland. *Bull. Bur. Miner. Resour. Geol. Geophys. Aust.* 116, 1–10.

Burger, D. 1973. Palynological observations in the Carpentaria Basin, Queensland. *Bull. Bur. Miner. Resour. Geol. Geophys. Aust.* 140, 27–44.

Burger, D. 1980. Palynological studies in the Lower Cretaceous of the Surat Basin, Australia. *Bull. Bur. Miner. Resour. Aust.* 189, 1–106.

Burger, D. 1981. Observations on the earliest angiosperm development with special reference to Australia. *Proc. 4th Int. Palynol. Conf. Lucknow* (1976–7) 3, 418–28.

Cooper, M. R. 1977. Eustacy during the Cretaceous: its implications and importance. *Palaeogeogr., Palaeoclimatol., Palaeoecol.* 22, 1–60.

Darwin, F. and A. C. Seward (eds) 1903. *More letters of Charles Darwin.* vol. 2. New York: Appleton.

Davis, P. H. and V. H. Heywood 1963. *Principles of angiosperm taxonomy.* Edinburgh: Oliver and Boyd.

Dettmann, M. E. 1963. Upper Mesozoic microfloras from southeastern Australia. *Proc. R. Soc. Vict.* 77, 1–148.

Dettmann, M. E. 1973. Angiospermous pollen from Albian to Turonian sediments of eastern Australia. *Geol. Soc. Aust. Spec. Publ.* 4, 3–34.

Dettmann, M. E. 1981. The Cretaceous flora. In *Ecological biogeography in Australia,* A. Keast (ed.), pp. 357–375. The Hague: W. Junk.

Dettmann, M. E. and G. Playford 1968. Taxonomy of some Cretaceous spores and pollen grains from eastern Australia. *Proc. R. Soc. Vict.* 81, 69–93.

Dettmann, M. E. and G. Playford 1969. Palynology of the Australian Cretaceous: a review. In *Stratigraphy and palaeontology – essays in honour of Dorothy Hill,* K. S. W. Campbell (ed.), 174–210. Canberra: Australian National University Press.

Dilcher, D. L. 1979. Early angiosperm reproduction: an introductory report. *Rev. Palaeobot. Palynol.* 27, 291–328.

Doyle, J. A. 1969. Cretaceous angiosperm pollen of the Atlantic coastal plain and its evolutionary significance. *J. Arnold Arbor. Harv. Univ.* 50, 1–35.

Doyle, J. A. 1973. Fossil evidence on early evolution of the monocotyledons. *Q. Rev. Biol.* 48, 399–413.

Doyle, J. A. 1977. Patterns of evolution in early angiosperms. In *Patterns of evolution as illustrated by the fossil record,* A. Hallam (ed.), pp. 501–46. Amsterdam: Elsevier.

Doyle, J. A. 1978. Origin of angiosperms. *Ann. Rev. Ecol. Syst.* 9, 365–92.

Doyle, J. A. 1984. Evolutionary, geographic, and ecological aspects of the rise of angiosperms. *Proc. 27th Int. Geol. Congr.,* 2, *Palaeontol.,* 23–33.

Doyle, J. A. and L. J. Hickey 1976. Pollen and leaves from the mid-Cretaceous Potomac Group and their bearing on early angiosperm evolution. In *Origin and early evolution of angiosperms,* C. B. Beck (ed.), pp. 139–206. New York: Columbia University Press.

Doyle, J. A., M. Van Campo and B. Lugardon 1975. Observations on exine structure of *Eucommiidites* and Lower Cretaceous angiosperm pollen. *Pollen Spores* 17, 429–86.

Doyle, J. A., S. Jardine and A. Doerenkamp 1982. *Afropollis,* a new genus of early angiosperm pollen, with notes on the Cretaceous palynostratigraphy and paleoenvironments of northern Gondwana. *Bull. Centres Rech. Explor.-Prod. Elf-Aquitaine* 6, 39–117.

Harland, W. B., A. V. Cox, P. G. Llewellyn, C. A. G. Pickton, A. G. Smith and R. Walters 1982. *A geologic time scale*. Cambridge: Cambridge University Press.

Hedlund, R. W. and G. Norris 1968. Spores and pollen grains from Fredericksburgian (Albian) strata, Marshall County, Oklahoma. *Pollen Spores* 10, 129–59.

Hickey, L. J. 1971. Evolutionary significance of leaf architectural features in the woody dicots. *Am. J. Bot.* 58, 469.

Hickey, L. J. 1978. Origin of the major features of angiospermous leaf architecture in the fossil record. *Cour. Forsch. Inst. Senckenberg* 30, 27–34.

Hickey, L. J. and J. A. Doyle 1977. Early Cretaceous fossil evidence for angiosperm evolution. *Bot. Rev.* 43, 3–104.

Hilu, K. W. 1983. The rôle of single gene mutations in the evolution of flowering plants. *Evol. Biol.* 16, 97–128.

Hopper, S. D. 1979. Biogeographical aspects of speciation in the southwest Australian flora. *Ann. Rev. Ecol. Syst.* 10, 399–422.

Hughes, N. F. 1961. Fossil evidence and angiosperm ancestry. *Sci. Prog.* 49(193), 84–102.

Hughes, N. F. 1976. *Palaeobiology of angiosperm origins*. Cambridge: Cambridge University Press.

Hughes, N. F. 1981. Palynological evidence and angiosperm ancestors. *Abstracts, 13th Int. Bot. Congr. Sydney, 21–28 August, 1981*, 193.

Hughes, N. F., G. E. Drewry and J. F. Laing 1979. Barremian earliest angiosperm pollen. *Palaeontology* 22, 513–35.

Jarzen, D. M. and G. Norris 1975. Evolutionary significance and botanical relationships of Cretaceous angiosperm pollen in the western Canadian interior. *Geosci. Man* 11, 47–60.

Kanis, A. 1981. An introduction to the system of classification used in the Flora of Australia. *Flora of Australia*, Vol. 1, pp. 77–111. Canberra: Australian Government Publishing Service.

Kemp, E. M. 1970. Aptian and Albian miospores from southern England. *Palaeontographica B* 131, 73–143.

Krassilov, V. A. 1973. Mesozoic plants and the problem of angiosperm ancestry. *Lethaia* 6, 163–78.

Krassilov, V. A. 1975. Dirhopalostachyaceae – A new family of proangiosperms and its bearing on the problem of angiosperm ancestry. *Palaeontographica B*, 153, 100–10.

Laing, J. F. 1975. Mid-Cretaceous angiosperm pollen from southern England and northern France. *Palaeontology* 18, 775–808.

Laing, J. F. 1976. The stratigraphic setting of early angiosperm pollen. In *The evolutionary significance of the exine*, I. K. Ferguson and J. Muller (eds), Linnean Society Symposium Series 1, 15–26. London: Academic Press.

Lange, R. T. 1981. Australian Tertiary vegetation. In *A history of Australasian vegetation*, J. M. B. Smith (ed.), pp. 44–89. Sydney: McGraw-Hill.

McGowran, B. 1978. Stratigraphic record of Early Tertiary oceanic and continental events in the Indian Ocean region. *Mar. Geol.* 26, 1–39.

Mildenhall, D. C. 1980. New Zealand late Cretaceous and Cenozoic plant biogeography: a contribution. *Palaeogeogr. Palaeoclimatol. Palaeoecol.* 31, 197–233.

Muller, J. 1970. Palynological evidence on early differentiation of angiosperms. *Biol. Rev.* 45, 417–50.

Muller, J. 1981. Fossil pollen records of extant angiosperms. *Bot. Rev.* 47, 1–142.

Niklas, K. J., B. H. Tiffney and A. H. Knoll 1980. Apparent changes in the diversity of fossil plants. *Evolutionary Biology* 12, 1–89.

Niklas, K. J., B. H. Tiffney and A. H. Knoll 1983. Patterns in vascular land plant diversification. *Nature* 303, 614–16.

Partridge, A. D. 1971. *Stratigraphic palynology of the onshore Tertiary sediments of the Gippsland Basin, Victoria.* Unpublished MSc thesis, University of New South Wales.

Partridge, A. D. 1976. The geological expression of eustacy in the Early Tertiary of the Gippsland Basin. *APEA J.* 16, 73–9.

Pierce, R. L. 1961. Lower Upper Cretaceous plant microfossils from Minnesota. *Minn. Geol. Surv. Bull.* 42, 1–86.

Pigram, C. J. and H. Panggabean 1984. Rifting of the northern margin of the Australian continent and the origin of some microcontinents in eastern Indonesia. *Tectonophysics* 107, 331–53.

Plumstead, E. P. 1956. Bisexual fructifications borne on *Glossopteris* leaves from South Africa. *Palaeontographica B* 100, 1–25.

Regal, P. J. 1977. Ecology and evolution of flowering plant dominance. *Science* 196, 622–9.

Retallack, G. J. and D. L. Dilcher 1981a. Arguments for a glossopterid ancestry of angiosperms. *Paleobiology* 7, 54–67.

Retallack, G. and D. L. Dilcher 1981b. A coastal hypothesis for the dispersal and rise to dominance of flowering plants. In *Paleobotany, paleoecology and evolution*, K. J. Niklas (ed.), pp. 27–77. New York: Praeger.

Scott, R. A., E. S. Barghoorn and E. B. Leopold 1961. How old are the angiosperms? *Am. J. Sci.* 258A, 284–99.

Shackleton, N. J. and J. P. Kennett 1975. Paleotemperature history of the Cenozoic and the initiation of Antarctic glaciation: oxygen and carbon isotope analyses in DSDP Sites 277, 279 and 281. *Initial reports of the deep sea drilling project*, No. 29, 743–55. Washington: U.S. Government Printing Office.

Simpson, G. G. 1953. *The major features of evolution.* New York: Columbia University Press.

Stewart, W. N. 1983. *Palaeobotany and the evolution of plants.* Cambridge: Cambridge University Press.

Stover, L. E. and P. R. Evans 1973. Upper Cretaceous – Eocene spore-pollen zonation, offshore Gippsland Basin, Australia. *Geol. Soc. Aust. Spec. Publ.* 4, 55–72.

Stover, L. E. and A. D. Partridge 1973. Tertiary and Late Cretaceous spores and pollen from the Gippsland Basin, southeastern Australia. *Proc. R. Soc. Vict.* 85, 237–86.

Takhtajan, A. L. 1969. *Flowering plants: origin and dispersal.* Washington: Smithsonian Institution.

Tiffney, B. H. 1977. Dicotyledonous angiosperm flower from the Upper Cretaceous of Marthas Vineyard, Massachusetts. *Nature* 265, 136–8.

Tiffney, B. H. 1981. Diversity and major events in the evolution of land plants. In *Paleobotany, paleoecology and evolution*, K. J. Niklas (ed.) pp. 193–230. New York: Praeger.

Truswell, E. M. and W. K. Harris 1982. The Cainozoic palaeobotanical record in arid Australia: fossil evidence for the origins of an arid-adapted flora. In *Evolution of the flora and fauna of arid Australia*, W. R. Barker and P. J. M. Greenslade (eds) 67–76. Adelaide: Peacock Publications.

Walker, J. W. and A. G. Walker 1982. Ultrastructural study of Lower Cretaceous angiosperm pollen and its evolutionary implications. *Abstracts, Bot. Soc. Am., Misc. Publ.* **162**, 66–7.

Ward, L. F. 1905. *Status of the Mesozoic floras of the United States.* U.S. Geol. Surv. Monogr. **48**, 1–616.

6

Selection or constraint?: a proposal on the mechanism for stasis[*]

P. G. WILLIAMSON

ABSTRACT

Though controversy over the rôle of stasis in the total evolutionary pattern continues, most workers now accept that many species persisted for long periods of time with little morphological change. Many extant species are relatively invariant over their geographical ranges. These two phenomena, which may be regarded as species-level integrity in time and space, have been considered to be the result of either 'stabilising selection' or 'constraint'. In the former explanation the environment plays the dominant rôle, but in the latter, intrinsic characteristics are dominant. It is now argued that these two mechanisms act in concert and that various genetic and developmental constraints drive and mediate stabilising selection.

INTRODUCTION

The 'punctuated equilibrium' mode for evolutionary change (Eldredge & Gould 1972, Gould & Eldredge 1977) proposes that significant evolutionary change is concentrated at rapid speciation events within small, geologically ephemeral, peripatric populations; apart from brief episodes of peripatric cladogenesis, species exhibit morphological stasis, i.e. no significant evolutionary change. The concept of temporal stasis is closely related to the idea of 'species integrity' (Mayr 1963, Dobzhansky 1970, Van Valen 1982) which maintains that many species are relatively invariant in

[*] *Editorial Note:* The paper presented at the symposium dealt with 'long term morphological analysis of species lineages: implications for stasis and cladogenesis', and used data on the fossil molluscs of the Lake Turkana Basin. The present contribution was prepared subsequently, and resulted from discussions at the meeting.

most basic respects across their geographic ranges (Van Valen 1982). The relative frequency of punctuationist versus traditional gradualistic modes of evolutionary change in the fossil record is widely debated (e.g. Levinton and Simon 1980, Hoffman 1982, Levinton 1983, Maynard Smith 1983a,b, Wake *et al.* 1983), but even critics of the punctuated equilibrium model concede that long-term morphological stasis is an important aspect of evolutionary change in many lineages (e.g. Maynard Smith 1980). Some proponents of the punctuationist view consider stasis to dominate the evolutionary patterns of most metazoan groups (e.g. Stanley 1979). In addition, most animal taxonomists acknowledge the reality of 'species integrity' – the comparative geographic invariance of most species: '. . . it would . . . seem important to stress the basic uniformity of most continuously distributed species. . . The fact that (every taxonomist) . . . can identify individuals as a species. . . regardless of where in the range of the species they come from is further illustration of this phenomenon' (Mayr 1963).

The two processes most frequently advocated as mechanisms are 'constraint' and stabilising selection. 'Constraint' would explain species integrity and stasis as a result of the fact that significant evolutionary change is restricted or prohibited by intrinsic features of the genetic and developmental architecture of species (Mayr 1970), whereas stabilising selection could promote integrity and stasis by maintaining a constant phenotype over the geographic and stratigraphic range of a given species.

In an important recent review, Van Valen (1982) identified no less than 11 mechanisms proposed for the maintenance of species integrity and stasis. Five of these clearly fall within the two general categories indicated above. In addition to 'constraint' and stabilising selection, Van Valen recognised six additional mechanisms none of which has been widely advocated as an important promoter of species integrity; Van Valen dismisses them, for a variety of good reasons, and they will not be considered further.

CONSTRAINT

A second class of explanations for morphological stasis and species integrity suggest that constraints inherent in the fundamental genetic and developmental architecture of species restrict or prohibit significant evolutionary change. The concept of genetic homoeostasis (Mayr 1963) falls into this category of 'constraint' and was initially advocated by Eldredge and Gould (1972) as the most likely explanation for stasis. On this view, the species genome is so coadapted that relatively small changes tend to disrupt it; coadapted genotypes are thus usually separated by 'non-adaptive valleys' (Mayr 1963), and change from one coadapted state to another may be very difficult. The species is therefore 'buffered' to resist significant evolutionary change. In addition to this 'balance' concept of genetic homoeostasis, other possible mechanisms have been suggested: Lerner (1954) emphasised the greater biochemical versatility of heterozygotes over homozygotes, and supposed that this might act to resist significant evolutionary change in natural populations (Mayr 1963). A concept closely related to genetic homoeostasis is that of developmental constraint; since genes express themselves developmentally, developmental systems may themselves be hard to change without major disruption (Mayr 1963, Frazetta 1975, Gould 1980). Developmental constraints may either prohibit change completely or restrict it to very few directions in character space; none of the limited options for change available to a species may be appropriate to the selective situation in which it finds itself, and evolution is therefore effectively blocked. The related concepts of genetic homoeostasis and developmental constraint are themselves clearly related to the phenomenon of canalisation (Waddington 1957, Rendel 1962, 1967). 'Canalisation' of characters is the phenomenon whereby the phenotypic expression of a trait is little affected by variation in environmental and genic influences within some range, which may be quite extreme. But, as Van Valen (1982) points out, canalisation, as usually conceived, is a relatively 'labile' restriction of evolutionary options which is not only maintained by selection but can also be modified by it. Canalisation acting alone is probably inadequate as a cause of species integrity, though it may be an important contributory factor.

The concept of 'constraint' has been severely criticised. One major problem is the observation that artificial human selection on numerous animal and plant species, involving a very large range of quantitative and meristic characters, has generally resulted in a very rapid shift of the mean phenotype by several phenotypic standard deviations, producing extreme forms not present in any original population (e.g. Falconer 1960, Lewontin 1974, pp. 86–94, Wright 1977, Chs. 7 and 8). The rapidity and magnitude of such changes in historical times, is a powerful argument against the supposition that species integrity and stasis is a result of an intrinsic blocking of evolutionary change by the fundamental genetic architecture of species (Charlesworth et al. 1982). An additional major problem is heritability. As Charlesworth and Lande (1982) pointed out, an appeal to developmental constraints as the primary 'blockers' of evolutionary response is equivalent to saying that the various morphological characters subject to selection lack genetic variability. However, an extensive body of work shows that wide ranges of phenotypic characters are highly heritable (e.g. Falconer 1960), as indeed they would have to be to accomplish the striking selective manipulation of agricultural stocks referred to above. In particular, allometric growth, an important determinant of adult shape (Huxley 1932, Gould 1966) has also been found to be highly heritable, genetically variable and subject to selective manipulation (e.g. Cock 1966, Atchley & Rutledge 1980). A third major problem involves the significant number of geographically widespread species, with continuous geographic ranges, that none the less exhibit pronounced geographical variability in fundamental aspects of the phenotype (Maynard Smith 1980). In summary, explanations for species integrity and stasis that rely largely or exclusively on genetic or developmental constraints do not appear at present to give a satisfactory explanation of this phenomenon, though, as discussed subsequently, constraint may be an important contributory factor.

STABILISING SELECTION

Stabilising selection – 'the elimination by selection of all phenotypes deviating too far from the population mean, and hence also of genes producing such deviating phenotypes' (Mayr 1963) – has been advocated by several authors as the primary mechanism for species integrity and stasis (e.g. Hoffman 1978, Lande 1980b, Charlesworth et al. 1982). Evidence for the ubiquity and potency of stabilising selection comes both from observations on natural populations and from experimental work. Comparisons of levels of phenotypic variation in different age cohorts – in natural populations of groups as diverse as gastropods (Weldon 1901, diCesnola 1907), reptiles (Dunn 1942, Mertens 1947, Camin & Ehrlich 1958), amphibia (Bell 1973), and insects (Dowdeswell 1961) – indicate that cohorts of increasing age often show decreasing phenotypic variance with little change in the mean phenotype. Selection acts towards an intermediate phenotype, and phenodeviants are preferentially eliminated. Direct observations on natural populations also indicate that phenodeviant individuals suffer disproportionate mortality rates (Karn & Penrose 1951, Jayant 1966, Hecht, 1952), particularly during catastrophic selection episodes (Bumpus 1896, Harris 1911, Johnson et al. 1972, O'Donald 1973). Experimental work has also demonstrated the reduced fitness of phenodeviant individuals compared to more modal phenotypes (Barnes 1968). Stabilising selection is generally considered to be primarily governed by extrinsic environmental factors: 'In the case of stabilizing selection the constancy of the environment regulates and maintains the constancy of the genetic makeup of the population' (Savage 1963); 'In these populations... we would expect stabilizing selection to be of principal importance... (this)... assumes that the environment remains reasonably constant' (Barnes 1968). As Van Valen (1982) noted, stabilising selection is viewed conventionally as a mechanism of 'external compulsion'. The species is essentially viewed as lying within an environmental 'Procrustean Bed', the extremes of its phenotypic distribution cut off by external environmental factors. On this view, should a given environment change, the previous stabilising selection regime will also change, the organism will make an appropriate evolutionary response, and a previously phenodeviant morphology will come to be the new modal form. The emphasis on extrinsic environmental factors as the primary agents for pruning phenodeviants and thus effecting stabilising selection leads proponents to stabilising selection as a primary mechanism for species integrity and stasis into a difficult position: environmental invariance must be a prerequisite for morphological invariance. 'We conclude that if prolonged morphological stasis exists in fossil populations, it must usually be caused by stabilizing selection rather than gene flow or developmental constraints. To the extent that morphological stasis can be documented, it is an important problem for paleoecology to explain why selective forces should often be so conservative over long periods of time' (Charlesworth et al. 1982). This reliance on environmental invariance had led many authors to regard conventional stabilising selection as an unlikely mechanism for the maintenance of species integrity and stasis. For example, Mayr explicitly considers that the wide range of environments experienced by many widely distributed but morphologically invariant species debars stabilising selection from any major rôle in the maintenance of species integrity (Mayr 1963). Many palaeontologists feel that the documentation of long-term morphological stasis in lineages that are known to have experienced extremely wide environmental fluctuations makes conventional stabilising selection an unlikely candidate for the promotion of stasis (e.g. Williamson 1981b, 1982). Indeed, in the original discussion of punctuated equilibrium, stabilising selection was not even considered as a candidate mechanism for the maintenance of morphological stasis (Eldredge & Gould 1972). However, given the apparent ubiquity and potency of stabilising selection, and its clear rôle in the short-term maintenance of species integrity, it seems clear that this process must play some rôle though not necessarily a major one.

DISCUSSION

This brief summary suggests that at present there is no general consensus on the mechanism or mechanisms responsible for the maintenance of species integrity and stasis (Maynard Smith 1983); neither of the two most widely advocated mechanisms – constraint and stabilising selection – offer a complete explanation of this phenomenon. Under these circumstances it is worthwhile pursuing Maynard Smith's (1980) suggestion that *both* constraint and stabilising selection, acting in some concerted manner, might provide a solution.

I suggest here that stabilising selection may not be primarily governed by extrinsic environmental factors, but may in large part operate by a process of intraspecific competition. As indicated above, conventional descriptions of stabilising selection quite explicitly consider that the fitness penalties attaching to phenodeviance derive primarily from some fundamental mismatch between extrinsic environmental factors and the phenodeviant individual. In the environmental 'Procrustean Bed', extrinsic environmental factors preferentially remove phenodeviant individuals. For convenience, I will term this model for stabilising selection 'Procrustean' stabilising selection. This conventional view of the process, that 'constancy of the environment regulates and maintains the constancy of the genetic makeup' (Savage 1963), carries two clear implications. First, changes in the environment and hence in the selective regime should elicit appropriate changes in the modal phenotype. Secondly, it is implicit that, even in the absence of the modal phenotype, phenodeviants should fare poorly, i.e. stabilising selection should be primarily a density-independent process. An alternative view of stabilising selection might be that the fitness penalties attaching to phenodeviance do not primarily derive from a fundamental mismatch between extrinsic environmental factors and the phenodeviant individual.

Rather, the mere fact of phenodeviance confers major selective disadvantages *vis-à-vis* the modal phenotype, and that these derive primarily from competition with modal phenotypes. On this view, phenodeviants might in fact 'manage' quite well in the absence of modal phenotypes, but in their presence are severely outcompeted or eliminated. This 'Saturnian' stabilising selection (Saturn, it will be recalled, ate his own children) seems likely to be a strongly density-dependent process.

An important observation relevant to this issue is that phenodeviant individuals are frequently characterised by pronounced developmental instabilities and abnormalities. As Lande (1980b) points out, the 'evolution of zones of phenotypic canalization around an optimum phenotype occurs because developmental stability imparts a selective advantage to the most common genotypes with mean phenotypes near the optimum, whereas there is less selection for phenotypic stability in extreme genotypes because of their rarity and preexisting disadvantage caused by deviation from the optimum'. The relationship between phenodeviance and developmental instability was first noted by Fisher (1930) who cited data on a herring population showing that individuals with the rarer vertebrae numbers have a pronounced tendency for abnormal development in other features; Wright (1978) gave other examples of meristic characters where the modal number in the population is the most canalized in its development. More recently, Soulé and Cuzin-Roudy (1982) summarised data from two insect genera, a reptile, three mammals and a bird, and demonstrated that extreme phenotypes are characterised by increased asymmetry, i.e. greater developmental instability. Various mechanisms have been suggested to account for this correlation between phenodeviance and developmental instability. Fisher originally suggested that the modal (i.e. optimal) phenotypic classes experience the strongest selection for standardised development; this process of 'canalisation' (Waddington 1957, Rendel 1962, 1967, 1979) ensures that the developmental mechanisms responsible for many characters of the optimal phenotype are governed by stable, coadapted gene complexes and are not responsive to variation in environmental and genic influences within some range of these influences. Phenodeviants would be relatively poorly 'balanced' on this view, and their development governed by poorly integrated gene complexes and hence poorly buffered to external influences (Thoday 1958). An alternative, though not unrelated, explanation for the correlation between phenodeviance and developmental instability, stresses the fact that such individuals are likely to be highly homozygous; their loss of heterozygosity and hence biochemical versatility would lead to an increase in the environmental component of variance during development (Soulé & Cuzin-Roudy 1982). Such an increase in the variance and asymmetry of characters in strongly inbred or selected, hence homozygous, strains has been described by various workers (e.g. Robertson & Reeve 1952, Lerner 1954, Waddington 1957, Thoday 1958, Reeve 1960). Whatever the precise mechanism responsible for the association between phenodeviance and instability, this association is well established. It is also well established that when the fine adjustment between developmental processes is disturbed, severe unfitness is generally introduced (Rendel 1968). This reduction in fitness is almost invariably seen in artificial selection experiments (Van Valen 1982); selection for extreme variants is accompanied by correlated genetic responses governed by the pleiotropic effects of modifiers and the epistatic interactions of linked loci (Levin 1970). Responses correlated with strong directional selection are usually deleterious (Haskell 1954, Lerner 1954, Latter & Robertson 1962, Spiess 1968). Strong directional selection often favours genes that relate poorly with other portions of the genotype, and which may in turn weaken the integration of the genetic system (Dun & Fraser 1959). This is presumably the basis for the common observation that directional selection weakens developmental canalisation and enhances phenotypic variability (Reeve & Robertson 1953, Guthrie 1965, Clayton & Robertson 1957). The correlations between phenodeviance, developmental instability and reduced fitness, which presumably reflect the fundamental genetic and developmental architecture of species, imply that phenodeviance confers important fitness penalties in most populations.

Given that both 'Procrustean' and 'Saturnian' stabilising selection mechanisms are probably acting on most populations, we might enquire as to their overall relative importance. Although conventional views of stabilising selection stress environmentally mediated 'Procrustean' selection, observations on natural populations strongly indicate that the 'Saturnian' mechanism – preferential elimination of phenodeviants due to their developmental instability and reduced fitness vis-à-vis the modal phenotype – in fact predominates. This suggestion is based on the common observation that relaxation of selection pressures generally leads to a major increase of phenotypic variation, i.e. an increase in the representation of phenodeviants in the population. In natural populations, such a relaxation of selection pressure is commonly experienced during periods of population 'flush' (Carson 1968, 1971, 1973), during periods when populations previously greatly reduced by an extrinsic agency such as disease or resource shortage are rapidly expanding due to the suddenly increased carrying capacity of the environment.

Several well-documented examples of the morphological consequences of such population flushes are available in the literature. For example, in a long-term study of a local population of the butterfly *Euphydras aurinia*, Ford and Ford (1930) demonstrated that a period of rapid population increase was accompanied by the appearance of large numbers of highly phenodeviant 'deformed' individuals. These unusual forms exhibited extreme variation in all aspects of morphology (Ford & Ford 1930, Plate XXX); some were even flightless. But as population numbers stabilised, these phenodeviants were eliminated and the population reverted to its previous monomorphic state. In a similar vein, Van Dalsum (1947) provided interesting documentation of the fact that extreme colour mutants in the freshwater snail *Planorbarius corneus* are restricted to populations kept at unusually low densities by annual mowing of reeds.

Population flushes also occur during the invasion of previously vacant habitats. The best documented of such 'introduction' flushes are those caused by human introductions, particularly of economic fish stocks. One of the best documented of these is the introduction of the gamefish *Bairdiella icistius* and *Cynoscion xanthulus* into the previously vacant habitat of the Salton Sea on the Colorado River delta. The initial rapidly increasing populations of these stocks were characterised by a very high incidence (up to 23%) of grossly abnormal individuals, phenodeviants displaying such unusual features as torsion of the vertebral column, lack of jaws, eyes and gill covers, and other unusual cranial configurations. However, after this initial flush phase, as population sizes became stable, these abnormal individuals were rapidly competitively eliminated by the 'typical' phenotype, and the population reverted to its previous monomorphic state (Whitney 1961a,b).

It is clear that under the permissive ecological conditions attending population flushes (Carson 1975), a reduction in density-dependent selection is accompanied by a marked increase in the representation of phenodeviant individuals. These are frequently extremely developmentally abnormal – flightless butterflies, fish lacking lower jaws and eyes – but these abnormalities often 'manage' quite well in the competitively permissive conditions of the flush phase; they survive to breeding age and are not eliminated by extrinsic 'Procrustean' environmental selection. As the population reaches carrying capacity, these phenodeviants are eliminated by competition with more phenotypically 'normal' conspecifics, and the population as a whole again becomes morphologically uniform. The importance of these flush phases in the context of the present argument is that they indicate quite unequivocally that the elimination of phenodeviants in most populations is primarily by intra-specific competition with modal phenotypes – 'Saturnian' stabilising selection – rather than by the action of extrinsic, environmentally mediated 'Procrustean' selection.

If 'Saturnian' stabilising selection is the major factor in maintaining the 'typical' modal phenotye in most species, the phenomenon of species integrity and stasis becomes much easier to understand. If stabilising selection were generally 'Procrustean', major shifts in environments should elicit appropriate shifts in modal morphology; on this view, the species is regarded as evolutionarily labile and responsive to selective change. As discussed previously, the apparent lack of response to major environmental perturbations exhibited by many lineages is the primary reason why many authors have rejected conventional stabilising selection as an adequate explanation for species integrity and stasis (Maynard Smith 1980). But if the 'Saturnian' model for stabilising selection suggested here in fact predominates, any shift in modal phenotype due to environmental changes will be hard to accomplish. Such shifts must involve movement of the phenotype into zones of pronounced developmental instability and reduced fitness; as Rendel (1968) points out:

> The unfitness introduced when the fine adjustment between developmental processes is destroyed will always be counteracting directional selec-

tion ... directional selection must introduce unfitness ... (this) antagonism between fitness and directional selection will have to be taken into account in interpreting outbursts of evolutionary change followed by long stability (and) failure of species to adapt to a change in their habitat.

The 'Saturnian' model of stabilising selection proposed here, which suggests that the constraints inherent in the genetic and developmental architecture of species are the primary 'motor' for stabilising selection, implies that populations are neither as labile in their potential evolutionary response to environmental change as implied by conventional 'Procrustean stabilising selection' models, nor as restricted as implied by conventional 'constraint' explanations. A labile adaptive response to changed environmental conditions – a prediction of conventional 'Procrustean' views of stabilising selection – is unlikely to be exhibited by most species on this 'Saturnian' view, as such a shift must involve movement into zones of developmental instability and reduced fitness. The high heritability of most traits is well established, but the constraints inherent in the genetic and developmental architecture of species, manifested in the reduced fitnesses of phenodeviants, will severely reduce the extent to which selection can actually translate genetic variance into substantial phenotypic change. But, given selection pressures extreme enough, the phenotypic mode may indeed be 'driven through' the zone of lowered fitness around the tails of the phenotypic distribution; selection pressures as extreme as those effected by human breeding programs may indeed be capable of deriving chihuahuas from wolves in a few thousand generations.

The 'Saturnian' model of 'constraint-driven' intraspecific stabilising selection advocated here appears to provide an intuitively appealing resolution of the current debate as to the mechanisms responsible for species integrity and stasis. But, if correct, it also has some important implications for the intriguing obverse of stasis – the rapid morphological transformations attending cladogenesis. First, it is clear that this model for stasis implies that the process of speciation must involve selection pressures severe enough to drive the modal phenotype through the 'reduced fitness' adaptive valleys of the tails of the phenotypic distribution (of course, other factors, such as isolation, may also facilitate this process). Secondly, it is clear that whatever the precise mechanism allowing the bridging of this adaptive valley, the process of significant phenotypic change must involve traversing the zone of developmental instability at the extremes of the phenotypic distribution; this difficult traverse must involve a period of extreme developmental instability and enhanced phenotypic variability for the entire population. Such a phase of enhanced phenotypic variability has been reported to attend cladogenesis in several palaeontological documentations of this process (e.g. Ovcharenko 1969, Sylvester-Bradley 1976, Williamson 1981a,b) although the claim that these palaeontological examples, particularly the last cited, genuinely document cladogenesis has been strongly questioned by some authors (Anon. 1982, Kat & Davis 1983, Cohen & Schwartz 1983,

Fryer *et al.* 1983). This phase of pronounced developmental instability may itself be an important contributor to the mobilisation of phenotypic diversity required for successful speciation (Levin 1970). Once a new and appropriate modal morphology is attained, the extreme selection pressures within the isolate will rapidly institute a new zone of canalisation and developmental constraint around the new modal phenotype. These genetic and developmental checks and balances, once instituted, will in turn serve to drive 'Saturnian' stabilising selection within the new species, and ensure its future geographic integrity and long-term morphological stasis.

SUMMARY

The current debate between 'constraint' and 'selection' explanations for species integrity and stasis represents an artificial polarity; the primary 'motors' and mediators of stabilising selection are the developmental and other constraints inherent in the genetic and developmental architecture of species, and both 'selection' and 'constraint' mechanisms actually act in concert to maintain the geographic and temporal integrity of species.

ACKNOWLEDGEMENTS

My thanks are due for the useful discussions and criticism of this chapter offered by Professors S. J. Gould, R. C. Lewontin, J. Maynard Smith, E. Mayr, G. L. Stebbins and L. Van Valen.

REFERENCES

Anonymous 1982. Punctuationism and Darwinism reconciled? *Nature* **296**, 608.
Atchley, W. R. and J. Rutledge 1980. Genetic components of size and shape. I. Dynamic components of phenotypic variability and covariability in the laboratory rat. *Evolution* **34**, 1161–73.

Barnes, B. W. 1968. Stabilizing selection in *Drosophila melanogaster*. *Hereditas* **23**, 433–42.
Bell, G. 1973. The reduction of morphological variation in natural populations of smooth new larvae. *J. Anim. Ecol.* **43**, 115–28.
Bumpus, H. C. 1896. The variations and mutations of the introduced sparrow *Passer domesticus. Biology lecture, marine biology lab., Woods Hole, (1896–1897)*, 1–15.

Camin, J. H. and P. R. Ehrlich 1958. Natural selection in water snakes (*Natrix sipedon* L.) on islands in Lake Erie. *Evolution* **12**, 504–11.
Carson, H. L. 1968. The population flush and its genetic consequences. In *Population biology and evolution*, R. C. Lewontin (ed.), pp. 123–38. New York: Syracuse University Press.

Carson, H. L. 1971. Speciation and the founder principle. *Stadler Genet. Symp.* 3, 51–70.

Carson, H. L. 1973. Reorganization of the gene pool during speciation. In *Genetic structure of populations*, N. E. Morton (ed.), pp. 274–80. Honolulu: University Press of Hawaii.

Carson, H. L. 1975. The genetics of speciation at the diploid level. *Am. Nat.* 109, 83–92.

Charlesworth, B. and R. Lande 1982. Morphological stasis and developmental constraint: no problem for neo-Darwinism. *Nature* 296, 610.

Charlesworth, B., R. Lande and M. Slatkin 1982. A Neo-Darwinian commentary on macroevolution. *Evolution* 36, 474–98.

Clayton, G. A. and Robertson, A. 1957. An experimental check on quantitative genetical theory. II. The long-term effects of selection. *J. Genet.* 55, 152–70.

Cock, A. G. 1966. Genetical aspects of metrical growth and form in animals. *Q. Rev. Biol.* 41, 131–90.

Cohen, A. S. and H. L. Schwartz 1983. Speciation in molluscs from Turkana Basin. *Nature* 304, 659–60.

diCesnola, A. P. 1907. A first study of natural selection on *Helix arbustorum* (Helicogena). *Biometrika*, 3, 387–99.

Dobzhansky, Th., 1970. *Genetics of the evolutionary process*. New York: Columbia University Press.

Dowdeswell, W. H. 1961. Experimental studies on natural selection in the butterfly, *Maniola jurtina*. *Heredity* 16, 39–52.

Dun, R. B. and A. S. Fraser 1959. Selection for an invariant character, vibrissae number, in the house mouse. *Aust. J. Biol. Sci.* 12, 506–23.

Dunn, E. R. 1942. Survival value of varietal characters in snakes. *Am. Nat.*, 16, 104–9.

Eldredge, N. and S. J. Gould 1972. Punctuated equilibria: An alternative to phyletic gradualism. In *Models in paleobiology*, T. J. M. Schopf (ed.), pp. 82–115. San Francisco: Freeman Cooper.

Falconer, D. S. 1960. *Introduction to quantitative genetics*. Glasgow: Oliver and Boyd.

Fisher, R. A. 1930. *The genetical theory of natural selection*. Oxford: Oxford University Press.

Ford, H. D. and E. B. Ford 1930. Fluctuation in numbers, and its influence on variation, in *Melitaea aurinia*, Rott. (Lepidoptera). *Trans. Ent. Soc. Lond.* 78(II), 345–53.

Frazetta, T. H. 1975. *Complex adaptations in evolving populations*. Sunderland, Mass.: Sinauer Associates.

Fryer, G., P. G. Greenwood and J. F. Peake 1983. Punctuated equilibria, morphological stasis, and the paleontological documentation of speciation: A biological appraisal of a case history in an African lake. *Biol. J. Linn. Soc. Lond.* 20, 195–205.

Gould, S. J. 1966. Allometry and size in ontogeny and phylogeny. *Biol. Rev.* 41, 587–640.

Gould, S. J. 1980. Is a new and general theory of evolution emerging? *Paleobiology* 6, 119–30.

Gould, S. J. and N. Eldredge 1977. Punctuated equilibria: the tempo and mode of evolution reconsidered. *Paleobiology* 3, 115–51.

Guthrie, R. D. 1965. Variation in characters undergoing rapid evolution, an analysis of *Microtus* molars. *Evolution* 19, 214–33.

Harris, J. A. 1911. A neglected paper on natural selection in the English sparrow. *Am. Nat.* 45, 314–8.

Haskell, G. 1954. Correlated responses to polygenic selection in animals and plants. *Am. Nat.* 88, 5–20.

Hecht, M. K. 1952. Natural selection in the lizard genus *Aristelliger*. *Evolution* 6, 112–24.

Hoffman, A. 1978. Punctuated-equilibrium evolutionary model and paleoecology. *Soc. Geol. Pologne, Ann.* 48, 327–31.

Hoffman, A. 1982. Punctuated versus gradual mode of evolution: a reconsideration, *Evol. Biol.* 15, 411–30.

Huxley, J. S. 1932. *Problems of relative growth*. London: Methuen.

Jayant, K. 1966. Birth weight and survival: A hospital survey repeat after 15 years. *Ann. Hum. Genet. (Lond.)* 29, 367–81.

Johnson, R. F., O. M. Niles and S. A. Rohwer 1972. Hermon Bumpus and natural selection in the house sparrow, *Passer domesticus*. *Evolution* 26, 20–31.

Karn, M. N. and L. S. Penrose 1951. Birth weight and gestation time in relation to maternal age, parity and infant survival. *Ann. Eugen. (Lond.)* 16, 147–64.

Kat, P. W. and G. M. Davis 1983. Speciation in molluscs from Turkana Basin. *Nature* 304, 660–1.

Lande, R. 1980a. Microevolution in relation to macroevolution. *Paleobiology* 6, 233–8.

Lande, R. 1980b. Genetic variation and phenotypic evolution during allopatric speciation. *Am. Nat.* 116, 463–78.

Latter, B. D. H. and A. Robertson 1962. The effects of inbreeding and artificial selection on reproductive fitness. *Genet. Res.* 3, 110–38.

Lerner, I. M. 1954. *Genetic homeostasis*. New York: Wiley.

Levin, D. A. 1970. Developmental instability and evolution in peripheral isolates. *Am. Nat.* 104, 343–53.

Levinton, J. S. 1983. Stasis in progress: The empirical basis of macroevolution. *Ann. Rev. Ecol. Syst.* 14, 103–37.

Levinton, J. S. and C. Simon 1980. A critique of the punctuated equilibrium model and implications for the detection of speciation in the fossil record. *Syst. Zool.* 29, 130–42.

Lewontin, R. C. 1974. *The genetic basis of evolutionary change*. New York: Columbia University Press.

Maynard Smith, J. 1980. Macroevolution. *Nature* 289, 13–14.

Maynard Smith, J. 1983a. Current controversies in evolutionary biology. In *Dimensions of Darwinism*, M. Grene (ed.), pp. 273–86. Cambridge: Cambridge University Press.

Maynard Smith, J. 1983b. The genetics of stasis and punctuation. *Ann. Rev. Genet.* 17, 11–25.

Mayr, E. 1963. *Animal species and evolution*. Cambridge, Mass.: Harvard University Press.

Mayr, E. 1970. *Populations, species and evolution*. Cambridge, Mass.: Harvard University Press.

Mertens, R. 1947. Studien zur Eidonomie und Taxonomie der Ringelnatter (*Natrix natrix*). *Abhandl. senckenberg. naturforsch. Ges.* 476, 1–38.

O'Donald, P. 1973. A further study of Bumpus' data: The intensity of natural selection. *Evolution*, 27, 398–404.

Ovcharenko, V. N. 1969. Speciation in Bathonian Brachiopoda from the Pamirs. *Paleont. J.* 57, 57–63.

Reeve, C. R. 1960. Some genetic tests on assymetry of sternopleural chaeta number in *Drosophila*. *Genet. Res.* 1, 151–72.

Reeve, C. R. and F. W. Robertson 1953. Studies in quantitative inheritance II. Analysis of a strain on *Drosophila* selected for long wings. *J. Genet.* 51, 276–316.

Rendel, J. M. 1962. The relationship between gene and phenotype. *J. Theor. Biol.* 2, 296–308.

Rendel, J. M. 1967. *Canalization and gene control*. London: Logos.

Rendel, J. M. 1968. The control of developmental processes. In *Evolution and environment*, E. T. Drake (ed.), pp. 34–9. New Haven: Yale University Press.

Rendel, J. M. 1979. Canalization and selection. In *Quantitative genetic variation*, J. N. Thompson and J. M. Thoday (eds) pp. 139–56. New York: Academic Press.

Robertson, F. W. and C. R. Reeve 1952. Heterozygosity, environmental variation, and heterosis. *Nature* 170, 286.

Savage, J. M. 1963. *Evolution*. New York: Holt, Rinehart and Winston.

Soulé, M. E. and J. Cuzin-Roudy 1982. Allometric variation. 2. Developmental instability of extreme phenotypes. *Am. Natur.* 120, 765–86.

Spiess, E. B. 1968. Experimental population genetics. *Ann. Rev. Genet.* 2, 165–241.

Stanley, S. M. 1979. *Macroevolution: pattern and process*. New York: W. H. Freeman.

Sylvester-Bradley, P. C. 1976. Speciation patterns in the Ostracoda. In 5th International symposium on *Evolution of post-Paleozoic Ostracoda*, G. Hartmann (ed.), pp. 29–37. Hamburg: Verlag Paul Parey; Naturwiss. Vereins Hamburg, Abhandl. Verhandl. (N. F.), Hamburg 18–19 (Suppl.) 29–37.

Thoday, J. M. 1958. Homeostasis in a selection experiment. *Heredity* 12, 401–15.

Van Dalsum, J. 1947. De kleurvarieteiten van *Planorbis corneus* (L.). *Basteria* 11, 100–9.

Van Valen, L. M. 1982. Integration of species: stasis and biogeography. *Evol. Theory* 6, 99–112.

Waddington, C. H. 1957. *The strategy of the genes*. London: Allen & Unwin.

Wake, D. B., G. Roth and M. H. Wake 1983. On the problem of stasis in morphological evolution. *J. Theor. Biol.* 101, 211–24.

Weldon, W. F. G. 1901. A first study of natural selection in *Clausilia laminata*. *Biometrika* 1, 109–24.

Whitney, R. R. 1961a. The bairdiellia, *Bairdiella icistius* (Jordan and Gilbert). In *The ecology of the Salton Sea, California, in relation to the sportfishery*, B. W. Walker (ed.), *Fish Bull.*, 113. State of California Dept. of Fish and Game.

Whitney, R. R. 1961b. The orangemouth corvina, *Cynoscion xanthulus* (Jordan and Gilbert). In *The ecology of the Salton Sea, California, in relation to the sportfishery*, B. W. Walker (ed.), *Fish Bull.*, 113. State of California Dept. of Fish and Game.

Williamson, P. G. 1981a. Paleontological documentation of speciation in Cenozoic molluscs from Turkana Basin. *Nature* 293, 437–43.

Williamson, P. G. 1981b. Morphological stasis and developmental constraint: real problems for neo-Darwinism. *Nature* **294**, 214–5.
Williamson, P. G. 1982. Reply to critics. *Nature* **296**, 608–12.
Wright, S. 1977. *Evolution and the genetics of populations*. Vol. 3. *Experimental results and evolutionary deductions*. Chicago: Chicago University Press.
Wright, S. 1978. *Evolution and the genetics of populations*. Vol. 4. *Variability within and among natural populations*. Chicago: Chicago University Press.

7

Developmental pathways and evolutionary rates

D. T. ANDERSON

ABSTRACT

For multicellular animals, the rate of evolutionary change is dependent, among other things, on the rate at which changes can be assimilated into the developmental system. Any proposal for rapid major change in morphology through time makes the assumption that the developmental system yielding the morphological organisation in question has been able to undergo major shifts in pathway rapidly and still remain functional. The question of whether the developmental system under consideration could have undergone the assumed changes is rarely explored. In most cases, it is extremely unlikely that function could have been sustained. The few examples among the modern fauna in which a major new adult form results from an 'instantaneous' developmental switch (e.g. axolotl, hexapods) are highly aberrant. In general, the complexities of the developmental system, including the need to maintain developmental function and physiological function throughout the process, dictate that evolutionary change in multicellular animals occurs as a consequence of the serial accumulation of small changes in development.

The rate at which this serial accumulation occurs can vary from zero (stasis) to rapid (small but stabilised change in each successive generation). Even stasis is a dynamic state from a developmental point of view, so the potential for change is constantly available. There is no problem in reconciling the developmental requirement of small serial changes with the fluctuations in rates of evolution through geological time.

The significant problem to be faced in evolutionary biology is the extraordinary conservatism of the developmental process. The causation of change in this process (evolution) will only be understood when the causation of stability in development (stasis) at all levels from the species to the phylum has been elucidated. Some progress is now being made towards the resolution of this problem.

FUNCTION IN DEVELOPMENT

The fact that changes in developmental pathways provide the nexus in animal evolution between genetic change and natural selection has been recognised for a long time. If adult populations are to change and, in doing so, to evolve as new species, there must be modifications of the developmental pathways through which adult structure is established. These developmental pathways are controlled and directed by the genome, hence the nexus (Katz 1982, Oster & Alberch 1982, Hall 1983, Horder 1983, Maynard Smith 1983).

What has taken longer to be appreciated is the fact that the process of development sets constraints to the potential for evolutionary change (Alberch 1982, Katz 1982, Oster & Alberch 1982, Hall 1983, Horder 1983, Maynard Smith 1983, Sander 1983). Only genetic changes that maintain the viability of the developmental process and the viability of the resulting adult organisation will survive natural selection. The vast range of genetic changes whose consequences are developmental abnormality (e.g. Hadorn 1961, Counce & Waddington 1973) serve only as tools in the demonstration that all components of the developmental process are under genetic control. In natural populations, such abnormalities are always fatal. Function must be maintained during the life of the individual, from egg to reproducing adult.

In this context, the concept of function is a complex one. Throughout life, physiological function must be maintained, and functional interaction with the environment must be maintained. Obviously the patterns of both of these will change during the course of development (compare any larva and ensuing adult). But the process of development itself adds another element to the system. Developmental processes function to generate further developmental processes, yielding emergent functional organisation. For viability, therefore, developmental function must also be maintained (Anderson 1979, Horder 1983, Maderson 1983, Maynard Smith 1983).

The complexity is amplified by the fact that in every animal there are two interwoven functional strands, the somatic system and the reproductive system. Of course, these are not separate entities. The somatic system is a vehicle without which the reproductive system cannot function; the reproductive system is a system without which the somatic system has no future; and each influences the other developmentally and physiologically (sex hormones!). Yet, in evolutionary terms, the somatic and reproductive systems come under somewhat different constraints. The soma is constantly subject to total functional integration of all of its parts at all levels (developmental, physiological and interactive with the environment). In somatic evolution through time, two things may happen, either a functional change of parts, commensurate with continued functional integration of the whole, resulting in new interactions, or a functional redundancy of parts, consequent on other changes in functional integration, leading to vestigiality or loss. Both, of course, may occur simultaneously. In the evolution of the reproductive system, in contrast, vestigiality of the gonads is not an option. Furthermore, the reproductive system may evolve in ways independent of the functional integration of the developing and mature soma, provided that in the end the reproductive function is maintained (or enhanced). A simple example is the egg enlargement and associated hypertrophy of one oviduct in birds. This has no bearing on the somatic integration of a bird as a functional feathered flyer, but a very significant impact on the mode of reproduction of this group of animals.

Because of these complicated interactions, the developmental constraints on evolution are very real. They impose directionality on the process (Maderson 1983, Langridge, this volume) and restrict the potential for change in such a way that the results are both directional and discontinuous. When new functional organisation evolves, it emerges from pre-existing functional organisation and does not arise *de novo*. At the same time, animal species are not a structural continuum from the simplest to the most complex, but a set of discrete functional entities, the gaps between which are the result of non-functionality of the somatic-reproductive variants that might fill them morphologically. We, for example, are an outcome of directional, discontinuous changes in pre-existing mammalian, reptilian, amphibian, choanichthyan, gnathostome and agnathan ontogenies. The parasitic barnacle *Sacculina*, in which the reproductive system is all and the somatic system almost nothing, is an outcome of directional, discontinuous change in pre-existing rhizocephalan, cirripede, crustacean, 'spiral cleavage worm' ontogenies. The timing is, of course, different. Rhizocephalans have probably been rhizocephalans since the Silurian, at which time we were still finny gnathostomes.

If we wish to understand the kinds and rates of change involved in the evolution of animal organisation, therefore, it is necessary to appreciate the kinds and rates of change that the developmental system allows. In view of the functional complexity of the system, gradual change seems to be a necessary postulate (Anderson 1965, Dullemeijer 1980, Maderson 1983, Maynard Smith 1983), and rapid major change a functional improbability. It is important to emphasise what is meant in the present context by gradual change as compared with rapid major change. The term gradual change is used here to imply a change accumulating progressively, in small increments, over a large number of generations. Major modifications consequent on such gradual change may emerge slowly, during millions of years, or more quickly, in perhaps fifty or a hundred thousand years. The change may proceed steadily as one generation succeeds another, or in bursts, with intervening periods of stasis. Whatever the pattern and rate, however, many generations and usually many species are involved. Some classical examples of this type of change have been registered among the ammonoids (Lehmann 1981). Ammonoid generation time is estimated at only a few years at most. Gradual changes in functional organisation in symmetrically coiled ammonoids can be traced through close series of fossils over many millions of years. More quickly accumulating changes are apparent in the repeated evolution of heteromorph ammonoids, expressing a modified functional pattern, but again million-year time spans and an implied multitude of generations are involved.

Rapid major change, in contrast, is taken here to refer to the evolution of radical modifications of structure and function taking place in a single or a few generations. Whether or not one needs to even contemplate the occurrence of rapid major change depends on the scale under consideration. No problem appears to attend the interpretation of specific, generic and familial divergences as gradualistic phenomena. At these levels of genetic relationship, the observed differences, and thus the inferred changes in development yielding the different adult configurations, are mainly quantitative, a consequence of genetically controlled changes in the rate and timing of developmental processes. At the level of orders, the differences are greater, due to the greater elapse of time since divergence, but the idea of gradual changes in developmental pathways is still applicable.

RAPID MAJOR CHANGE

In certain favourable combinations of genetic–epigenetic expression, quite major morphological modifications can result from simple genetic changes even in natural populations. Echinoderms provide a good example (Raff & Kaufmann 1983). In echinoderm development, pentamerous symmetry is secondarily imposed on bilateral symmetry, presumably reflecting an evolutionary sequence. The pentamerous symmetry of the adult is determined during development by a counting process, based on the development of five successive diverticula from the growing water vascular ring. Studies of mutants have shown that only a few genes are involved in controlling the counting system. In echinoids, pentamerous symmetry is highly regulated. The counting system is constrained by the fact that for symmetries other than five, Aristotle's lantern shows reduced functional efficiency. In asteroids, in contrast, the number of ambulacra varies greatly between species. Here the counting system is much less constrained, since differences in the number of ambulacra do not functionally disadvantage either locomotion or feeding. A simple control mechanism acting early in development thus offers, in this case, a potential for major morphological changes in short evolutionary time at the familial level of divergence.

In contemplating the problem presented by the evolution of the more basic organisational levels, such as the distinct body plans characteristic of phyla or classes, the gradualistic interpretation has often failed to satisfy. Special problems are posed by the sudden emergence of major new body forms in the Cambrian fossil record (Campbell & Marshall, this volume, Runnegar, this volume). The gaps are real; they can be filled if one accepts that small genetic changes can yield large morphological changes (see above), and the opportunity is often presented to tender an imaginative proposal for a sudden evolutionary leap. Rapid evolutionary rates, in other words, are more often inferred for the big changes than for the small ones. This is an interesting paradox, because the proposed big changes actually involve far more fundamental functional redirection of developmental pathways than any of the small changes. Little changes can, of course, have big results, e.g. insectivore to man, but these are always recognised to have involved long times (e.g. 70 million years) and gradual changes at slow rates.

All of the big-event hypotheses in which the need for a suitable developmental mechanism has been recognised are based on the concept of heterochrony – the idea that changes in major functional organisation can result from changes in the relative timing of developmental events (De Beer 1958, Maderson 1983, Raff & Kaufmann 1983). Four general categories of heterochronous change are defined, of which two retain a gradualistic flavour while the other two are big-bang propositions. The gradualistic categories are acceleration, in which somatic development is accelerated relative to gonad maturation and thus proceeds further before sexual maturation is attained; and hypermorphosis, the reverse phenomenon, in which gonad maturation is retarded while somatic development continues to a new (usually larger) adult end point. Evolutionary trends towards larger size can often be identified as consequences of these changes in relative timing, but neither acceleration nor hypermorphosis lead to radical oganisational change. Major somatic change relative to reproductive maturation can be envisaged, however, in the categories of neoteny and paedogenesis (Maynard Smith 1983, Raff & Kaufmann 1983).

In neoteny, the development of the somatic organisation is retarded relative to development of the reproductive system, such that the animal attains sexual maturity at normal size but retains a juvenile aspect in its soma. This new end point of development must obviously be associated with a different configuration of functions and a different mode of life.

Many examples ascribable to this pattern of change in developmental pathways have been identified (Raff & Kaufmann 1983). Neoteny is common among urodele amphibians, yielding aquatic end points such as the axolotl, and specialised sidelines such as arboreal salamanders. Neoteny also appears to underlie the repeated evolution of flightless birds, involving the failure of development of the adult features associated with the flight mechanism and the retention of many chick-like juvenile features as components of the functional running animal. It is interesting that this modification has always occurred in bird species in which sternum development begins after hatching, so that the deletion of the flight mechanism does not interfere with general somatic development of the embryo, but is a post-hatching event. The association of large size with flightlessness in birds is a secondary phenomenon involving acceleration. Neoteny is also evident in the evolution of the genus *Homo*, especially in *H. sapiens*.

In the amphibian examples of neoteny, the developmental causal system whose modification brings about the changed result can be recognised. It is the system of interaction of the developing tissues with those few hormones that normally cause amphibian metamorphosis. Here a large difference *can* result from a small developmental change. In the evolution of flightless birds and of man, on the other hand, the developmental changes underlying neoteny are more subtle shifts in the balance of rates and timing between a multitude of integrated developmental processes, so that a gradual sequence of evolutionary change is required. Human evolution, if we retain the concept Hominoidea for the time being, provides an example of the timing involved – from forest dwelling ape to bipedal hominoid over 8 million years, and from basic hominoid progressively to *Homo sapiens* over 4 million years, the latter through a succession of at least two genera and several species (Leakey 1981). This is not unusually fast (see Templeton, this volume, for supporting genetic evidence indicating gradual evolution in hominoids), so that neoteny as a mode of evolution through developmental change does not necessarily take us from a gradualistic to an instantaneous rate of change.

In the case of paedogenesis, there is again one proposal for rapid major evolution that must be seriously entertained. This is the evolution of hexapods from myriapodous ancestors (De Beer 1958, Anderson 1973, Manton 1977, Manton & Anderson 1979). It involves the postulation of a group of hypothetical early Devonian myriapods hatching as hexapodous juveniles, completing the development of a further 11 leg-bearing segments post-hatching, and reaching sexual maturity as adults with 14 pairs of trunk limbs. These animals also need to have been trignathous labiate at the front end and opisthogoneate at the back end (Anderson 1973, Manton & Anderson 1979). The latter suggests that the hatching juvenile already had the full complement of trunk segment rudiments, but with functional leg pairs only on the front three. This is a perfectly satisfactory kind of developmental sequence for a hypothetical herbivorous myriapod, but how much developmental change is required to transform such an animal into a functional hexapod? Paedogenesis requires that the development of the reproductive system should be accelerated relative to that of the somatic system, bringing the reproductive system to sexual maturity and function when the animal is still small and hexapodous (previously juvenile). Since growth, moulting and sexual maturation in hexapods are under hormonal control, it is possible that our hypothetical myriapod species could suddenly spawn a sprinkling of sexy little hexapods through a small change in the genetic control of the hormonal control of post-hatching development. Given the right circumstances and antecedents, then, paedogenesis has a lot of potential for major modification of developmental pathways and functional end points at a fast evolutionary rate.

Once the basic hexapod organisation had evolved, its diversification in subsequent evolution can be explained almost entirely on the basis of gradually evolved modifications of development (French 1983), but with one exception. There is another event in hexapod evolution that requires the concept of an instantaneously introduced developmental change. I refer to the origin of endopterygote insects from their exopterygote forebears. In the first instance, exopterygotes became endopterygotes because wing buds invaginated instead of evaginating. All other aspects of development, structure and function could have remained the same, except for a subsequent evagination of the grown buds at the penultimate moult. A change from evagination to invagination of a small patch of growing epithelium is a very minor developmental event, something which could evolve in consequence of a single mutation and be immediately of positive selective value. A small, ground-dwelling juvenile hexapod without extraneous knobs is obviously more efficient than one with them. Subsequently, the endopterygote developmental system has undergone very elaborate modifications of pathway, leading to the extremes of dual body form developed in the higher Diptera (Anderson 1972, 1973, Raff & Kaufmann 1983); but all of this, once again, like the diversification of hexapods in general, required the evolution of gradual changes in a complex of integrated developmental pathways, occurring at moderate rates.

The exopterygote/endopterygote case is analogous to another evolutionary event involving a change from epithelial evagination to epithelial invagination. Oster and Alberch (1982) have demonstrated that in the development of amniote scales and feathers, the epidermis evaginates over each dermal papilla, whereas in the development of mammalian hairs, the homologous epidermis invaginates. It is obvious that the initiation of hair evolution could thus have been rapid, but this does not belie the gradual evolution of the complex mammalian organisation over many millions of years during the Mezozoic, after proto-hair had evolved.

Another famous proposal for major evolution by paedogenesis is the origin of vertebrates from tunicate larvae. It is now realised that the development of the ascidian tadpole is extremely specialised and unlikely to provide a suitable functional basis for the evolution of segmental chordates; however, the general idea of a paedogenetic origin of vertebrates lingers on (e.g. McFarland *et al.* 1979). Interestingly, the concept, which initially postulated a big-bang event (Berrill 1955), has become remarkably watered down. In recent versions, we find only paedogenetic unsegmented tunicates being postulated as the first pelagic chordate adults. The hypothetical sequence of developmental changes leading from this beginning to the first segmented vertebrates then has a strongly gradualistic flavour. Perhaps the time has come to abandon the notion of paedogenesis as a significant factor in early chordate evolution, and to develop a wholly gradualistic interpretation of this phenomenon.

DEVELOPMENTAL CONSERVATISM

Although phenomena such as the origin of hexapods and the origin of endopterygotes provide examples of major evolutionary events consequent on instantaneous deviations in development, such processes are extremely rare. On the whole, a gradual assimilation of developmental changes is usually essential if developmental pathways are to change in ways that yield the evolution of new types of functional organisation. In fact, the highly integrated stepwise nature of animal development causes it to be in many respects an extremely conservative process. Basic developmental events established during the early evolution of a group are maintained repetitively over hundreds of millions of years, since any change in them would spell extinction. Some obvious examples of this, selected here because the developmental pathways involved have been studied in detail by embryologists, are the development of the segmental compartments in pterygote insect embryos (French 1983, Sander 1983), the notochord–neural tube induction system of chordates (Hall 1983), and the development of the tetrapod pentadactyl limb (Maderson 1983, Wolpert 1983). But one could, in fact, nominate any of the bauplan pathways that result in the development of the basic organisation in any phylum or class. For these basic developmental processes, once established, the rate of evolution has been minimal. We shall see below that such developmental conservatism rests with highly conserved components of the genome controlling the developmental pathways involved. The ancient and conservative nature of these systems can also be demonstrated epigenetically. The notochord–neural tube induction system of chordates, for example, is operational with combinations of tissues from any of different chordate groups (Hall 1983). A Cambrian or Precambrian origin of this inductive interaction can be inferred. Similarly, a basic single pattern of development, presumably evolved in the Devonian, pertains to all pentadactyl limbs. Variations in rate and timing account for the ways in which this pattern yields a wide range of functional end products, all without disruption of the original pattern (Maderson 1983).

At the same time, it is possible to identify the time of origin of a small change in the basic pattern of development of the pentadactyl limb, obviously resulting from a functional genetic change. In amphibians, the formation of the digits during limb-bud development is effected by differential mitosis. In reptiles, the same process results from differential cell death. The evolution of this change in a basic epigenetic process must have stemmed from a genetic change during the amphibian–reptilian transition in the Carboniferous. We have no precise knowledge of when it arose or of how long the transition took, but we can say that, once evolved, the reptilian mechanism of basic digit formation has persisted throughout amniote evolution. The functional significance of the change is not immediately apparent, but can perhaps be understood in the context of the rôle of cell death in morphogenesis. As pointed out by Katz (1982), selective cell death depends on the interaction of many genes. A morphogenetic process involving controlled cell death therefore has greater developmental plasticity than a process based on controlled mitosis. One can speculate that the greater digit diversity potentiated by this plasticity might have been of considerable functional consequence in the early evolution of fully terrestrial tetrapods.

Added to the overt developmental conservatism of animals is the phenomenon of hidden conservation of developmental mechanisms (Hall 1983). The unexpected re-emergence of these mechanisms is the underlying cause of atavisms. They can also be revealed experimentally. For example, chick mandibular arch epithelium, when combined with mouse dental mesenchyme, forms dental enamel; yet birds have not had teeth since the Cretaceous. In spite of the long-standing deletion of the dental inducer in bird development, the epithelial response mechanism is still conserved.

We can discern a whole series of levels at which the conservation of developmental control mechanisms operates, with the general rule that later developmental events are more open to the functional divergence of developmental pathways than earlier ones (Horder 1983). Von Baer's rules, and the palaeontological history of any multidivergent group, give conceptual and factual expression to this aspect of the relationship between development and evolution. In special circumstances, certain early events in development may be subject to extreme functional divergence (e.g. events related to the evolution of large yolky eggs, or placentation, or polyembryony), but even in these situations the essential bauplan pathways leading to adult structure are still conserved.

On a shorter time scale, developmental conservatism is the basis of the stability of a species through time (Maderson 1983, Maynard Smith 1983). At this level, all developmental pathways are held to a minimum of variation. When a species is in stasis, the cumulative rate of change in developmental pathways, and hence in the control systems generating these pathways, is zero. This is, of course, a dynamic condition. The variance is there genetically (Carson, this volume), and to some extent it is expressed, but mainly it is held in check by the processes subsumed under natural selection.

VESTIGIALITY

Sometimes, as mentioned above, the divergent pathways of evolution lead animal populations into modes of life in which previously functional components of integrated adult organisation become redundant. Such redundancy always leads to the vestigiality, and sometimes to the complete disappearance, of the component. As pointed out by Carson (this volume), if selection is removed, characters decay back to the level at which selection again operates. From a developmental point of view, the fact that vestigiality precedes disappearance implies a more gradual loss of developmental function than of physiological and interactive function (Horder 1983, Raff & Kaufmann 1983), and is another manifestation of the integrated nature of development.

Usually the vestigiality of redundant components is so far advanced in contemporary organisms that the developmental changes that led to the vestigial condition cannot be analysed; but in a few cases, as the eyes of cave-dwelling fishes and amphibians, this has been possible. What happens here is very interesting. Once the functional pressure is off, the developmental pathways begin to go into random deviations, the only common feature of which is a reduction in the resources expended on the redundant pathway. For example, in a cave fish with reduced eyes, the development of the optic cup is delayed in time, the rate of retinal mitosis is slowed, the inductive capacity of the optic cup is reduced in consequence, and only a vestigial lens is developed. Perhaps the most interesting thing about this example is the fact that these developmental processes persist at all. From an interactive functional point of view, they could easily be deleted totally. Evidently, the conservatism of developmental control mechanisms is such that the natural breakdown of a component of the control system can be tolerated only through gradual change over many millions of years. In a cave salamander, the breakdown of the eye-development system is much less advanced and has taken a different route. Here, a normal optic cup is still developed but lens induction fails, and no lens is formed.

EVOLUTIONARY RATES

Although it is possible to identify various ways in which the rate of evolution is related to the rate of change of developmental pathways over time, the topic suffers from an almost complete absence of hard data. No contemporary natural system of development is undergoing change at any recognisable rate. The time scale is too short. Changes in developmental pathways over time can be identified as a universal feature of historical evolution, but how fast or how slowly did they occur? More importantly, how fast did the underlying genomic modifications that brought about these changes evolve? The general considerations outlined above lead to the conclusion that almost all such changes have come about gradually and that the developmental system has a very high level of in-built conservatism. This is not really surprising when one recognises that developmental conservatism is an essential prerequisite of species stability. Whether this interpretation is applicable to the early phase of evolutionary diversification of the animal phyla can only be surmised. Organisations were simpler and developmental pathways concomitantly simpler during that phase of animal evolution. But changing them still involved the same set of complex functional problems, so probably was constrained to similar rates. Exceptions in which major changes have occurred quickly always involve special circumstances. For example, the evolution of large animals has often taken place in a relatively short time, as with the Pleistocene megafaunas. Developmentally, this is not a major problem. Accelerated growth rates yielding large size have little impact on any other aspect of functional developmental integration. The initial evolution of hexapod uniramians may have occurred almost instantaneously, as discussed above, but a whole concatenation of favourable circumstances of developmental and functional organisation was required for this to be possible. Perhaps the message of developmental biology for evolutionary studies, as already suggested by Hall (1983), is that we should be investigating not the causation of divergence, but the mechanisms that conserve and constrain developmental pathways from the level of species stasis to the level of persistence of phyla for hundreds of millions of years. Only when the capacity to hold the status quo is understood in genetic/developmental terms will the process underlying evolutionary change over time, and the factors that govern the rates of this change, begin to be identified. For the palaeontologist, too, developmental biology has a message. The interpretation of palaeontological data must take account of a dictate of living systems – that gradual functional change is required – but can recognise that the dynamics of the developmental system are able to accommodate rates of change that are geologically fast, and are not at odds with the concept of punctuated equilibrium any more than with the concept of gradualism. Hopeful monsters and dragons teeth are not required.

GENETIC CONTROL SYSTEMS

If developmental systems, as is argued here, are generally conservative and tend to impose slow rates of change on evolution, is there any corroboratory evidence bearing on the rate of change of the genetic control systems that underlie them. Severe limitations attend this topic because of the present lack of understanding of how the higher levels of developmental expression are causally genetically controlled (Horder 1983, Miklos & John, this volume). There is some evidence about the evolution of developmentally active genes, however, and it does tend to indicate that even small changes take a long time to evolve. By judicious selection of echinoderm material in relation to known events of the fossil record, Raff *et al.* (1984) have been able to elucidate differences in gene expression by regulatory genes during early development, in cidaroids as compared with more advanced echinoids. The results of this investigation indicate:

(a) that these differences evolved during a period of 5–10 million years of the Early Triassic,
(b) that they have remained stabilised since that time,
(c) that the ancestral (cidaroid) condition, shared with asteroids and holothuroids, had been a stable component of echinoderm development since the Cambrian.

Unfortunately, the developmental differences examined in this study by Raff and his colleagues are very minor; a question of whether α-histone genes are transcribed either during early cleavage (the cidaroid condition) or during oogenesis (the more advanced condition). The functional significance of the difference is not known. If minor modifications at this level of the genetic control of development took several million years to establish themselves, there is every reason to believe that anything more developmentally complex, involving functional interactions between many genetic components, might normally take considerably longer to undergo significant evolutionary change.

This view is sustained by recent advances in molecular genetics that have revealed an astonishing complexity in the organisation and functioning of eukaryote genomes. As pointed out by Campbell (this volume), and Miklos and John (this volume), developmental processes are controlled by highly organised but dynamically structured multigene families. The manner in which these complex components of the genome encode and express the information required for the development of complex phenotypic traits is far from clear. It also seems likely (Campbell, this volume) that the genetic system itself imposes further direction and constraint on the process and rate of evolutionary change. Indeed, the conservatism so evident in developmental systems has proved to be more accessible to molecular genetic analysis than the genetics of change. The 'homoeo-box' component of the eukaryote genome discussed by Thomson (this volume), which plays a fundamental controlling rôle in the establishment of segmentation in the embryos of metameric animals, provides an example of a highly conserved genetic mechanism that directly controls the delimitation of groupings of cells. The recognition of gene complexes that are so directly involved in developmental control is certainly a breakthrough, but there is still a long way to go. The genetic mechanisms controlling the further development of these defined cell groupings into supracellular organisation, for example, are still unknown. It is also now known that the genes for development reside in only a small part (Miklos & John, this volume, Templeton, this volume) of a highly dynamic, constantly changing genome replete with redundancy and comprising a number of semi-independent subsystems. A considerably greater understanding of structure and function at the genomic level will need to be achieved before the elusive functional nexus of genomic change, epigenetic change and evolutionary change is effectively gaffed and landed.

REFERENCES

Alberch, P. 1982. Developmental constraints in evolutionary processes. In *Evolution and development*, J. T. Bonner (ed.) pp. 313–32. Berlin: Springer Verlag.

Anderson, D. T. 1965. Morphological integration and animal evolution. *Scientia* 102, 1–6.

Anderson, D. T. 1972. The development of holometabolous insects. In *Developmental systems – Insects*, S. J. Counce and C. H. Waddington (eds), pp. 165–242. London: Academic Press.

Anderson, D. T. 1973. *Embryology and phylogeny in annelids and arthropods*. Oxford: Pergamon Press.

Anderson, D. T. 1979. Embryos, fate maps and the phylogeny of arthropods. In *Arthropod phylogeny*, A. P. Gupta (ed.) pp. 59–105. New York: Van Nostrand Reinhold.

Berrill, N. J. 1955. *The origin of vertebrates*. Oxford: Claredon Press.

Counce, S. J. and C. H. Waddington, 1973. *Developmental systems – Insects*, Volume 2. London: Academic Press.

De Beer, G. R. 1958. *Embryos and ancestors*. Oxford: Clarendon Press.

Dullemeijer, P. 1980. Functional morphology and evolutionary biology. *Acta Biotheoretica* 29, 151–250.

French, V. 1983. Development and evolution of the insect segment. In *Development and evolution*, B. C. Goodwin, N. Holder and C. C. Wylie (eds), pp. 161–93. Cambridge: Cambridge University Press.

Hadorn, E. 1961. *Developmental genetics and lethal factors*. New York: Wiley.

Hall, B. K. 1983. Epigenetic control in development and evolution. In *Development and evolution*, B. C. Goodwin, N. Holder and C. C. Wylie (eds), pp. 353–80, Cambridge: Cambridge University Press.

Horder, T. J. 1983. Embryological bases of evolution. In *Development and evolution*. B. C. Goodwin, N. Holder and C. C. Wylie (eds), pp. 315–52. Cambridge: Cambridge University Press.

Katz, M. J. 1982. Ontogenetic mechanisms: the middle ground of evolution. In *Evolution and development*, J. T. Bonner (ed.), pp. 207–12. Berlin: Springer Verlag.

Leakey, R. E. 1981. *The making of mankind*. London: Michael Joseph.

Lehmann, U. 1981. *The ammonites*. Cambridge: Cambridge University Press.

McFarland, W. N., F. H. Pough, T. J. Cade and J. B. Heiser 1979. *Vertebrate life*. New York: Macmillan.

Maderson, P. F. A. 1983. The rôle of development in macroevolutionary change. In *Evolution and development*, J. T. Bonner (ed.), pp. 279–312. Berlin: Springer Verlag.

Manton, S. M. 1977. *The arthropoda: habits, functional morphology and evolution*. Oxford: Clarendon Press.

Manton, S. M. and D. T. Anderson 1979. Polyphyly and the evolution of arthropods. In *The origin of major invertebrate groups*, M. R. House (ed.), pp. 269–321. London: Academic Press.

Maynard Smith, J. 1983. Evolution and development. In *Development and evolution*, B. C. Goodwin, N. Holder and C. C. Wylie (eds), pp. 33–45. Cambridge: Cambridge University Press.

Oster, G. and P. Alberch 1982. Evolution and bifurcation of developmental programs. *Evolution* 36, 444–59.

Raff, R. A. and T. C. Kaufmann 1983. *Embryos, genes and evolution*. New York: Macmillan.

Raff, R. A., J. A. Anstrom, C. J. Huffman, D. S. Leaf, J-H. Loo, R. M. Showman and D. E. Wells 1984. Origin of a gene regulatory mechanism in the evolution of echinoderms. *Nature* 310, 312–14.

Sander, K. 1983. The evolution of patterning mechanisms: gleanings from insect embryogenesis. In *Development and evolution*, B. C. Goodwin, N. Holder and C. C. Wylie (eds), pp. 137–59. Cambridge: Cambridge University Press.

Wolpert, L. 1983. Constancy and change in the development and evolution of pattern. In *Development and evolution*, B. C. Goodwin, N. Holder and C. C. Wylie (eds), pp. 47–58. Cambridge: Cambridge University Press.

8

Population biology and evolutionary change

I. R. FRANKLIN

ABSTRACT

Evolutionary biology, traditionally the domain of the naturalist, the palaeobiologist and the theoretical geneticist, is now dominated by molecular clocks, molecular drives, selfish DNA and the reconstruction of phylogenetic relationships from molecular data. The theoretical basis of evolutionary theory, population biology, has declined in importance due to failure of theory to cope with observation.

This chapter examines the current structure of theoretical population genetics and ecology and their contribution to contemporary evolutionary biology.

POPULATION BIOLOGY – OBSOLETE, IRRELEVANT OR PREMATURE?

> In many ways the lot of the theoretical population geneticist ... is a most unhappy one. For he is employed, and has been employed for the last 30 years, in polishing with finer and finer grades of jeweller's rouge, those three colossal monuments of mathematical biology *The causes of evolution*, *The genetical theory of natural selection* and *Evolution in Mendelian populations*. (Lewontin 1964)

In 1979, Bentley Glass presented a list of the major advances in genetics since the rediscovery of Mendel in 1900. Among nearly 200 citations, population genetics was credited with only 11. The most recent was the 1966 finding that the genome is highly polymorphic. The latest theoretical contribution listed was from the mid-1930s. The most recent Cold Spring Harbour Symposium on Quantitative Biology to discuss population genetics or ecology was the Darwin Centennial issue 25 years ago. These observations are symptomatic of population biology's declining image. Molecular biology's dominant rôle in science is now firmly established, but many problems that are not expressed simply in molecular terms have been relegated to the background.

However, there has been no shortage of debate in the last 25 years on the nature of genetic loads, on neutralism, on kin selection and, most recently, on punctuated equilibria. Also, we now have information on genetic variability in natural populations which should by now have marked the beginning of new insights into the theory of evolutionary change. Unfortunately, the results have been disappointing.

> Quite suddenly, the situation has changed. The mother-lode has been tapped and facts in profusion have been poured into this theory machine. And from the other end has issued – nothing. (Lewontin 1974)

Lewontin goes on to point out that the machinery does indeed work. The problem is the inability to discriminate between opposing hypotheses. In practice, and sometimes in principle, much of population genetics theory fails the test of falsifiability. Even where there exist null hypotheses, such as those provided by neutral theory, attempts to apply statistical tests have lacked the power to reject the hypothesis under test. Attempts to generate null models for testing hypotheses in community ecology have also encountered difficulties (Quinn & Dunham 1983).

What, then, is the rôle of population genetics and ecology in modern biology? Kimura (1983) is optimistic. Flushed with the success of the neutralist interpretation of variability and evolutionary change at the molecular level, he sees a 'new uncharted territory awaiting exploration'. Ewens (1983) looks to the future from a viewpoint strongly influenced by the analysis of neutral differences at the genic level. He argues for a shift in emphasis in theoretical genetics from a primarily deductive science to an inductive one. The reason, he suggests, is in part because deductive theory has 'largely fulfilled its rôle in describing evolution as a genetic process'.

There is still a place for traditional deductive analysis. Almost daily, molecular biology reveals new features of the genetic material which call for evolutionary explanations. We are still ignorant of the genetics of quantitative variability, the genetics of speciation and the relationship between complexity and community stability.

Nevertheless, I agree with Ewens. Theories that are directed at the interpretation of data, based around hypothesis testing and estimation, are crucial if population biology is to occupy more than a peripheral rôle.

A digression on prematurity and irrelevance

> A discovery is premature if its implications cannot be connected by a series of simple logical steps to canonical, or generally accepted, knowledge. (Stent 1980)

The theory of evolution links all of biology. Consequently, findings in molecular biology should, through evolutionary considerations, have strong implications for population biology, and vice versa. Part of this influence is already evident. New branches of population genetics have sprung up as theory struggles to cope with molecular data and with evolutionary explanations for newly revealed complexities in the genome.

However, population genetics is basically a theory of rates of change of heritable and mutable entities and often it matters little whether these are DNA sequences or beads on a string. Even today, most of the theory established in the 1930s by Fisher (1930), Wright (1931) and Haldane (1932) remains the cornerstone of population genetics. In some respects, population genetics has been revitalised by molecular biology, but the relationship is not mutual. The situation is even worse for ecology and molecular biology – the two disciplines scarcely intersect.

Does this lack of connection mean that the findings of population biology are premature to molecular and cellular biology? We need, I think, to draw a distinction between irrelevancy and prematurity. Prematurity implies that we have some reason to expect that a discovery will eventually be connected to the discipline in question. However, we do not expect new estimates of the Hubble constant to have much impact on population biology. Cosmology and population biology are largely, but not entirely, irrelevant to each other and will remain so for some time to come.

Will population biology become more strongly integrated with the rest of biology, or are the disciplines destined to remain separate, requiring a different level of description? Even worse, is the current theory so misdirected that it will be replaced eventually by something of greater heuristic and empirical value?

CONFLICT IN POPULATION BIOLOGY

In any dispute between A and B, there are four possible outcomes: A can be right and B wrong; B right and A wrong; both can be wrong; or both can be right. . . . Seemingly, the fourth case of disputes should not occur at all. (Salt 1983)

Population biology has a history of thesis and antithesis but rarely synthesis. The resolution of conflict between the Mendelians and the Biometricians (Fisher 1918 – a reference not in Glass's list) is perhaps an exception. Fisher and Wright clashed over the relative importance of subdivision and genetic drift. The theoretical and experimental schools argued about mutational versus segregational loads. Heated debates followed between neutralists and selectionists, between sociobiologists and everyone else, and, most recently, between the advocates of punctuated equilibria and those supporting the more traditional 'phyletic gradualism'. None of these issues have been resolved to the satisfaction of the participants. Eventually it seems the antagonists become bored with the debate and search for new shibboleths. In fact most of these issues are straw men; if it were available I would expect critical evidence to reveal much truth on both sides.

Population ecologists are not immune to these squabbles – the arguments that preceded the general acceptance of the logistic curve of population growth were bitter and divisive. Even here the case was not decided on its merits – acceptance had more to do with the strong will of Raymond Pearl than anything else (Kingsland 1982). There remains, as in population genetics, a long standing mistrust of mathematical modelling by experimental biologists (see Roughgarden 1983, Simberloff 1983).

In the physical sciences, such a polarisation of viewpoints is common and considered healthy. In population biology I consider these conflicts a sign of malaise. The rôle of debate in science is to clarify the issues and to stimulate the collection of data designed to discriminate between opposing theses. Such is rarely the case in population biology. For a variety of reasons the mutual stimulation of theory and experiment (or observation) is inadequate.

The nexus between theory and experiment

Evolutionary processes involve slow rates of change. Even industrial melanism, an extraordinarily fast process by evolutionary standards, involved a period of more than half a century. Evolution at higher levels, such as those involving processes of speciation and extinction, are many orders of magnitude slower still. These long timescales make observation difficult and the experimental study of evolution even more so. However, this is only part of the problem. The major difficulty lies in the paucity of falsifiable theory.

The mathematical equations of population genetics and population ecology have not culminated in a dynamical description; the equations are, for the most part, rearrangements of kinematic variables. Current theory, therefore, has a strong tautological component. Dynamics are necessary for prediction, and prediction is necessary to stimulate experiment and observation. In fact, the purpose of the early work on the foundations of theoretical population genetics and ecology was not to produce falsifiable hypotheses but to deduce the properties of evolutionary systems. In doing so, it mattered little whether the analysis involved variables that make biological sense or those that are measurable in other than a tautological fashion.

Finally, experimental biologists are not always noted for their numeracy, nor for their understanding of the analytical goals of theory. Some, unfortunately, enter biology for that very reason. Conversely, some theoreticians, especially those entering from physics or mathematics, have a poor experimental knowledge of biology. Misunderstanding is commonplace. Indeed many of the debates, both in genetics and ecology, have been fueled by the differing aims and perspectives of experimental and theoretical biology. Roughgarden (1983), reacting to some recent criticisms of mathematical ecology, suggested that the disagreement between theoreticians and empiricists springs from the ethic of perceptiveness – the attention to detail – of the naturalist and the need of the theoretician to simplify.

There is now a strong need to integrate theory and experiment in population biology. I believe that our first step should be to eradicate sloppy and ambiguous definition of important evolutionary variables. Also, I suggest, many of the problems in communication and co-ordination among biologists result from a failure to specify clearly the biological level appropriate to the evolutionary system under study.

ON THE DEFINITION OF TERMS

Let us borrow, if we can the method of the physicist: He discovers a quantity $1/2mv^2$ possesses certain important properties. Then, he proceeds to name it: Energy, in particular kinetic energy. But biologists have been disposed sometimes to adopt the reverse procedure: they have named a vital force, a mental energy, and what not, and now they entertain the pious hope that in due time they may discover these 'things'. That there is something radically at fault with such terms is evident from the fact that forces and energies are magnitudes, and 'to define a magnitude and to say how it is measured is one and the same thing'. (Lotka 1925)

It is not mere pedantry to expect that variables in biology be defined unambiguously and, preferably, in such a way that their measurement is implicit. Much of the debate mentioned above has been obfuscated by ambiguous definition of important biological concepts. I will illustrate this with some definitions of three important notions in evolutionary theory, namely *fitness*, *polymorphism* and *niche*.

Fitness

Fitness enters biology as a vague heuristic notion, rich in metaphor but poor in precision. (Levins 1968)

Fitness is perhaps the most abused term in population biology. It is used by various population biologists in at least three major ways, and there are variants within each category.

The number m... measures the relative rate of increase or decrease of a population when in a steady state appropriate to any such system. In view of the emphasis laid by Malthus upon the 'law of geometric increase' m may appropriately be termed the Malthusian parameter of population increase. m measures fitness by the objective fact of representation in future generations. (Fisher 1930)

The fitness of any particular genotype is half the mean number of progeny left by an individual of that genotype. (Haldane 1937)

These two definitions are similar conceptually but different in fact. At the genic level there is a close correspondence, but since sexually reproducing individuals do not reproduce their own kind, the expected number of offspring of a genotype is not the same as its intrinsic rate of increase.

Haldane's definition is similar to that used by Sewall Wright (1968) for a quantity which he called '*selective value*'. Subsequently, the term fitness and selective value have become synonomous for most population geneticists, although Wright never equated the two.

The fitness of a unit (of evolution) is its probability of leaving descendents after a given long period of time. (Thoday 1953)

Here Thoday introduces another notion: the long-term survival of an evolutionary unit. No matter how sympathetic one might be to the concept, this is the kind of definition that we should seek to eradicate. It defines a quantity that is unmeasurable in principle and thus is empirically empty.

The fourth, which I believe to be the appropriate historical usage, relates fitness to adaptation, i.e. the 'fit' of an organism to its environment.

Fitness is a measure of the quality of the relationship between an organism and its environment. (Williams 1970)

However, though adaptation is a term widely used in biology and is fundamental to modern evolutionary theory, it too has never been adequately defined; this most important Darwinian concept, together with the fitness concept, remains inaccessible to the theorist. The above definition, for example, provides us with no guidelines for measurement. Descriptive biology is not so constrained; lyrical and metaphorical accounts of adaptations abound in the evolutionary literature.

Polymorphism

Polymorphism is the occurrence together in the same locality of two or more discontinuous forms of a species in such proportions that the rarest of them cannot be maintained merely by recurrent mutation. (Ford 1940)

Ford's definition of *polymorphism* is a curious one; it excludes continuous variation and calls for a conclusion about the selective forces involved. These qualifications render the definition operationally useless. There is a need to quantify the amount of variability in a population without reference to the causes of that variability. At the genic level, average heterozygosity is such a measure and is generally used in preference. The quantitative geneticist uses genetic and phenotypic components of variance to measure variability, and the ecologist has chosen a number of *ad hoc* measures such as *diversity* and *abundance*.

Niche

For any species S_1 .. an n-dimensional hypervolume is defined, every point in which corresponds to a state of the environment which would permit the species to exist indefinitely . . . This hypervolume will be called the *fundamental niche* of S_1. (Hutchinson 1958)

MacArthur (1968) suggested that this definition allowed the term *niche* to enter into falsifiable statements for the first time. In ecology it marks the beginning of theoretical modelling of competition, character displacement and island biogeography. However, as an operational definition of a niche it is completely open ended. It allows an observer to decide that two niches are different but can never serve to define a niche for a single species.

THE HIERARCHICAL STRUCTURE OF EVOLUTIONARY BIOLOGY

> Are the different levels of organisation such as atom, molecule, cell, organism, species, and community only the epiphenomena of underlying physical principles of are the levels separated by real discontinuities? (Levins & Lewontin 1980)

Evolutionary change occurs at a number of biological levels. Three common examples are change in gene frequency, change in the average phenotype of a population of organisms and changes in the composition of biological communities. All of these units of evolution exhibit the properties of heredity, multiplication and variation (Maynard Smith 1983), and at any time natural selection may be operating at any one of or at all three levels. But this is far from an exhaustive list. Transposable elements, plasmids, chromosomes, demes and species are all self-reproducing entities, and natural selection can occur at any of these levels. Each of these entities is, biologically, a subset of another; therefore, they can be ordered hierarchically (Arnold & Fristrup 1982).

It is vital when attempting to model evolution, either mathematically or linguistically, to draw a distinction between the level at which natural selection is operating, the unit of selection and the level at which we wish to describe the change, viz. the unit of inheritance.

For the geneticist the most important unit of inheritance is the haploid genome, not the gene, since the bulk of the genome is replicated co-ordinately and transmitted to the next generation as a unit. Also, the genome is an important functional unit since genes are not transcribed and translated independently of one another. For these reasons, and because it is the diploid organism that is most commonly exposed to the environment, the most important focus for natural selection is the individual.

However, the major contribution of population genetics is to show that changes that occur at one level (say, the genotype) can be expressed at a different level (such as the gamete or the gene) *if and only if* certain interaction terms are assumed to be unimportant. Modelling in population genetics has depended almost entirely on this reductionist approach, and much effort has gone into analysing the consequences of, and justification for, various simplifying assumptions. Unfortunately, this work is greatly misunderstood by those who adopt a holistic view of biology. At the other extreme, these explorations have led some to suppose that the gene is not only a unit of inheritance but, predominantly, a unit of selection as well. This view permeates sociobiology and is exemplified, in its extreme, in *The selfish gene* (Dawkins 1976). However, there are circumstances where genes can be fixed by mechanisms other than natural selection operating at the organismic level. Genetic drift and meiotic drive are two examples well covered in the Neo-Darwinian synthesis. The discovery of entities that vary and are capable of differential multiplication does not lead to new principles in evolution, as authors such as Dover (1982) would have us believe.

Any change at a particular level, for whatever reason, results in change at all higher levels in the hierarchy. Also, if a change is observed, say in gene frequency, there can be no grounds for attributing that change to natural selection at that level without ancillary biological information. The partitioning of the genome, at least notionally, into non-interacting subsets has allowed an exploration of the genetic basis of evolutionary change in a way that would be impossible if we were forced to treat the problem in its full complexity.

CONDITIONS FOR EVOLUTIONARY CHANGE

> Can we doubt ... that individuals having any advantage, however slight, over others, would have the best chance of surviving and procreating their kind? On the other hand, we may feel sure that any variation in the least degree injurious would be rigidly destroyed. This preservation of individual differences and variations, and the destruction of those which are injurious, I have called Natural Selection, or the Survival of the Fittest. (Darwin 1859)

Natural selection is a logical consequence of the following conditions (Lewontin 1970).
(a) A population, composed of discrete entities of different types.
(b) Competition between these entities to survive and reproduce.
(c) Correlation between parent and offspring for some attribute related to survival or fertility.

An often cited paradigm for evolution by natural selection is industrial melanism. This is perhaps atypical of most evolutionary change, for the genetic basis of most traits is multifactorial. Consider, instead, the more abstract example outlined below.

A population, at time t, has a mean for some characteristic X_t. This might be the proportion of dark-coloured butterflies, the average value of a quantitative trait such as body weight, or some measure of the abundance of a community. A subset of individuals survive and reproduce and the mean of these is X_t' ($\neq X_t$). In the following generation the mean value is X_{t+1}. Let $S_t = X'_t - X_t$ and $R_t = X_{t+1} - X_t$. $R_t \neq 0$, corrected for environmental differences in the two generations, indicates evolutionary change. Quantitative geneticists would be familiar with this representation. S and R are known as the selection differential and the response to selection. The ratio $R : S$ is known as the heritability – a measure of the correlation between parent and offspring. (Even the term heritability does not have a universal meaning amongst biologists (Jacquard 1983).) By definition, if the heritability is zero, there is no evolutionary change.

This general formulation makes no assumptions about the biology of the response of the nature of the attributes involved, but I believe is a much more apposite description of the phenomenology of evolutionary change, and one around which we need to develop predictive equations.

DEFINITION OF STATE VARIABLES

Physics-envy is the curse of biology. (Cohen 1971)

Past attempts to develop an evolutionary dynamics system have been strongly influenced by mechanics, where all laws are differential equations with respect to time. The aim has been to achieve a description of an evolving system in terms of laws of transformation operating on a set of dynamically sufficient variables.

> Evolution of the system is movement of the point through the space, tracing out a trajectory, and the laws of transformation are the equation whose solution is that trajectory. The axes of the space are the *state variables*, and the space they span is the *state space* of our system. (Lewontin 1974)

Here we encounter our first major problem in the development of a mathematical theory of evolutionary change. The state variables of the evolutionary process cannot be defined *a priori*, for without dynamics we cannot say which variables are sufficient to describe the system. A second difficulty to be faced is the representation of the dimensions of time and space, particularly the former. In mechanics, time is a primitive term (see below). In biology, the use of time in this way presents a major epistemological difficulty. The appropriate scale for most evolutionary processes is not elapsed time but generation time. However, generation time is a biological variable, and subject to modification by environment and by the evolutionary process itself. (This is particularly evident in plants where flowering or germination often occurs in response to environmental stimuli.)

Furthermore, as discussed above, the change in frequency of an evolutionary unit depends on the contribution of selective and random processes at a number of biological levels. Each of these has its own unit of time, as determined by birth and death rates of the unit of inheritance at that level. The specification of time, implicit in any discussion of evolutionary rates, presents real difficulties. The same difficulties apply to any measure of distance, for organisms differ in their vagility. Movement through space, like time, depends on biological variables which are subject to environmental and genetic modification. Attempts to describe evolution through the use of differential equations is less appropriate than in mechanics (Williams 1977).

Is there a systematic way to develop an appropriate mathematical framework? An approach taken by Williams (1970), among others, is to attempt a formal axiomatisation of the evolutionary process in which fundamental concepts of the theory are designated primitive terms. The aim is to deduce a set of generalised theorems about evolutionary systems and then to give these meaning to the real world by interpreting the primitive terms biologically. I shall not discuss this approach in detail but use some of the principles of axiomatisation to highlight the deficiencies of much of our current theory.

THE KINEMATICS OF EVOLUTIONARY CHANGE

Part of classical population genetics is just an elaborate restatement of the fact that if X_1 and X_2 are populations of two alleles, and if:

$$\frac{1}{X_1}\frac{dX_1}{dt} > \frac{1}{X_2}\frac{dX_2}{dt}$$

then X_1 is gaining on X_2. If further,

$$\frac{1}{X_1}\frac{dX_1}{dt}$$

stays greater than

$$\frac{1}{X_2}\frac{dX_2}{dt}$$

X_1 will eventually form all of the combined population. These results are routine because all of the interesting biology is hidden in the expression

$$\frac{1}{X}\frac{dX}{dt}$$

The population geneticist is content to say,

$$\text{'Let fitness} = \frac{1}{X}\frac{dX}{dt},$$

and he proceeds in a purely mathematical program. (MacArthur 1972)

Despite the cautionary comments about the use of time as a primitive term, it is impossible to discuss most of the basic mathematical theory, especially that concerned with evolutionary rates, without accepting it as primitive. I shall be concerned in this section with five primitive terms. These are entity, attribute, population, population size and time. The purpose of the exercise is to define some variables that are simple functions of these primitive terms, and to derive relationships that will be true for any evolving system. It should be apparent that theorems in population biology involving only functions of primitive terms have little meaning unless they can be interpreted biologically.

An *entity* is an identifiable unit of inheritance, such as a nucleotide sequence, gene, gamete, genotype, mating type, deme etc. It is assumed that each of these entities can exist in a variety of alternative forms, discrete or continuously varying, which I shall call its *attribute*. A set of entities, which in turn form an entity at the next higher level, I shall call a *population*. The number of entities in the population is the *population size*.

I will avoid the use of the term fitness. There seems little hope that the world's biologists will cease to equate either Fisher's Malthusian parameter or Wright's selective value with fitness, just because I, among others, feel that this is inappropriate. However, I can but try. If one were forced to accept one or the other, precedent must go to Fisher, and this usage accords better with mathematical developments in ecology. Consequently, I shall proceed from the following definition: *The Malthusian parameter of an entity is its proportional rate of change in number.* There is nothing in this definition that explicitly ties rates of change to natural selection, let alone to adaptation. The quantity defined above is (tautologically) a measure of evolutionary change, regardless of the cause. Mutation, migration and sampling effects all contribute.

Following Fisher, I shall denote the Malthusian parameter by the letter M. Consider a population comprising $N(t)$ entities at time t. If $b(t)$ and $d(t)$ are instantaneous birth and death rates:

$$N(t) = b(t)N(t) - d(t)N(t)$$
$$= M(t)N(t)$$

Writing this difference equation as a differential, and dropping the (t)s for convenience, then:

$$M = \frac{1}{N}\frac{dN}{dt} = \frac{d\log N}{dt}$$

Suppose now that the population can be subdivided into subclasses, and that the number of representatives in the ith class is N_i. Then:

$$M_i = \frac{1}{N_i}\frac{dN_i}{dt}$$

The change in frequency (p_i) of the ith attribute is:

$$d\log p_i/dt = d\log(N_i/N)/dt$$
$$= d\log(N_i)/dt - d\log(N)/dt$$
$$= M_i - M$$

therefore:

$$dp_i/dt = p_i(M_i - M)$$

This formula, or something like it, such as:

$$\Delta p = \frac{p(1-p)}{2W}\frac{\partial W}{\partial p}$$

where W is Wright's selective value, often appears in the genetical literature purporting to be some sort of dynamical equation for gene frequency change. It is, of course, nothing of the kind.

The proportional rate of change in frequency, $m_i = M_i - M$ we can call the *relative Malthusian parameter* of the ith attribute. Relative and absolute values are statistically uncorrelated. Malthusian parameters have useful properties. For example, for two alternatives A_1 and A_2, if $m_1 > m_2$ then A_1 is replacing A_2. If $m_1 = m_2$ then A_1 and A_2 are in equilibrium.

It is often desirable to consider the joint frequency of evolutionary units. For example, the genotype at a locus can be redefined, at a lower level, as a pair of alleles. Then, the frequency of the kth genotype becomes:

$$p_k = p_{ij} = p_i p_j \theta_{ij}$$

and:

$$m_{ij} = m_i + m_j + \varepsilon_{ij}$$

where:

$$\varepsilon_{ij} = \mathrm{d} \log \theta_{ij} / \mathrm{d}t$$

This is a mathematical formalisation of the principle discussed above, namely that frequencies and rates of change can always be partitioned into frequencies and their rates of change at lower levels. The new variables θ and ε are interaction terms, and almost all of theoretical population biology has been developed around assumptions of additivity ($\theta = 1$, $\varepsilon = 0$) or that θ and ε have a simple functional form.

It is not appropriate here to discuss the theory derived from simple rearrangements of primitive terms. Indeed, whole books have been written on the subject. A recent example is Ginzberg (1983). However, I will discuss one aspect of this theory, namely the theory of rate of change of M, which is a measure of the rate of evolutionary change.
From:

$$M = \sum_i p_i M_i$$

we have

$$\frac{\mathrm{d}M}{\mathrm{d}t} = \sum p_i \frac{\mathrm{d}M_i}{\mathrm{d}t} + \sum M_i \frac{dp_i}{dt}$$

$$= \frac{\overline{\mathrm{d}M}}{\mathrm{d}t} + \sum M_i p_i (M_i - M)$$

$$= \frac{\overline{\mathrm{d}M}}{\mathrm{d}t} + \mathrm{Var}\,(M)$$

The rate of change in M is, therefore, separable into two components, the first of which ($\overline{\mathrm{d}M/\mathrm{d}t}$) is the rate of change of the M_i averaged over all subpopulations, and the second is the population variance of M. This formula can be simply extended to lower levels. For subpopulations of size two:

$$\frac{dM}{dt} = \sum_{i,j} p_{ij} \frac{dM_{ij}}{dt} + \sum_{i,j} p_{ij} (M_{ij} - M)^2$$

$$= \frac{\overline{dM}}{dt} + \text{Var}_1(M) + \text{Var}_2(M) - \sum p_{ij} \frac{d\varepsilon_{ij}}{dt}$$

The extension to many factors is straightforward. In general:

$$\frac{dM}{dt} = \frac{\overline{dM}}{dt} + \sum \text{Var}_i(M) - \frac{\overline{d\varepsilon}}{dt} \tag{1}$$

Population geneticists will recognise in this formulation something similar to Fisher's (1930) fundamental theorem of natural selection, which states that 'the rate of increase in fitness of any organism is equal to its genetic variance (in fitness) at that time'. This theorem is an interpretation of the simple formulation above, in which fitness is defined as the difference in birth and death rates for individuals, not as rates of increase for genotypes, and the variance in fitness is defined by least squares. A rigorous proof and further discussion can be found in Kimura (1958).

Other interpretations are possible. The Red Queen's Hypothesis (Van Valen, 1973) proposes that in a stable ecosystem, where dM/dt is zero, by definition, each increase in fitness in one species is distributed as a loss in fitness over the rest of the community. If the ecosystem is stable, the variance of the rates of increase of the component species is zero. Under these circumstances any tendency of individual species to increase can be balanced only by changes in the interaction terms.

The quantities $d \log (N)/dt$ and $d \log (p)/dt$ clearly have useful properties and it is reasonable that these be named. I have already indicated that I am reluctant to call these 'fitness'. The term *Malthusian parameter* is, unfortunately, cumbersome. Too often evolutionary theory is left at this point. There is no biology in any of the above, but often an illusion is generated that something profound has been revealed. Insight can only come from biological interpretation and modelling of these kinematic relationships.

Before leaving this topic, consider a simple corollary. Suppose we have a measure for a trait X such that the value of the ith class is X_i. Then:

$$X = \sum p_i X_i$$

and

$$\frac{dX}{dt} = \sum p_i \frac{dX_i}{dt} + \sum X_i \frac{dp_i}{dt}$$

$$= \frac{dX}{dt} + \text{Cov}(X, M)$$

$$= \frac{dX}{dt} + b_{XM} \text{Var}(X)$$

where b_{XM} is a measure of the association between the mean value of the trait and its rate of change. The rate of change in the mean value of a trait is discussed in greater detail in the next section.

POPULATION BIOLOGY AND EVOLUTIONARY RATES

Evolutionary rates can be inferred from data on extant species or estimated directly from the fossil record. I have gone to some lengths to indicate that population biology has little to contribute to the analysis of these data. Even the molecular clock owes little to theory; its usefulness is essentially empirical. Neutral theory proposes that the rate of substitution is constant for a given protein, but that the actual rate depends on functional constraints on the molecule. As natural selection improves the function and adaptation of the encoded protein these constraints should increase, thereby reducing the neutral mutation rate. This is a possible explanation for faster substitutions in early lineages. A major difficulty of the theory is that mutation rates, and hence the rate of substitution, depends on the generation time, whereas empirically, substitution rates appear constant over elapsed time.

An example of the ambiguity of theory is illustrated by considerations of the relationship between population size and evolutionary rates. The answer depends on whether one believes that evolutionary rates are limited by mutation rates. The rate of fixation of neutral alleles is simply the neutral mutation rate, and hence is independent of population size. On the other hand, the rate of fixation of favourable mutations is directly proportional to the population size, the mutation rate to favourable mutations, and the selective advantage of the mutation. This view predicts that adaptive change is favoured by large populations, whereas neutral substitutions are largely independent of size (cf. Templeton, this volume).

However, the proposition that evolution is limited by mutation rates is questionable. Most morphological change is polygenic, and for such traits it appears that ample genetic variability exists in natural populations. If this view is correct, population size also is largely irrelevant to the rate of adaptive change.

The fossil record provides two kinds of information, namely, estimates of morphological change in a lineage and the rates of taxonomic origination and extinction. There is little that can be said regarding taxonomic extinction, and even less about origination, from the point of view of population genetics theory. It is commonly assumed that species become extinct because they cannot adapt fast enough to cope with a changing physical or biotic environment. The other side of the coin is stasis. A possible insight into the phenomenon of long periods of stasis followed by mass extinction, at the community level, lies in formula (1). Any tendency for population increase $(\overline{dM/dt})$ when resources are limiting will tend to favour increased species interaction $(\overline{d\varepsilon/dt})$. This means that the community becomes more and more connected. The result is that the loss of one species becomes more likely to affect all others adversely.

One contribution of population biology to the analysis of fossil data is the quantification of evolutionary change at the phenotypic level. Haldane (1949) offered two suggestions. One was to consider the proportional rate of change over the lineage. This measure can be approximated by $(\log_e X_0 - \log_e X_t)/t$, where X_t is the mean value of the trait t years after the initial measurement. A unit change in this measure in 10^6 years is called a *Darwin*. The alternative suggestion is that changes in a trait over time be measured in standard deviations. Haldane considered this a more difficult option since 'standard deviations are not so accurately known as means'. In the following section I review these alternatives from the point of view of our current knowledge of response to selection for continuously varying traits.

Quantitative genetics and morphological change

Quantitative genetics is a highly developed empirical discipline, and it offers to evolution the potential for a predictive theory. In experiments with quantitative characters it is possible to predict the consequences of selection over many generations from a knowledge of only three state variables, namely the mean of the population at generation zero, the heritability, and the accumulated selection differential. Population geneticists have not seriously considered using this simple theory to describe long-term evolutionary change.

There is no theoretical reason to suppose that genetic variances and genetic correlations remain constant in time. However, variances remain approximately constant over many generations of very intense selection, thousands of times more intense than occur in nature. The genetic component of variance is a product of evolution, and is presumably the result of an equilibrium between generation of new variation and loss through drift and selection (Fisher 1930, Lande 1976). It seems reasonable to suppose that these components of variance do not change rapidly – after all, heritabilities are similar across species that have been separated for millions of years. The existence of allometric relationships suggest that constancy may also be a property of genetic correlations.

Consider now the elementary theory of response to selection for quantitative variables. Almost any elementary textbook tells us that the selection differential S, can be written as $i\sigma$, hence:

$$\Delta X = h^2 i \sigma$$

where i is the standardised selection differential, h^2 is the heritability, and σ is the phenotypic standard deviation. Approximating by the differential, where time is measured in generations, Haldane's second measure becomes:

$$\frac{1}{\sigma}\frac{dX}{dt} = h^2 i \sigma$$

Since we assume that h^2 is approximately constant over an evolutionary lineage, the standardised rate of change depends only on the selection differential. Haldane's first measure is less satisfactory, for if we were to express the changes scaled against the mean, we have the less elegant expression:

$$\frac{1}{X}\frac{dX}{dt} = h^2 i\sigma / X$$

I suggest that, rather than express morphological change in standard deviations, we go one step further, and express all phenotypic change scaled by the phenotypic standard deviation and the heritability. That is, our measure should be:

$$\frac{1}{\sigma h^2}\frac{\Delta X}{\Delta t}$$

Phenotypic standard deviations should be estimated from the fossil record if possible. Heritabilities must be obtained from extant species. It may be necessary to transform the observations to remove dependencies of variances on the mean.

Morphological change, as observed in the fossil record, is very slow compared with what is thought to be possible. Simpson's (1944) measurements of ectoloph length and paracone height in the lineage of horses indicates

Table 8.1 Minimum culling rates for morphological change.[a]

Change per generation	Selection differential	Proportion culled
10^{-5}	0.00004	2.0×10^{-5}
10^{-4}	0.0004	1.0×10^{-4}
10^{-3}	0.004	1.1×10^{-3}

[a] heritability is assumed to be 0.25.

rates of change of no more than 10^{-6} standard deviations per year. In order to appreciate just how little selection this entails, Table 8.1 shows the proportion of the population that needs to be culled per generation in order to achieve rates of one standard deviation per 10^5, 10^4 and 10^3 years. Assuming that the coefficient of variation is 20%, and the generation interval is 5 years, these rates are 1, 10 and 100 Darwins respectively. Human selection in domesticated species can achieve rates that are many orders of magnitude faster than has ever been observed in the fossil record. Clearly, periodic bursts of rapid morphological change present no theoretical difficulty.

CONCLUSIONS

> When you coin a term, it ought to mark a real species, and a specific difference; otherwise you get empty, frivolous, verbiage. (Aristotle, *The Rhetoric*)

I have briefly described some of the theory of evolutionary change, and pointed to some of the epistemological difficulties of the theory. In particular, the use of time as a primitive variable, and hence the tendency to use differential equations, has serious difficulties. I have also discussed the concept of levels of description, an important property of evolutionary models and one that is implicit in most theories but not always clearly defined.

As population genetics and ecology move closer together it is time for a careful examination of the general structure of evolutionary theory. Some of my colleagues may feel that I have been overly critical. The major point of this chapter has been to attempt constructive comments in the hope that the deficiencies and strengths of population biology will be more clearly understood. In particular, we need:

(a) to agree on the definition of evolutionary variables;
(b) to be cautious about the use of time and space as primitive terms;
(c) to distinguish clearly between a theory that is simply a rearrangement of axiomatic terms and a theory that has real biological content;
(d) to develop a better understanding of the biological rules that connect one level of description with another.

A primary function of mathematical population biology is to provide methods for the rigorous analysis of the evolutionary process. The field has provided a number of strong assertions; for example, that evolution will, in general, favour the close linkage of interacting loci. This theory clearly demonstrates the Markovian nature of evolutionary change and provides an effective counter to teleological arguments in biology. The discoveries in molecular biology have spawned a new discipline based on the mathematical and statistical analysis of DNA sequence data. The newer theory can only become connected, and hence relevant, to the more traditional fields of population biology through a greater understanding of the biological relationship between genes and phenotypes. If population biology is premature to current activities in molecular biology, the key seems to lie with developmental genetics.

ACKNOWLEDGEMENTS

I am grateful to John Sved, Oliver Mayo and John Black for reading and criticising earlier drafts of the manuscript. My greatest debt lies with Richard Lewontin, who stimulated my interest in these matters.

REFERENCES

Arnold, A. J. and K. Fristrup 1982. The theory of evolution by natural selection: a hierarchical expansion. *Paleobiology* 8, 113–29.

Cohen, J. E. 1971. Mathematics as a metaphor. *Science* 172, 674–5.

Darwin, C. 1859. *The origin of species*. London: John Murray.

Dawkins, R. 1976. *The selfish gene*. London: Oxford University Press.

Dover, G. 1982. Molecular drive: a cohesive mode of species evolution. *Nature* 299, 111–16.

Ewens, W. J. 1983. Inference problems in population genetics: DNA sequences, restriction endonucleases and ascertainment sampling. *Proc. R. Soc. Lond. B* 219, 223–39.

Fisher, R. A. 1918. The correlation between relatives on the supposition of Mendelian inheritance. *Trans. R. Soc. Edinb.* 52, 399–433.

Fisher, R. A. 1930. *The genetical theory of natural selection*. Oxford: Clarendon Press.

Ford, E. B. 1940. Polymorphism and taxonomy. In *The new systematics*, J. Huxley (ed.), pp. 493–513. Oxford: Clarendon Press.

Ginzberg, L. R. 1983. *Theory of natural selection and population growth*. California: Benjamin/Cummings.

Glass, B. 1979. Milestones and rates of growth in the development of biology. *Q. Rev. Biol.* 54, 31–53.

Haldane, J. B. S. 1932. *The causes of evolution*. New York: Harper & Row.

Haldane, J. B. S. 1937. The effect of variation on fitness. *Am. Nat.* 71, 337–49.

Haldane, J. B. S. 1949. Suggestions as to the quantitative measurement of rates of evolution. *Evolution* 3, 51–6.

Hutchinson, G. E. 1958. Concluding Remarks. *Cold Spring Harbour Symp. Quant. Biol.* 22, 415–27.

Jacquard, A. 1983. Heritability: one word, three concepts. *Biometrics* 39, 465–77.

Kimura, M. 1958. On the change of population fitness by natural selection. *Heredity* 12, 145–67.

Kimura, M. 1983. *The neutral theory of molecular evolution*. Cambridge: Cambridge University Press.

Kingsland, S. 1982. The refractory model: the logistic curve and the history of population ecology. *Q. Rev. Biol.* 57, 29–52.

Lande, R. 1976. The maintenance of genetic variability by mutation in a polygenic character with linked loci. *Genet Res.* 26, 221–35.

Levins, R. 1968. *Evolution in changing environments*. Englewood Cliffs, N. J.: Princeton University Press.

Levins, R. and R. C. Lewontin 1980. Dialectics and reductionism in ecology. *Synthese* 43, 47–78.

Lewontin, R. C. 1964. The role of linkage in natural selection. In *Genetics Today*, *Proceedings XI Int. Congr. Genetics*, pp. 517–25. Oxford: Pergamon Press.

Lewontin, R. C. 1970. The units of selection. *Ann. Rev. Ecol. Syst.* 1, 1–18.

Lewontin, R. C. 1974. *The genetic basis of evolutionary change*. New York: Columbia University Press.

Lotka, A. J. 1925. *Elements of physical biology*. Baltimore: Williams and Wilkins.

MacArthur, R. H. 1968. The theory of the niche. In *Population biology and evolution*, R. C. Lewontin (ed.), pp. 159–76. New York: Syracuse University Press.

MacArthur, R. H. 1972. *Geographical ecology: patterns in the distribution of species*. New York: Harper & Row.

Maynard Smith, J. 1983. Models of evolution. *Proc. R. Soc. Lond. B.* 219, 315–25.

Quinn, J. F. and A. E. Dunham 1983. On hypothesis testing in ecology and evolution. *Am. Nat.* 122, 602–17.

Roughgarden, J. 1983. Competition and theory in community ecology. *Am. Nat.* **122**, 583–601.

Salt, G. W. 1983. Roles: their limits and responsibilities in ecological and evolutionary research. *Am. Nat.* **122**, 697–705.

Simberloff, D. 1983. Competition theory, hypothesis testing and other community ecological buzzwords. *Am. Nat.* **122**, 626–35.

Simpson, G. G. 1944. *Tempo and mode in evolution*. New York: Columbia University Press.

Stent, G. S. 1980. Prematurity and uniqueness of scientific discovery as illuminated by the history of molecular biology. In *Life: hierarchies of interacting molecules*, W. J. Peacock (ed.), pp. 7–15. Canberra: Australian Academy of Science.

Thoday, J. M. 1953. Components of fitness. *Symp. Soc. Exp. Biol.* **7**, 96–113.

Van Valen, L. 1973. A new evolutionary law. *Evol. Theory* **1**, 1–30.

Williams, M. B. 1970. Deducing the consequences of evolution: a mathematical model. *J. Theor. Biol.* **29**, 343–85.

Williams, M. B. 1977. Needs for the future: radically different types of mathematical models. In *Mathematical models in biological discovery*, D. L. Solomon and C. Walter (eds), pp. 225–40. Berlin: Springer-Verlag.

Wright, S. 1931. Evolution in Mendelian populations. *Genetics* **16**, 97–159.

Wright, S. 1968. *Evolution and the genetics of populations*, vol. 1. Chicago: University of Chicago Press.

9

Comparative rates of molecular, chromosomal and morphological evolution in some Australian vertebrates

P. R. BAVERSTOCK and M. ADAMS

ABSTRACT

Rates of molecular evolution were assessed by two approaches, allozyme electrophoresis and immunology (microcomplement fixation). The former gives a measure of the proportion of genes that differ between taxa and the latter gives a measure of the proportion of DNA triplets that differ between taxa for a single gene. We have both electrophoretic and immunological data for dasyurid marsupials, Australian rodents and Australian *Rattus* snakes. The estimates of evolutionary rates for each group, based on the two data sets, are comparable.

Rates of chromosome evolution were assessed from G-banding analyses. Comparisons between rates of chromosome evolution and rates of molecular evolution were available for the family Dasyuridae (marsupial mice), Australian *Rattus* (rodents), and Australian Hydromyinae (rodents). The data show virtually no correlation between chromosomal and molecular evolution. Moreover, cladistic analysis of the data shows that rates of chromosome evolution vary enormously between different lineages not only between groups but also within groups.

Although rates of electrophoretic, immunological and chromosomal evolution can be measured objectively, the same cannot be said of morphological evolution. We have attempted to overcome this problem by choosing extremes of morphological divergence, from sibling species (which by definition are morphologically very similar) to species that are clearly very different morphologically even to the casual observer. Examples of sibling species that nevertheless differ greatly at the molecular level include bats of the *Eptesicus pumilis* complex in Australia and marsupial mice of the *Sminthopsis murina* complex. The other extreme is represented by

Australian cockatoos of the genus *Cacatua* and Australian elapid snakes, where species that are very similar at the molecular level are nevertheless very different at the morphological level.

The data show that molecules, chromosomes and morphology evolve at very different rates. A large body of evidence suggests that, of the three, molecular evolution is the closest to being constant with time. If this is true, then rates of morphological and chromosomal evolution vary enormously with time.

INTRODUCTION

The purpose of the symposium was to bring together workers studying rates of evolution from various points of view, including morphology, genes, chromosomes and developmental biology. In this chapter we compare and contrast rates of evolution at the genetic, chromosomal and morphological levels using various Australian vertebrates as examples. We show that, although different measures of genetic divergence correlate closely with each other, chromosomal and morphological divergence bear little relationship to genetic divergence.

These results have important implications for palaeontologists, many of whom seem to accept that morphological stasis in the fossil record implies genomic stasis, and that rapid morphological shifts imply rapid genomic evolution. The present results show that this is not necessarily so.

THE GENETIC LEVEL

Genes can be studied directly in various ways such as DNA sequencing and DNA–DNA hybridisation, or indirectly via their protein products. Although the technology exists for direct DNA sequencing, it is nevertheless an expensive and time-consuming process and is limited at this stage to very few genes. Most of the information available on rates of evolution at the gene level has therefore been derived from studies of proteins. Protein evolution can be best assessed by direct amino acid sequencing. This also is an expensive and time-consuming process. Two techniques that are relatively inexpensive and rapid are microcomplement fixation (MC'F) and allozyme electrophoresis.

Microcomplement fixation

In this technique (Champion *et al.* 1974), a protein is purified from one species and injected into rabbits which raise antibodies against the protein. The antisera are then cross-reacted with the same protein in related taxa, and the 'strength of cross-reaction' is measured to yield an immunological distance (ID). The ID is directly proportional to the number of differences in amino acid content between the protein used to raise the antisera and the protein of the test species (Maxson & Wilson, 1974). Hence it is a measure of the divergence between the taxa for the gene encoding that protein. Provided that convergences do not exceed divergences at the amino acid level, the ID between two taxa will be a monotonically increasing function of time since divergence.

Allozyme electrophoresis

The electrophoretic technique detects different mobility rates of proteins when individual samples are loaded on a support medium (usually a gel) and subjected to an electric current. These different rates are a consequence of charge differences resulting from amino acid substitutions. A typical study will use allozyme electrophoresis to compare about 30 gene loci.

In the vast majority of cases, two taxa that possess electrophoretically distinguishable proteins for a particular gene will have different structural alleles for that gene. However, two taxa that possess the same electrophoretic form of a particular protein do not necessarily possess the same structural allele for that protein since not all amino acid substitutions are electrophoretically detectable; not all nucleotide substitutions in the coding regions yield amino acid substitutions (i.e. the code is redundant); and some base substitutions occur in non-coding regions.

Nevertheless, it seems reasonable to assume that the genetic distances assessed by electrophoresis are proportional to the total genetic divergence for the suite of proteins studied.

The correlation between MC' F and electrophoresis

Here we examine whether there is a correlation between these two methods of measuring genetic divergence between taxa. That is, do taxa that differ greatly at one gene (the gene encoding the protein being studied by MC' F) differ at a high proportion of genes as studied electrophoretically? Before addressing this question, however, it will be necessary to consider the limits of the two techniques.

There is an upper limit to the extent of genetic divergence that can be estimated by the two techniques. For MC' F, the limit is a technical one; at IDs much above 100, high antibody concentrations inhibit the reactions being measured (see Champion *et al.* 1974). For allozyme electrophoresis the upper limit will theoretically occur when the taxa share no alleles at any loci, i.e. they are 100% different. In practice, saturation occurs at a lower level because of convergences and parallelisms. Therefore a positive correlation between genetic distances as estimated by MC' F and electrophoresis will be expected only for closely related taxa.

There is a second statistical problem that also needs to be considered when searching for such correlations. While for MC' F distances, reciprocal data are independent (i.e. the distance from taxon A to taxon B is measured independently of the distance from B to A), the same is not true for electrophoretic data. Moreover, the data are not totally independent in the sense that if the distance from A to B is small, and A is distant from C, then B will also be distant from C. To obviate these potential statistical artifacts, we have chosen to plot a subset of the data, i.e. where A is close to B and distant from C, the A–B and A–C distances are plotted but not the B–C distance.

Both electrophoretic and MC′ F data are available for three groups of Australian vertebrates, the rodent genus *Rattus*, the marsupial family Dasyuridae and the rodent subfamily Hydromyinae. The electrophoretic data are given by Baverstock *et al.* (1985) for *Rattus*, by Baverstock *et al.* (1982) for Dasyuridae and by Baverstock *et al.* (1981) for the Hydromyinae; the M′ CF data are unpublished (P. R. Baverstock). In the case of the Dasyuridae and Hydromyinae, only one member of each genus was used for the intergeneric comparisons to avoid the statistical problem outlined above. In the case of the *Rattus* data, *R. rattus*, *R. norvegicus* and the Australian *Rattus* were treated as three lineages (Baverstock *et al.* 1985) and only one member of each lineage was used for interlineage comparisons.

The plot of genetic distance (measured as Nei D) based on 55 electrophoretic loci against albumen immunological distance (AID) for the Australian *Rattus* is shown in Figure 9.1. There is a strong correlation between the two of 0.91 ($P < 0.001$). The slope of the line through the origin is 0.074. Note that in this case the maximum AID is 13. Figure 9.2 shows a similar plot for the dasyurid marsupials, the electrophoretic data being based on 32 loci. Again there is a strong correlation between the two of 0.69 ($P < 0.001$). The points are scattered more widely in this case than in the *Rattus*, especially at higher AIDs. However, it is clear that we are here seeing the effects of saturation in the electrophoretic data, since the maximum AID is 34. Figure 9.3 depicts a similar plot for the Australian Hydromyine rodents. The electrophoretic data are based on 20 loci. The scatter of points is considerable in this case, but this is clearly due to the

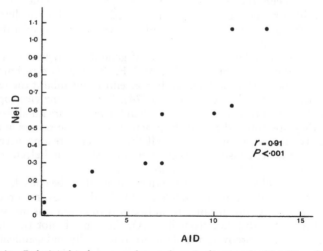

Figure 9.1. Relationship between electrophoretic distance (as Nei D) and albumen immunological distance (AID) for Australian *Rattus* (see text for detail).

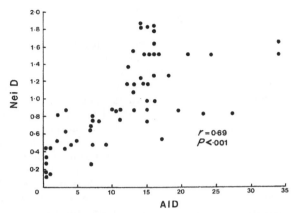

Figure 9.2. Relationship between electrophoretic distance (as Nei D) and albumen immunological distance (AID) for Dasyurid marsupials (see text for detail).

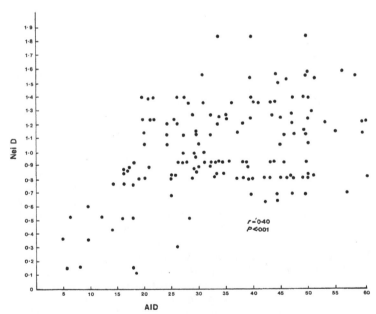

Figure 9.3 Relationship between electrophoretic distance (as Nei D) and albumen immunological distance (AID) for Australian hydromyine rodents (see text for detail).

saturation effect for the electrophoretic data since we are here dealing with AIDs of up to 60.

For these three groups of Australian vertebrates, there is a strong correlation between genetic distance estimated from a single gene and genetic distance estimated from a suite of genes. A similar strong correlation between AID and Nei D has been reported for frogs, salamanders, lizards and mammals (summarised in Wyles & Gorman 1980).

We emphasise that electrophoresis and MC' F measure genetic divergence between taxa in what would appear to be totally independent ways: electrophoresis measures the proportion of electrophoretic loci that differ between taxa, and MC' F measures the proportion of amino acids that differ between taxa for a particular protein. That they do correlate suggests that rates of evolution at the genome level are the same whether they are viewed from the number of differences in a single gene, or from the proportion of genes different over a set sample of structural genes. Moreover, Maxson and Maxson (1979) have shown a strong correlation between AID and DNA–DNA hybridisation data for *Plethodon* salamanders, suggesting that the correlation also applies to the whole genome.

Thus if the immunological distance between two taxa is known, the electrophoretic distance can be predicted, at least approximately. As will become clear, this is not true for either chromosomal differences or morphological differences.

THE CHROMOSOME LEVEL

A number of studies have attempted to assess rates of chromosomal evolution from the differences seen in gross karyotypes (e.g. Wilson *et al.* 1975, Schnell & Selander 1981, Gothran & Smith 1983). There are two general criticisms that may be made of such studies. First, some chromosomal rearrangements do not affect the gross karyotype, e.g. paracentric inversions, equal reciprocal translocations, independent fusions between chromosomes of the same size, etc. The gross karyotype can also be affected by modifications that do not involve rearrangement of the structural genome, e.g. addition or deletion of heterochromatin. Both these limitations can be overcome by using the techniques of G-banding and C-banding. We will therefore restrict ourselves to groups whose chromosome evolution has been assessed by these techniques. The second problem is to relate the chromosome differences to specific branches of the tree leading to extant taxa. Approaches that summarise karyotypic diversity of a group are fraught with statistical problems. Thus a speciose group may appear to have high karyotypic diversity simply because of the high number of species. Simply dividing by the number of species, however, overcompensates – if a high proportion of the chromosomal rearrangements have occurred in a branch leading to a highly speciose group, the amount of chromosome evolution will be underestimated. One way out of this dilemma is to work with groups whose cladistic relationships are fairly well established independently of the chromosome data, and plot the proposed chromosomal changes on the branches in a parsimonious fashion.

Thus an appropriate analysis requires groups whose chromosomes have been analysed by G-banding and C-banding and whose cladistic relationships are established. Three such groups of Australian vertebrates are again the Australian *Rattus*, the dasyurids and the hydromyine rodents.

Cladistic relationships among the Australian *Rattus* are now firmly established. The group has been the subject of an intensive morphometric analysis (Taylor & Horner 1973), as well as an electrophoretic analysis based on 55 loci (Baverstock *et al.* 1985), and the two studies are in the main concordant. The group has also been the subject of intensive chromosomal analyses using G-banding and C-banding (Baverstock *et al.* 1977, 1983c).

The cladistic relationships among the Australian *Rattus*, along with the minimum number of chromosomal rearrangements that have occurred along each branch, are shown on Figure 9.4. The rates of chromosome evolution along various branches are highly disparate. For example, from the common ancestor of the Australian *Rattus*, no chromosomal rearrangements have occurred in the line leading to *R. lutreolus*. However, the lineages leading to members of the *R. sordidus* group (*R. sordidus, R. villosissimus* and *R. colletti*), have undergone considerable chromosome evolution. Yet at the genetic level, these three taxa are very similar – *R. colletti* and *R. villosissimus* share alleles at all 55 loci screened and *R. sordidus* differs at only two of these loci. Clearly, in this case, extent of genetic divergence is a very poor predictor of the extent of chromosomal divergence.

Of the 35 or so species of Australian dasyurids, about half have been chromosomally analysed by G-banding and C-banding (Rofe 1979, Young

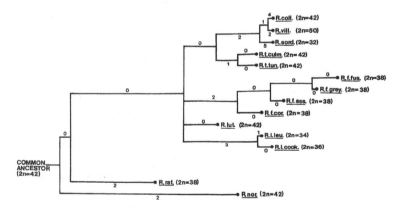

Figure 9.4 Cladistic relationships among Australian *Rattus* showing minimum amount of chromosome evolution along each branch (in Arabic numerals) and 2*n* is the diploid number. Branch lengths shown are proportional to the amount of electrophoretic evolution among each branch. (See Baverstock *et al.* (1985) for species and subspecies abbreviations).

et al. 1982, Baverstock *et al.* 1983b). The cladistic relationships among the species for which such chromosome data are available are shown in Figure 9.5. These relationships are based primarily on electrophoretic analysis of 32 loci (Baverstock *et al.* 1982) and MC′F data (P. R. Baverstock, P. A. Woolley & M. Archer unpublished work), but are supported on morphological criteria (M. Archer, personal communication).

The same karyotype is retained in 11 of the 13 species representing all major lineages. This karyotype is therefore the presumed ancestral state. *Antechinomys spenceri* possesses a unique paracentric inversion (Rofe 1979), and *Ningaui yvonneae* a unique pericentric inversion (Baverstock *et al.* 1983b). The rates of chromosome evolution within the Dasyuridae are therefore seen to be fairly uniform, with no lineage showing a spurt of chromosome evolution. Moreover, compared with the Australian *Rattus*, the chromosomes of the dasyurids are highly conservative. This is all the

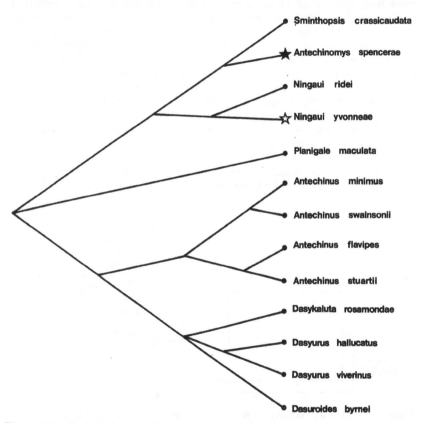

Figure 9.5 Cladistic relationships among Australian dasyurid marsupials whose chromosomes have been G-banded: species with proposed ancestral complement – •; species with single paracentric inversion – ★; species with single pericentric inversion – ☆.

more remarkable when it is remembered that the dasyurids are about three times as diverse at the molecular level.

The third group for which extensive G-banding data are available are the Australian hydromyine rodents. Of the 35 species in 12 genera, 25 species in 10 genera have had their chromosomes G-banded (Baverstock *et al.* 1983a). Although the cladistic relationships among these 25 species are not known precisely, sufficient is known for the rates of chromosome evolution to be interpreted from the available data.

It is clear, for example, that there are three major lineages referable to the tribal level (Lee *et al.* 1981). A single G-banded karyotype occurs in all three tribes, suggesting that this is the ancestral karyotype. This ancestral karyotype is retained unmodified or only slightly modified in the vast majority of taxa studied. Thus the hydromyines are generally characterised by karyotypic conservatism. Yet at the molecular level, the hydromyines are about five times as diverse as the Australian *Rattus*.

One genus (*Zyzomys*) in one tribe has, however, undergone a high degree of karyotypic modification, to the extent that at least seven chromosomes show partial or no homology with the ancestral complement. Moreover, the rearrangements that characterise the *Zyzomys* complement are not explicable by simple inversions of translocations. Clearly the genus *Zyzomys* has undergone rapid chromosome evolution, whereas all other lineages in the subfamily have remained chromosomally conservative.

The above three examples illustrate two points. First, rates of chromosome evolution can vary enormously within groups, and secondly, that chromosomal diversity does not correlate with molecular diversity. Therefore karyotypic data cannot be used to predict genetic diversity.

THE MORPHOLOGICAL LEVEL

One of the problems in assessing rates of morphological evolution both between and within groups is to establish an objective measure of morphological diversity (see Cherry *et al.* 1982, Wyles *et al.* 1983). As these authors point out, morphological diversity is too often assessed subjectively, and they have developed an objective measure for the intuitive concept of overall morphological difference.

Unfortunately we do not have such measures for any Australian vertebrate. We have attempted to overcome this problem by taking two extremes of morphological diversity, and assessing the molecular diversity for these cases. We consider a group to show very little morphological diversity if it consists of cryptic species, i.e. species that are so similar morphologically that taxonomic experts with the group have failed to recognise the species on morphology alone. We consider a group to show extensive morphological diversity if a lay person could instantly recognise the different taxa. As examples of the former, we will use members of the bat genus *Eptesicus* in Australia and members of the *Sminthopsis murina* complex (dasyurid marsupials). As examples of the latter, we will use the

'white' cockatoos of Australia (Aves), and snakes of the family Elapidae in
Australia.

Bats of the genus Eptesicus

Species of this genus in Australia are so similar morphologically that Ride
(1970) recognised only one species. McKean *et al.* (1978) subsequently
recognised four species separable on the shape of the penis bone in males.
Discriminant function analysis was needed to separate males on other
skeletal characters, but these same characters failed to discriminate females.
Kitchener (1976) subsequently recognised a fifth species. Electrophoretic
analysis has revealed, however, that there are at least nine species of
Eptesicus in Australia (Tidemann *et al.* 1981, Adams *et al.* 1982, M.
Adams, P. R. Baverstock, C. H. S. Watts & T. Reardon unpublished obser-
vations), some of which are only differentiable electrophoretically (D. J.

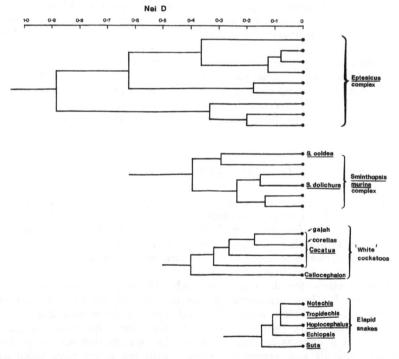

Figure 9.6 Phenograms expressing genetic relationships (as Nei Ds) among
members of four groups of Australian vertebrate. The *Eptesicus* complex consists
of nine sibling species; the *Sminthopsis murina* complex consists of six sibling
species (*S. ooldea* and *S. dolichura* are morphologically indistinguishable in
sympatry); the 'white' cockatoos include the genera *Cacatua* and *Callocephalon*: the
Elapid snakes shown are five genera of the '*Notechis*' lineage.

Kitchener, personal communication). Clearly, by any criterion, the nine species of this genus in Australia are very similar at the morphological level.

The genetic relationships among these nine species based on 36 loci are shown phenetically in Figure 9.6. Nei D values range up to 0.9, this being higher than the genetic distances typifying species within a genus (e.g. Avise & Aquadro 1982).

The Sminthopsis murina complex

The members of the complex are so similar morphologically that Troughton (1965) recognised only the species *S. murina*. In an extensive review of the genus, Archer (1981) recognised *S. ooldea* as distinct from *S. murina*, and provided evidence that *S. leucopus* also represented a distinct species. Subsequent electrophoretic analysis revealed that there were at least six species in the complex (Baverstock *et al.* 1984), which proved to be distinguishable by discriminant function analysis (Kitchener *et al.* 1984). However, it has recently become apparent that the characters used in this discriminant analysis fail to distinguish *S. ooldea* from *S. dolichura* where they are sympatric on the Nullabor Plain (D. J. Kitchener, personal communication). Clearly the members of this group satisfy the criterion of cryptic species and are morphologically very similar.

The genetic distances based on 30 loci (Baverstock *et al.* 1984) among six of the species of the complex are shown in Figure 9.6. Nei D values range up to 0.4. Indeed, this is the genetic distance between *S. ooldea* and *S. dolichura*, the two species most difficult to separate on morphology alone.

The 'white' cockatoos of Australia

This group of cockatoos includes the corellas (which are all white and no prominent crest), the sulphur crest (mainly white with a prominent yellow crest), the pink cockatoo (pink with a prominent pink and white crest), the galah (deep pink, white cap, all grey wings, no prominent crest), and the gang-gang (dark grey with a red head and medium sized crest). Excellent colour photographs of these birds appear in *The Reader's Digest Complete Book of Australian Birds* (1983). Other than the two species of corellas, all species are so different morphologically that anyone can distinguish the species even at long range. The two corellas are clearly distinguishable close up.

An electrophoretic analysis based on 28 loci (Adams *et al.* 1984) showed maximum Nei D values within *Cacatua* of 0.35, and even the gang-gang had a Nei D of only 0.42 to the species of *Cacatua* (Fig. 9.6). The mainly white corellas were electrophoretically most similar to the pink and grey galah, which were separated by a Nei D of only 0.15. We thus see here a very strong contrast with the sibling species of *Eptesicus* and *Sminthopsis*, in that morphologically very divergent forms are less divergent at the genetic level than are the sibling species of bats and marsupial mice.

Elapid snakes of the Notechis *lineage*

The *Notechis* lineage of Australian snakes include the genera *Tropidechis*, *Hoplocephalus*, *Echiopsis* and *Suta*. Colour photographs of all but *Notechis* can be found in Cogger (1975). They vary considerably in size, head shape and colour to the extent that none could be confused by a lay person. Indeed these taxa are so distinct morphologically that they have traditionally been placed in separate genera (Cogger 1975).

Unfortunately, we do not have electrophoretic data for these snakes. However, we have made an analysis based on MC′ F using the protein transferrin (Schwaner *et al.* 1985). It is known that the molecular diversity of transferrin is roughly twice that of albumen (Sarich 1977). Using the slope of the line relating Nei D to AID established earlier, it is possible to relate these transferrin IDs (TID) to expected Nei D. The result is shown in Figure 9.6. Although it must be admitted that the proposed relationship between TID and Nei D may be tenuous, these different genera are obviously very similar at the molecular level. Thus the snakes, like the cockatoos, are highly divergent at the morphological level, but at the genetic level are less divergent than the sibling species of *Eptesicus* and *Sminthopsis*.

The foregoing analysis demonstrates quite clearly that the extent of morphological diversity bears little relationship to the extent of diversity of the whole genome. This is not to deny the rôle of genes in determining morphological diversity. Clearly the morphological diversity exhibited among the cockatoos and among the snakes has a genetic basis. Rather, the proportion of the genome that is reflected in gross morphology may be a very small proportion of the total. Therefore morphological similarity cannot be taken to indicate genomic similarity.

CONCLUSIONS

In this analysis, we have compared data from a wide range of Australian vertebrates that illustrate (a) that independent measures of molecular genetic diversity (MC′ F and electrophoresis) give concordant results, and (b) that molecular, chromosomal and morphological evolution can occur at enormously discordant rates in different lineages.

A large body of evidence suggests that of the three, molecular evolution is closest to being constant with time (Wilson, this volume). If this is true, then rates of morphological and chromosomal evolution vary enormously with time.

The data also emphasise that rates of morphological evolution cannot be equated with rates of total genomic evolution. This result has important implications for the palaeontologist, namely that morphological stasis in the fossil record cannot be equated to total genomic stasis, and conversely rapid morphological shifts in the fossil record cannot be equated to rapid genomic shifts.

ACKNOWLEDGEMENTS

Much of the work reported here was done with the able technical assistance of J. Birrell, M. Krieg and T. Reardon. We thank Drs R. Andrews and C. H. S. Watts for their critical comments on the manuscript and P. Kidd for typing it.

REFERENCES

Adams, M., P. R. Baverstock, C. R. Tidemann and D. P. Woodside 1982. Large genetic differences between sibling species of bats, *Eptesicus*. *Heredity* **48**, 435–8.

Adams, M., P. R. Baverstock, B. A. Saunders, R. Schodde and G. T. Smith 1984. Biochemical systematics of the Australian cockatoos (Psittaciformes:Cacatuinae). *Aust. J. Zool.* **32**, 363–7.

Archer, M. 1981. Results of the Archbold Expeditions No. 104. Systematic revision of the marsupial genus *Sminthopsis* Thomas. *Bull. Am. Mus. Nat. Hist.* **168**, 61–224.

Avise, J. C. and Aquadro, C. F. 1982. A comparative summary of genetic distances in the vertebrates. Patterns and correlations. *Evol. Biol.* **15**, 151–85.

Baverstock, P. R., C. H. S. Watts, A. C. Robinson and J. T. Robinson 1977. Chromosome evolution in Australian rodents. II. The *Rattus* group. *Chromosoma (Berl.)* **61**, 227–41.

Baverstock, P. R., C. H. S. Watts, M. Adams and S. R. Cole 1981. Genetical relationships among Australian rodents (Muridae). *Aust. J. Zool.* **29**, 289–303.

Baverstock, P. R., M. Archer, M. Adams and B. J. Richardson 1982. Genetic relationships among 32 species of Australian dasyurid marsupials. In *Carnivorous marsupials*, M. Archer (ed.), pp. 641–50. Sydney: Royal Zoological Society of New South Wales.

Baverstock, P. R., C. H. S. Watts, M. Gelder and A. Jahnke 1983a. G-banding homologies of some Australian rodents. *Genetica* **60**, 115–17.

Baverstock, P. R., M. Adams, M. Archer, N. McKenzie and R. How 1983b. An electrophoretic and chromosomal study of the dasyurid marsupial genus *Ningaui* Archer. *Aust. J. Zool.* **31**, 381–92.

Baverstock, P. R., M. Gelder and A. Jahnke 1983c. Chromosome evolution in Australian *Rattus* – G-banding and hybrid meiosis. *Genetica* **60**, 93–103.

Baverstock, P. R., M. Adams and M. Archer 1984. Electrophoretic resolution of species boundaries in the *Sminthopsis murina* complex (Dasyuridae). *Aust. J. Zool.* **32**, 823–32.

Baverstock, P. R., M. Adams and C. H. S. Watts 1985. Biochemical differentiation among karyotypic forms of Australian *Rattus*. *Genetica*. (in press).

Champion, A. B., E. M. Prager, D. Wachter and A. C. Wilson 1974. Microcomplement fixation. In *Biochemical and immunological taxonomy of animals*, C. A. Wright (ed.), pp. 397–416. London: Academic Press.

Cherry, L. M., S. M. Case, J. G. Kunkel, J. S. Wyles and A. C. Wilson 1982. Body shape metrics and organismal evolution. *Evolution* **36**, 914–33.

Cogger, H. G. 1975. *Reptiles and amphibians of Australia*. Sydney: Reed.

Gothran, E. G. and M. H. Smith 1983. Chromosomal and genic divergence in mammals. *Syst. Zool.* **32**, 360–8.

Kitchener, D. J. 1976. *Eptesicus douglasi*, a new vespertilionid bat from Kimberley, Western Australia. *Rec. West. Aust. Mus.* 4, 295–301.

Kitchener, D. J., J. Stoddart and J. Henry 1984. A taxonomic revision of the *Sminthopsis murina* complex (Marsupialia, Dasyuridae) in Australia, including descriptions of four new species. *Rec. West. Aust. Mus.* 11, 201–48.

Lee, A. K., P. R. Baverstock and C. H. S. Watts 1981. Rodents – the late invaders In *Ecological biogeography in Australia*, A. Keast (ed.), pp. 1523–48. The Hague: Junk.

Maxson, L. R. and A. C. Wilson 1974. Convergent morphological evolution detected by studying proteins of tree frogs *Hyla eximia* species group. *Science* 185, 66–8.

Maxson, L. R. and R. D. Maxson 1979. Comparative albumen and biochemical evolution in plethodontid salamanders. *Evolution* 33, 1057–62.

McKean, J. L., G. C. Richards and W. J. Price 1978. A taxonomic appraisal of *Eptesicus* (Chiroptera:Vespertilionidae) in Australia. *Aust. J. Zool.* 26, 529–37.

Ride, W. D. L. 1970. *A guide to the native mammals of Australia*. Melbourne: Oxford University Press.

Rofe, R. H. 1979. *G-banding and chromosomal evolution in Australian marsupials.* PhD thesis, University of Adelaide.

Sarich, V. M. 1977. Rates, sample sizes, and the neutrality hypothesis for electrophoresis in evolutionary studies. *Nature* 265, 24–8.

Schnell, G. D. and R. K. Selander 1981. Environment and morphological correlates of genetic variation in mammals. In *Mammalian population genetics*, M. H. Smith and J. J. Joule (eds), pp. 60–99. Athens: Georgia University Press.

Schwaner, T. D., P. R. Baverstock, H. C. Dessauer and G. A. Mengden 1985. Immunological evidence for the phylogenetic relationships of Australian elapid snakes. In *Biology of Australian frogs and Reptiles*, G. Grigg, R. Shine and H. Ehmann (eds), pp. 177–84. New South Wales: Royal Zoological Society.

Taylor, J. M. and B. E. Horner 1973. Results of the Archbold expeditions. No. 98. Systematics of native Australian *Rattus* (Rodentia, Muridae). *Bull. Am. Mus. Nat. Hist.* 97, 183–430.

Tidemann, C. R., O. P. Woodside, M. Adams and P. R. Baverstock 1981. Taxonomic separation of *Eptesicus* (Chiroptera:Vespertilionidae) in south-eastern Australia by discriminant analysis and electrophoresis. *Aust. J. Zool.* 29, 119–28.

Troughton, E. 1965. A review of the marsupial genus *Sminthopsis* (Phascogalinae) and diagnoses of new forms. *Proc. Linn. Soc. New South Wales*, 89, 309–21.

Wilson, A. C., G. L. Bush, S. M. Case and M. C. King 1975. Social structuring of mammalian populations and rate of chromosomal evolution. *Proc. Natl Acad. Sci. USA* 72, 5061–5.

Wyles, J. S. and G. C. Gorman 1980. The albumin immunological and Nei electrophoretic distance correlation: a calibration for the Saurian genus *Anolis* (Iguanidae). *Copeia* 1980, 66–71.

Wyles, J. S., J. G. Kunkel and A. C. Wilson 1983. Birds, behaviour and anatomical evolution. *Proc. Natl Acad. Sci. USA* 80, 4394–7.

Young, G. J., J. A. Graves, I. Barbieri, P. A. Woolley, D. W. Cooper and M. Westerman 1982. The chromosomes of dasyurids (Marsupialia). In *Carnivorous marsupials*, M. Archer (ed.) pp. 783–95. Sydney: Royal Zoological Society of New South Wales.

10

Evolution of gene structure in relation to function

JOHN A. THOMSON

ABSTRACT

Recent studies of gene expression in higher organisms highlight the possible significance of change in control sequences for the evolution of new phenotypes in which co-ordinated activity of a number of gene-action systems is required. Three examples are briefly reviewed: (1) a set of genes involved in the control of transcriptional activity in embryogenesis and under stress in adult life; (2) a gene complex involved in a co-ordinated behavioural activity in a mollusc; and (3) the genes involved in segmental organisation and compartmentalised gene function in determination and differentiation during insect development. These three examples illustrate the significance of regulatory sequences in allowing for the evolution of gene function to cover more than one rôle, and also show how the time of activation of a set of genes and their relative productivity may be co-ordinated.

In seeking to explain the basis of major evolutionary divergence of form and function, it is necessary to analyse the mode of action of genes which have an especially significant rôle in mediating morphological and/or functional (including behavioural) organisation. Such genes include those affecting transcriptional patterns at developmental stages crucial for morphogenesis (as in early cleavage and during metamorphosis in holometabolous insects), and those influencing alternative states of cells compartmentalised in cellular fields.

The basic processes involved in body organisation of the segmented animal phyla may have been highly conserved through evolution. Homoeotic mutations take on new significance as showing how compartmentalisation of gene action allows major change in phenotype without concomitant disruption of body function. It has been suggested that rapid evolutionary change may have been favoured in primitive Metazoa by a higher degree of focusing of gene action at the compartmental level than is observed in more advanced groups. The contention in this chapter is that gene action is focused at the compartmental level in advanced as well as in primitive groups. What has changed in the course of evolution of at least the segmented Metazoa is the imposition of tagmatal and other higher order functional groupings of body structures on a primitive, compartmentalised body plan, resulting in an apparent slowing down of the rate of appearance of 'large-scale' morphological variations in later geological time.

INTRODUCTION

One approach to analysis of evolutionary processes is to start by identifying the time of origin and mechanismal basis of innovations in the phenotype, especially where the appearance of a new higher taxon is involved. The first step, then, is to define the new gene functions. The second stage involves tracing back to establish the changes in gene structure, the alterations of nucleotide sequence and organisation that brought about the new genotype, relating the new to the old. The third step in the process is the elucidation of the mechanism whereby the genetic reorganisation took place, while a fourth concerns how and why the new genotype came to be perpetuated in the population, or participated in speciation. It is with the first two of these stages, i.e. with the delimitation of kinds of gene function and then with the relation between function and structure of genes that this chapter is concerned. The third and fourth stages suggested are covered by other contributions to this book.

Where there is a simple and direct relationship between gene, protein and phenotype, and especially where the protein product is the phenotype studied, it is often reasonably clear from knowledge of gene structure at the nucleotide level how a new or altered function was derived and what structural changes in the genome were needed to produce it. For instance, the source in the progenitors of the Triassic mammalian radiation of the α-lactalbumin subunit of lactose synthetase needed in milk production was presumably an ancestral lysozyme gene (reviewed by Raff & Kaufman 1983).

It is, however, the appearance of new phenotypes involving complex, integrated patterns of interplay between a number of nucleotide sequences and their products (Fig. 10.1), between a number of cell types and between discrete organs, that constitutes a more interesting and challenging problem. This is nearer the idea of 'macroevolution', of the origin of distinctive new life forms, of new species and, by accumulation, new higher taxa. We have only just begun the task of analysing the ways in which such integrative control of, and evolutionary change in, interacting gene-action systems (Waddington 1962) can be brought about.

Recognition of the importance of cellular compartments as units of genetic control in development has led to the conclusion that they are also significant units with respect to evolutionary change. The need is now to establish in detail how the early cleavage products of the zygote come to receive differing signals which operate control sequences that bring about co-ordinated, localised and differential patterns of gene activation. When this mechanism is fully analysed, many of the major difficulties in the way

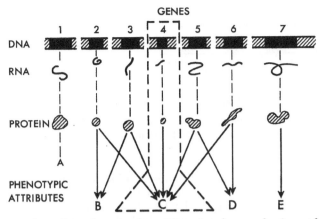

Figure 10.1 Co-ordinated gene interaction in the production of complex phenotypic attributes (modified after Waddington 1962). A direct link between gene and phenotype is readily evident only when the protein product of the gene itself comprises the aspect of phenotype under study (gene 1). In most instances many genes will contribute to a specific phenotypic attribute.

of deeper understanding of evolutionary change in body architecture, as well as in development, will have been solved.

The idea that '. . . evolutionary change in regulatory systems accounts for evolution at and beyond the anatomical level' (Wilson *et al*. 1974) has been increasingly emphasised in the last decade (e.g. Raff & Kaufman 1983, Erwin & Valentine 1984). Regulatory controls must determine the timing of developmental events, the on–off binary choice of developmental path according to whether a gene product is made or not, and the integrated switching of banks of genes in the co-ordinated production of a spectrum of proteins. For an understanding of these processes it is necessary to review our present knowledge of what regulatory sequences are, what they can do, and how such sequences are put together in the evolution of gene control systems.

PATTERNS OF GENE CONTROL IN EUKARYOTES

Gene structure in eukaryotes

The organisation of a typical eukaryotic gene is summarised in Figure 10.2. Control of the time and amount of transcription from the structural protein-coding nucleotide sequence of the gene is mediated by at least three kinds of untranscribed promoter sequence upstream from the transcription start-point. The 'TATA box' sequence is the site of binding of one or more proteins recognising this nucleotide sequence and facilitating the attachment of RNA polymerase II for transcription. Some such proteins appear to recognise selectively duplex consensus sequences downstream (3') to the

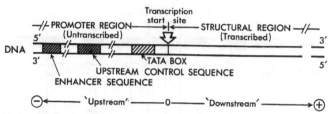

Figure 10.2 Organisation of a typical eukaryote gene. Nucleotides are numbered relative to the transcription start point: to the right, positive (downstream); to the left, negative (upstream).

TATA nucleotides, forming a pre-initiation complex. Other general transcription factors also associate with TATA sequences along with RNA polymerase II. Upstream from the TATA box, promoter regions typically contain a control sequence recognised by DNA binding proteins specific to that gene or to a set of genes and capable of controlling the level of transcription. Even further upstream, enhancer sequences capable of modulating transcription provide yet another level of control (Fig. 10.2). The first two of the regulatory elements are well exemplified by the heat shock genes discussed below.

Further levels of transcriptional control may be mediated by downstream sequences within the transcribed structural region of the gene (briefly reviewed by Reudelberger 1984); such control may be one function of the intron sequences that are transcribed but not represented in the mRNA produced from the transcript. The introns of certain co-ordinately controlled, neurone-specific genes active in the rat brain, for example, contain sequences in common that may reflect this function (Milner *et al.* 1984).

Many lines of evidence contribute to the present understanding of the rôle of the promoter sequences is gene control. One of the most striking examples is provided by the now well known experiments on enhanced growth in mice which incorporated in their genomes an artificially constructed gene consisting of the strong promoter from the mouse metallothionein gene with the structural region of the rat growth hormone gene. Enhanced activity of this artificial gene compared with that of the normal mouse growth hormone gene led to growth hormone levels 100 to 800 times that of control mice, and to the increased size of the affected mice (Palmiter *et al.* 1982).

The detailed consideration of three genetic systems will illustrate the rôle of regulatory sequences in the control of gene action affecting metabolism, development and morphogenesis as well as behaviour.

Three case histories

The heat shock system   The eukaryotic heat-shock protein (*hsp*) genes were first identified in studies of the effect of elevated temperature on puffing in salivary polytene chromosomes of *Drosophila*. The *hsp* genes fall into three distinct groups within each of which the protein products, the heat-shock proteins (HSPs) are homologous; the genes are named from the molecular mass of their products in kilodaltons (kd; Table 10.1) The *hsp* loci puff rapidly when the insect, or isolated polytene tissues *in vitro*, are exposed to a range of stresses including elevated temperature, ethanol or a variety of other chemicals, certain viral infections, radiation, anoxia and so on (reviewed by Schlesinger *et al.* 1982). Except for the product of *hsp-82*, which is detected in the cytoplasm, especially in association with the Golgi apparatus, the HSPs are largely concentrated in the nucleus. Two major functions of these genes are to modify the pattern of transcription in cells under stress, thus presumably suppressing the formation of abnormal mRNAs, and to modify translational activity during stress to an extent and by mechanisms that vary at least quantitatively from species to species. In some cases, polypeptide chain elongation is blocked (Ballinger & Pardue 1982), in others most of the mRNAs on ribosomes at the time of stress are discharged, but may be re-engaged when the normal pattern of protein synthesis resumes after recovery from the stress (Lindquist *et al.* 1982), at which time the nuclear HSPs return to the cytoplasm.

Cloning of the *D. melanogaster hsp-70* gene allowed the homology of the *hsp-70 and hsp-68* genes to be established, and revealed the general existence of homologous stress-inducible loci in other eukaryotes from yeast to man (Schlesinger *et al.* 1982 for overview). Very importantly, however, use of the cloned *hsp-70* gene as a probe established the existence in many organisms, including *Drosophila*, of genes related to the *hsp* loci but not inducible by heat – the *heat shock cognates* (e.g. Ingolia & Craig 1982). Presumably it is these cognate genes that are differentially activated in the early *Drosophila* zygote (Zimmerman *et al.* 1983) and in imaginal disc cells

Table 10.1 Three groups of *hsp* genes are defined by the homologies of the protein products of the varius loci, all of which are inducible by heat shock (and certain other stresses). Co-ordinate regulation in response to heat shock is mediated by a common regulatory sequence of nucleotides 'upstream' (5′) to the nucleotides forming the transcription start signal in each of the genes. This regulatory sequence differs in nucleotide composition in different *hsp* genes relative to the consensus sequence CTNGAATNTTCTAGA and also differs in position relative to the transcription start point. (After Davidson *et al.* 1983.)

Product group	*hsp* locus	Heat-shock protein (mol.wt $\times 10^{-3}$)	Percentage homology with consensus sequence	Position relative to transcription start
1	82	82	92	−74
2	70	70	85	−62
	68	68	69	−61
3	27	27	54	−68
	26	26	85	−62
	23	23	62	−75
	22	22	85	−147

(Ireland *et al.* 1982, Cheney & Shearn 1983) to produce the 'heat shock' peptides found at these stages and which seem likely to be related to regulation of transcription in normal development (Bensaude *et al.* 1983).

Correlation of the activation of the *hsp* gene set with induced protection from subsequent stress is evident in two Australian drosophilids feeding on the bracken fern, *Pteridium*. *Drosophila megagenys* and *D. notha* have quite different thermotolerance patterns, but both species are readily protected from otherwise lethal short-term temperature stress by prior exposure to a sub-lethal temperature shock (illustrated for *D. megagenys*; Table 10.2).

Using *D. melanogaster*, Lindquist (1980) showed that production of HSP-82 is induced at a lower temperature than that of HSPs 68 and 70. Relative levels of induction of the individual *hsp* loci are highly dependent on conditions of treatment. *Drosophila megagenys* and *D. notha*, when compared under standardised heat stress, consistently show species differences in the temperature pattern of induction of the HSPs, most markedly of HSP-82, which is readily induced as in *D. melanogaster*, and in HSPs 68 and 70 considered together (Fig. 10.3).

The regulation of the *Drosophila hsp* genes and their cognates illustrates many features of selective and integrated gene-control systems in eukaryotes and of their evolution. First, all the *hsp* genes are induced in response to a common stimulus derived from certain kinds of stress, so these loci must share a common control element. But, as we have seen, each heat-shock gene has a characteristic and individual pattern of switch-on in any species so that regulatory controls show discrimination between the genes of the *hsp* set. Further, the heat-shock cognates are activated only at specific life stages, for specific times. Production of the heat proteins outside that

Table 10.2   Results of typical experiments showing survival to adult emergence of *Drosophila megagenys* exposed as third instar, post-feeding larvae to the same total thermal loads administered in contrasting sequence as periods of relatively low (32.5°C) and relative high (37.5°C) temperature at 100% relative humidity. Treatment sequences are indicated as hours at the lower (L) or higher (H) temperature e.g 2L + 2H indicates 2 hours at 32.5°C followed by 2 hours at 37.5°C before transfer to the rearing temperature (20°C). Exposure to 37.5°C for 2 hours or more is normally lethal to *D. megagenys*. The significance of the difference between treatments was assessed using Fisher's exact test. (J. A. Thomson and C. M. Shearer, unpublished observations.)

Treatment received by larvae	Adult flies emerged	Deaths in larval or pupal stage	Probability of difference between treatments
2L + 2H	11	9	<0.0002
2H + 2L	0	20	
2L + 3H	5	15	0.047
3H + 2L	0	20	

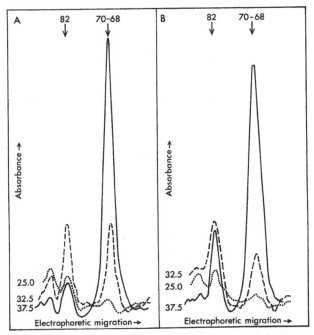

Figure 10.3   Densitometer tracings of fluorographs showing the pattern of incorporation of ^{35}S-methionine into two major heat-shock proteins of larval ganglia of (a) *Drosophila megagenys* and (b) *D. notha* following heat shock for 30 minutes *in vivo* at 32.5°C and 37.5°C (J. A. Thomson and C. M. Shearer, unpublished observations). Incorporation into 68–70 kd proteins follows a similar pattern in the two species, but in *D. notha* proportionately more 82 kd HSP is synthesized at 37.5°C than in *D. megagenys*.

developmental range, as when the *hsp* loci are induced under conditions of stress, leads to changes of normal transcription and translational patterns, so breaking the continuity of normal metabolic functions. Once the heat-shock cognate genes have played their early developmental rôle, whether in embryogenesis or in imaginal disc differentiation, there is a cost to the organism in producing HSPs. A trade-off is necessary in evolving gene controls maximising survival chances. Triggering of *hsp* genes at too low a stress threshold will lead to unnecessarily frequent and disadvantageous disruption of normal function. Triggering at too high a threshold will leave many individuals unprotected against a potentially lethal stress by delaying the reprogramming of transcription and translation in favour of synthesis of gene products needed to counteract the effects of the stress. Detoxifying enzymes in the case of chemical stress, and repair enzymes in the case of radiation exposure, for instance, may be produced too late.

The work of Wu (1984a,b) and of Parker and Topol (1984a,b) has recently provided evidence of how these various controls are achieved. These studies employed the ability of DNA binding proteins to protect a double-stranded DNA sequence from exonuclease digestion, permitting control sequences covered by regulatory proteins to be separated from the rest of the genome. Wu's model of the function of control sequences identified in *hsp* genes upstream (5′) from their transcription start points is summarised in Figure 10.4.

Wu's model suggests that a TATA box binding protein (TAB) must complex with the TATA sequence before the upstream controlling sequence common to the *hsp* genes can become inducible by heat shock. According to Wu's proposal heat shock or other stress involves binding of a second protein, heat-shock activator protein (HAP) to the upstream control element. HAP could not on this model bind to the promoters of heat shock cognate loci, for which a separate inducing signal would be required (Fig. 10.4).

Within the *hsp* gene set, Wu ascribes the variation of response to stress of differing kind and intensity to the differing affinity of HAP for the specific control sequences (Table 10.2) of each gene. Wu (1984b) has extended analysis of the control sequence to 28 bases. The control sequence of *hsp-82*, induced at the lowest temperature and showing the highest affinity

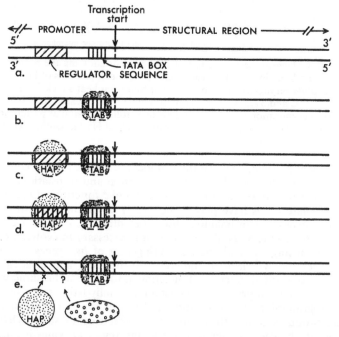

Figure 10.4 Model of proposed control of heat-shock and cognate genes in *Drosophila melanogaster* (after Wu 1984a,b). Details as in text: TAB, TATA (box) binding protein; HAP heat-shock activated protein. (a–c), *hsp-82*; in (a) uninducible, (b) inducible and (c) activated states; (d) *hsp-70*, activated state; (e) heat-shock cognate, not heat-shock inducible.

for HAP, has almost perfect internal dyad synmmetry:

GAAGCCTCTAGAAG.TTTCTAGAGACTTC

The less easily induced *hsp-70* gene has nucleotide substitutions destroying the near perfect internal symmetry characterising *hsp-82*, and shows a lower affinity for HAP. Thus Wu's model proposes that each gene in the co-ordinately controlled set is either on or off, but each has its own threshold of sensitivity to the inducing stimulus mediated by its binding affinity for a regulatory protein (Fig. 10.5).

In the heat-shock gene system of *Drosophila* we see the re-use in evolution of a set of genes, the original function of which seems to have been set embryonic transcription patterns, to control transcription under stress conditions in later life. The latter function evolved in such a way to maintain compatibility with the former. The protein products of the two systems remain highly conserved, but separate control systems permit realisation of the two functional rôles.

Other examples of shared 5' control sequences in the co-ordinate regulation of gene sets are well documented, for instance, in setting transcription patterns required for reproduction. Conalbumin, lysozyme and ovalbumin synthesis in the chicken appear to be co-ordinated in this way, as are the late and middle functioning chorion genes in the silkmoth, *Antheraea* (reviewed by Davidson *et al.* 1983).

Genetic control of a molluscan behaviour sequence   Elucidation of the nature and structure of a gene complex mediating innate behaviours in the tectibranch mollusc *Aplysia*, results largely from work of Scheller, Kandel, Axel and colleagues at the Centre for Neurobiology and Behavior at Columbia University. This study provides an exciting insight into one

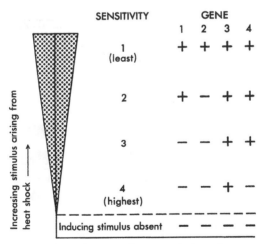

Figure 10.5   General scheme for co-ordinate regulation of a gene set as the *hsp* and cognate loci. Gene 3 is most responsive to heat stress, gene 2 is least sensitive.

mechanism of physiological co-ordination in the production of an integrated, in this case behavioural, phenotype.

Egg laying in *Aplysia* involves a set sequence of behaviours. First, movement and feeding of the animal is inhibited, then characteristic head waving takes place as the mollusc grasps the emerging egg string and assists its expulsion from the genital aperture. Finally, the egg string is attached to a solid substratum with the aid of a mucoid buccal secretion (Scheller *et al.* 1982).

One neuroactive peptide, the egg laying hormone (ELH), is involved in eliciting some, but not all, of this behavioural sequence. Strumwasser *et al.* (1980) showed that ELH and three companion peptides, the alpha and beta bag-cell factors and an acidic peptide, are released from specialised neurons known as bag cells attached to the abdominal ganglion into which these peptides move (for details see Scheller *et al.* 1982, 1983). ELH is an excitatory transmitter selectively augmenting the firing of a specific neuron in the abdominal ganglion, while the beta factor excites other neurons and the alpha factor inhibits a third group of neurons (Fig. 10.6). In addition to the effect of ELH as a neurotransmitter, this peptide diffuses into the ovotestis to act hormonally on the smooth muscle of follicles in the initiation of egg laying.

A number of ELH-producing gene clones were identified by screeening a genomic library of recombinant clones, using cDNA from mRNAs of bag cell and non-neuronal tissue to locate clones which hybridised with bag-cell cDNA but not that from non-neuronal tissue (Scheller *et al.* 1982). Sequencing suggested that the mRNA from the ELH-producing gene would code for a precursor polypeptide of 271 amino acids of which 36 amino acids are

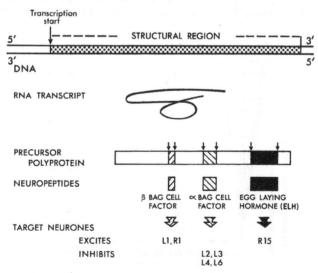

Figure 10.6   The *Aplysia* ELH gene, its initial polyprotein product and its functional neurotransmitter polypeptides (based on Scheller *et al.* 1982, 1983).

those of ELH. Detailed examination of the ELH sequence enabled Scheller *et al.* (1983) to deduce that the precursor is better regarded as a polyprotein. It contains 10 presumptive sites for endopeptidase cleavage, four of the fragments produced being ELH itself, alpha and beta bag-cell factors and the acidic bag-cell peptide. Further, it became evident that the ELH and the bag-cell factors are related and presumably derived by duplication and subsequent divergence of a common ancestral sequence. These sequences are also related to those of the A and B neuropeptides produced by genes active in atrial neurons: the A and B neuropeptides control the initiation in the bag cells of ELH production and thus the egg-laying sequence (Scheller *et al.* 1982).

These genetic studies reveal the key to the evolution of an extraordinarily elegant system of co-ordinate control for a physiological and behavioural phenotype. Related genes coding for similar but selective neurotransmitters have been linked into a linear series within one transcription unit, so that the synthesis and subsequent processing of a single polyprotein co-ordinates the timing of related behaviours in the array. Further, no single behaviour will be released without the related elements preceding and subsequent to it (Scheller *et al.* 1982).

From the point of view of evolution of a co-ordinated system of gene activities, the ELH case may be considered a highly sophisticated system in which the production of a set of gene products from separate structural sequences is integrated by splicing them together under the control of a single regulatory sequence, rather than by separate *cis*-acting control sequences in front of each. Control under a single promoter, with a single transcript, processed to give a message subsequently translated into a polyprotein, may be evolutionarily preferable when the emphasis on the activities of a gene set is on distinctive quality of their products, on their relative amounts, and on the time of release. Certainly similar mechanisms have evolved repeatedly as in the cases of the amphibian and avian yolk proteins, phosvitin and lipovitellin, processed from the precursor vitellogenins (Wahli *et al.* 1981); the production of insulin A and B chains: and the corticotropin (β-lipoprotein-related peptides, α-melanotropin, endorphin and methionine–enkephalin (reviewed by Herbert & Uhler 1982).

Control of the segmental body plan in animals In contrast to integration through linear control, co-ordinated regulation of gene activities mediated by separate *cis*-acting control sequences may most reliably produce qualitatively and quantitatively distinct spectra of proteins. The genetic analysis of embryogenesis and morphogenesis in *Drosophila* illustrates the significance of such controls in eukaryotic development and differentiation, as well as the probable evolutionary significance of genetic change in the regulatory mechanisms involved.

The body plan of *Drosophila* is laid down by genes referable to several classes according to their mode of action. The first of four broad groups of genes considered here controls the axis of the embryo, determining in the oocyte before fertilisation the future orientation of the zygote. Presumably such genes control the distribution in the egg of morphogens specifying subsequent developmental processes. Females homozygous for *bicaudal* produce embryos with posterior structures at each end; homozygotes for *dorsal* give rise to embryos with two dorsal surfaces (Nusslein-Volhard 1979).

Subsequent to specification of polarity and orientation, segment number and pattern is defined in the zygote by a second class of genes, of which some 20 have so far been identified, and which fall into two groups: 'gap' loci and 'pair-rule' loci. Mutants of the gap group produce changes in the number of segments in the body tagmata. Homozygotes for *knirps*, for instance, have a normally segmented head and thorax, but only one large 'segment' in place of abdominal segments 1 to 7 (Nusslein-Volhard & Wieschaus 1980). Mutants of the pair-rule group show phenotypes with half the normal number of apparent segments, reflecting organisation of the germ band into units two segments wide. These double units are defined by boundaries in the normal segmental positions, although in different mutants the pairing of segments into these double units may be out of phase (Nusslein-Volhard & Wieschaus 1980)

A third group of genes appears to be involved in specifying the polarity of segments as anterior or posterior half-segment units. Evidence from the fate of cells marked genetically by induced mutation shows that, at the blastoderm stage, particular cells are already determined as belonging to anterior or posterior compartments in specific segments (Garcia-Bellido & Ripoll 1978). The segment polarity loci confirm that the anterior and posterior half-segment units participate in delimitation of the ultimate compartmental building blocks of *Drosophila* (Lawrence 1981). Homozygotes for *gooseberry*, for instance, showed structural alterations limited to the posterior half of every segment (Nusslein-Volhard & Wieschaus 1980).

Genes affecting the kind of tissue organisation in the subsegmental compartments comprise a fourth group. Included here are the homoeotic loci, mutations in which lead to the appearance in one segment of structures characteristic of another segment. Genes of this group control which of several possible pathways a given lineage of cells will take, so producing the special structures typical of a particular segment rather than of segments in general. Examples include the *Antennapedia* (*Antp*) complex that affects primarily cephalic and thoracic segments (Kaufman *et al.* 1980) and the *bithorax* (*bx*) complex, a group of some eight mutationally distinct sites that determines features of thoracic and abdominal segments (Lewis 1978, Bender *et al.* 1983).

The general features of the *bithorax* gene complex are summarised in Figure 10.7. Evidence from mutation studies, and especially from deletion mapping, shows that a series of related and closely linked sites in the cluster specify segmental characteristics in similar but distinctive ways, so that these genetic units evolved by duplication and later divergence of a common ancestral sequence (Lewis 1978). The nucleotide sequence in the *bx* complex is generally colinear with the segmented pattern of expression: the left to right order of the genetic material largely corresponds with the anterior to posterior order of the body structures affected. Loss of function mutations (typically recessive) shift the phenotype of an adult segment

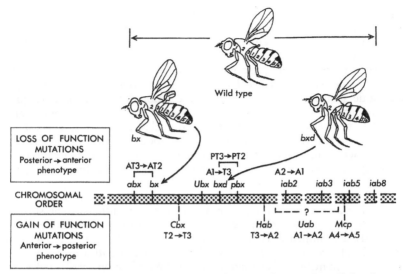

Figure 10.7 Organisation of the *bithorax* (*bx*) complex in *Drosophila*, showing colinearity of chromosome position to phenotypic effect (after various sources including Lewis (1978) and Bender *et al.* (1983)). (A)T, (anterior) thoracic: (P)T, (posterior) thoracic: A, abdominal: numbers refer to segments of each tagma.

towards a more anterior type. Gain of function (typically dominant) mutations in this series shift the phenotype of an adult segment towards a more posterior type.

Struhl (1984) briefly summarises the many similarities between the *Antp* amd *bx* homoeotic complexes. The transcripts of both complexes appear at the same early stage of embryogenesis (cellular blastoderm) in a pattern showing tight localisation to particular presumptive segments. The initial transcripts of each are large and then processed into smaller mRNAs. Most importantly both homoeotic complexes contain highly homologous copies of a short nucleotide sequence of 180 bases now known as the homoeo 'box' or domain, representing an open reading frame corresponding to a basic polypeptide of 60 amino acids (McGinnis *et al.* 1984a; Scott & Weiner 1984). Not only is the homoeo box near the 3' end of the mature transcript of two homoeotic sequences in *Antp* and three in *bx*, but it is also repeated at least twice further in the genome of *D. melanogaster*, one of these copies being at the 3' end of the transcript of the developmentally significant gene *fushi tarazu* (Laughon & Scott 1984). The hypothesis of Garcia-Bellido (1975) that homoeotic genes specify segment-specific patterns of cell differentiation by their activity as 'selector' genes controlling expression of sets of 'realisator' structural genes seems close to being proved.

Homoeo boxes highly homologous with those of *Drosophila* have recently been discovered (McGinnis *et al.* 1984b, Carrasco *et al.* 1984) in segmented animals ranging from the earthworm, a beetle, chicken and clawed toad to the human. This finding focuses attention on the idea of unitised segmental structure as an homologous building plan common to all these groups. Sequences homologous to the homoeo box have not yet been found in the nematode or sea urchin genomes, but this negative evidence is difficult to evaluate until more non-segmented animals have been examined.

The genetic control of development, as now understood, involves sequential binary decisions specified by the on or off condition of the relevant nucleotide sequence in terms of its transcriptional and translational products (Kauffman *et al.* 1978). In succession these alternative gene states must determine anterior *versus* posterior, dorsal *versus* ventral, then a series of successive transverse partitionings of the body down to subsegmental compartments, followed by the control of tissue type and organisation involving the homoeotic series of genes (see Raff & Kaufman 1983, for discussion).

Three additional points are relevant to considerations of gene switching in bringing about the developmental sequence just outlined. The sharply delimited and non-overlapping boundaries between compartments of differently determined cells arise from two features. First, the initial event setting the choice along one or other of the alternative pathways for a given gene is deterministic rather than stochastic (Waddington 1962), reflecting the ability of each progenitor cell to distinguish sharply between levels of the controlling morphogen, i.e. in a gradient of morphogen concentration, each gene seems likely to respond according to a clearly defined threshold for its induction or repression. A series of duplicate genes with different thresholds, mediated for instance by different affinities for an inducer as in the model of Figure 10.5, could then control a compartment-by-compartment pattern of determination and differentiation (Fig. 10.8).

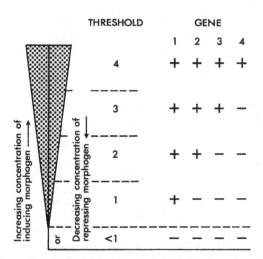

Figure 10.8 General scheme for co-ordinate regulation of a gene set by hierarchical switching of a series of loci recognising distinct threshold levels in a gradient of an inducing or repressing morphogen.

Secondly, cells of different compartments do not normally mix in later development (Garcia-Bellido 1975, Lawrence 1981). Thirdly, gene settings are fixed early in development when distances are small (Crick 1975), and the contrast in the level of signal received by two adjacent cells along a gradient may be relatively great. In *Drosophila*, at the cellular blastoderm stage when the crucial events of determination are already complete, each presumptive segment is only three to four cells wide (Scott & Weiner 1984).

Genetic studies of development, such at those centering on *Drosophila*, now enable formulation of models of gene control that seem to offer new insight into the evolution of change in developmental pattern, and hence into change of phenotype with respect to morphological attributes of considerable scale.

EVOLUTIONARY ASPECTS OF GENE STRUCTURE IN RELATION TO FUNCTION

Rapid growth in our understanding of the structure and function of genes coding for the proteins of eukaryotic cells has highlighted the many processes of evolution that are still far from adequately explained. Nucleotide substitutions leading to change in amino acid sequence resulting in altered kinetic properties of enzymes, for instance in substrate affinity, thermostability or relation to inhibitors, are clearly relevant to adaptation and to change in allele frequences in populations. So are the intron – exon structure of the eukaryotic transcription unit, the conservation and re-use of domain sequences in families of proteins, duplication and subsequent mutational divergence of nucleotide sequences to provide new genes coding for new products meeting new needs in the evolutionary sequence, or gene rearrangement providing for immunoglobulin or trypanosome cell-surface protein diversity. Although these processes undoubtedly play important rôles in evolution, such features of evolutionary history as rapid bursts of speciation and the appearance of new taxa of quite radically altered body plan and behaviour seem inadequately explained in terms of such phenomena. Nor do theories of horizontal transfer of genetic information, for instance via RNA viruses (Erwin & Valentine 1984), contribute significantly to solution of the paradox. Such mechanisms may increase available variability by pooling components from the contributing genomes in a manner analogous to occasional events of introgressive hybridisation, but it is difficult to see here any special mechanisms for radical change in body form without disruption of function.

Increasingly, attention has recently been turned to the idea that change in regulatory elements controlling genes, and integrating groups of genes, may have a particular significance in evolutionary diversification of form and function (reviewed by Raff & Kaufman 1983). In ascribing evolutionary significance to regulatory components in the genome two problems need consideration. The first concerns the danger of merely pushing back the key events by one step to invoke yet another (hypothetical) gene. Many contemporary discussions of genome function in relation to evolution seem to employ the terms regulatory and structural in an exclusive dichotomous classification of genes. But the regulator of one locus is likely to be itself the product of a structural gene or to require complexing with one or more such products to become functional in the induction or repression of a gene activity, so that there is here a semantic difficulty. Thus genes producing the components specifying the anterior–posterior axis and dorsal–ventral surfaces of the *Drosophila* embryo are structural genes, coding for products that subsequently interact with the regulatory sequences of compartment-specific genes to control the programming of later development. It is exciting that, for *Drosophila*, we seem now to have identified most of the genetic loci involved, so that postulation of additional hierarchies of genes is increasingly unnecessary.

The second problem to be considered in evaluating the rôle of regulatory elements in evolutionary change is the perception that life processes will necessarily be adversely disrupted by change in co-ordinated gene activities. If a gene set is integrated by way of a common regulatory element, does it allow that mutational change in that regulator sequence bringing about a substantial shift in phenotype will generally cause chaotic disorganisation? Erwin and Valentine (1984) write that:

'Viable mutations with major morphological or physiological effects are exceedingly rare and usually infertile: the chance of two identical rare mutant individuals arising in sufficient propinquity to produce offspring seems too small to consider as a significant evolutionary event'.

Homoeotic mutants of *Drosophila* and other species have often been regarded as genetic peculiarities of limited developmental interest. It is now emerging that homoeotic mutations may provide us with two of the keys to analysis of the evolution of at least the segmented animal phyla. Firstly, these mutations provide evidence of how change in nucleotide sequences regulating developmental patterns can produce gross phenotypic change. Secondly, such mutations commonly bring about quite extreme changes in body structure which are not inconsistent with viability and even fertility.

Under the subtitle 'Homoeology recapitulates phylogeny', Raff and Kaufman (1983) outline a speculative scheme of arthropod evolution based on the kinds of genetic step compatible with homoeotic gene functions identified in the analysis of *Drosophila* development. One example used by Raff and Kaufman to illustrate these ideas is the onychophoran to myriapod to apterygote to pterygote series. A deletion of bx^+ and pbx^+ in *Drosophila* transforms the fly to a condition mimicking the more primitive four-winged condition. The loss of the $Antp^+$ function leads to a *Drosophila* phenotype resembling the apterygote condition; the three thoracic segments are similar and non-wing bearing. If the entire *bx* complex is deleted, the posterior segments in the *Drosophila* embryo develop as a series of similar thoracic or trunk segments to give a body organisation mimicking that of the trignathous myriapods. Deletion of both the *Antp* and *bx* complexes produces an embryo with three head segments and a series of similar trunk segments on the onychophoran pattern. This genetic series in *Drosophila* appears interpretable as providing a model in reverse of the major features resulting from successive steps of gene duplication and acquisition that may have produced the original phylogenetic sequence.

Lesions in the genetic material of homoeotic loci producing extreme phenotypic effects are certainly often lethal in homozygous embryos. But what is really surprising is that a simple genetic change such as homozygosity for three mutations *abx*, *bx* and *pbx*, converts the *Drosophila* phenotype into one in which the third thoracic segment is a fully formed wing-bearing segment duplicating the second thoracic segment. Some of these grossly reorganised animals can survive to emergence from the puparium, as do some *bx* mutants with four pairs of legs, although these flies are of poor viability and often show postural and behavioural abnormalities. The important point is that extreme changes in phenotype of this kind do not necessarily disrupt either the larval or adult functions essential to life. Body organisation on a unitary compartmentalised basis protects the organism from developmental chaos that would otherwise result from mutations; the alimentary and nervous systems, musculature and so on are maintained as well as possible, consistent with the scale of the disruption experienced.

Another situation in which a single gene may cause substantial and concerted phenotypic change in many characters is exemplified by sex-change phenomena seen in many animals, including mammals. In *Drosophila*, the *transformer* mutation converts the female phenotype into that of the male with respect to genitalia, body pigment patterns, sex-combs and behaviour, although such individuals have small testes and are sterile (Sturtevant 1945).

Several recent authors have drawn attention to the idea that rapid evolution in the early Metazoa was favoured by a higher degree of compartmentalisation of gene action than that seen in later metazoans (Erwin & Valentine 1984). It is difficult to see how any higher degree of focusing of gene action could have existed in more primitive groups, to be later lost. A more plausible view appears to be that animals of the present day have in fact retained their ancestral patterns of compartmentalisation, but have increasingly superimposed on them a higher order of organisation, both of complex behaviour patterns and of tagmata as functional morphological and physiological units. The apparently higher rate of appearance of larger scale morphological variants in Palaeozoic metazoans would thus reflect restrictions imposed on developmental flexibility in more evolutionarily advanced groups, rather than a higher degree of compartmentalisation of gene action in more primitive groups. A corollary of this argument is that the viable homoeotic mutants seem in present day groups, such as the higher arthropods, may represent a subset of the *viable* phenotypes which could have arisen in the absence of these constraints.

Compartmentalisation of gene action in morphogenesis as a means of permitting phenotypic flexibility, without chaotic phenotypic disorganisation, may also have a logical counterpart in the evolution of integrated behaviour patterns. The whole set of egg-laying behaviours in *Aplysia*, for example, occurs as a systematic repertoire related specifically to egg laying because of the unitary control leading to multiple neuropeptide synthesis and release. A genetic change in any one part of the complex gene producing the polyprotein precursor molecule may lead to behavioural change, but that change will be manifested within the context in time and space of the normal behavioural series to which it relates (unless the regulatory element controlling the gene complex is involved). Perhaps such behavioural compartmentalisation plays a part in the rapid evolution of ethological barriers acting at the pre-mating level to reinforce reproductive isolation during the speciation process.

ACKNOWLEDGEMENTS

I thank D. T. Anderson and J. A. Sved for helpful comments on the manuscript.

REFERENCES

Ballinger, D. G. and M. L. Pardue 1982. The subcellular compartmentalization of mRNAs in heat-shocked *Drosophila* cells. In *Heat shock from bacteria to man*, M. J. Schlesinger, A. Ashburner and A. Tissieres (eds), pp. 183–90. Cold Spring Harbor: Cold Spring Harbor Laboratory.

Bender, W., M. Akam, F. Karch, P. A. Beachy, M. Peifer, P. Spierer, E. B. Lewis and D. S. Hogness 1983. Molecular genetics of the bothorax complex in *Drosophila melanogaster*. *Science* 221, 23–9.

Bensaude, O., T. C. Babinet, M. Morange and F. Jacob 1983. Heat shock proteins, first major products of zygotic gene activity in mouse embryo. *Nature* 305, 331–3.

Carrasco, A. E., W. McGinnis, W. J. Gehring and E. M. De Robertis 1984. Cloning of an *X. laevis* gene expressed during early embryogenesis coding for a peptide region homologous to *Drosophila* homoeotic genes. *Cell* 37, 409–414.

Cheney, C. M. and A. Shearn 1983. Developmental regulation of *Drosophila* imaginal disc proteins: synthesis of a heat-shock protein under non-heat shock conditions. *Dev. Biol.* 95, 325–30.

Crick, F. 1975. Diffusion in embryogenesis. *Nature* 225, 420–2.

Davidson, E. H., H. T. Jacobs and R. J. Britten 1983. Very short repeats and coordinate induction of genes. *Nature* 301, 468–70.

Erwin, D. H. and J. W. Valentine 1984. 'Hopeful monsters', transposons and Metazoan radiation. *Proc. Natl Acad. Sci. USA* 81, 5482–3.

Garcia-Bellido, A. 1975. Genetic control of wing disc development in *Drosophila*. *Ciba Found. Symp.* 27, 161–82.

Garcia-Bellido, A. and P. Ripoll 1978. Cell lineage and differentiation in *Drosophila*. In *Results and problems in cell differentiation*, W. J. Gehring (ed.), Vol. 9, pp. 119–56. Berlin: Springer-Verlag.

Herbert, E. and M. Uhler 1982. Biosynthesis of polyprotein precursors to regulatory peptides. *Cell* 30, 1–2.

Ingolia, T. D. and E. A. Craig 1982. *Drosophila* gene related to the major heat shock induced gene is transcribed at normal temperatures and not induced by heat shock. *Proc. Natl Acad. Sci. USA*, *525–9*.

Ireland, R. C., K. Berger, M. A. Sirotkin, D. Osterbur and J. Fristrom 1982. Ecdysterone induces the transcription of four heat shock genes in *Drosophila* S3 cells and imaginal discs. *Dev. Biol*. **93**, 498–507.

Kauffman, S., R. Shymko and K. Trabert 1978. Control of sequential compartment formation in *Drosophila*. *Science* **199**, 259–70.

Kaufman, T. C., R. Lewis and B. Wakimoto 1980. Genetic analysis of chromosome 3 in *Drosophila melanogaster*: The homoeotic gene complex in polytene chromosome interval 84A-B. *Genetics* **94**, 115–33.

Laughon, A. and M. P. Scott 1984. Sequence of a *Drosophila* segmentation gene: protein structure homology with DNA binding proteins. *Nature* **310**, 25–31.

Lawrence, P. A. 1981. The cellular basis of segmentation in insects. *Cell* **26**, 3–10.

Lewis, E. B. 1978. A gene complex controlling segmentation in *Drosophila*. *Nature* **276**, 565–70.

Lindquist, S. 1980. Varying patterns of protein synthesis in *Drosophila* during heat shock: Implications for regulation. *Dev. Biol*. **77**, 463–79.

Lindquist, S., B. Didomenico, G. Bugaisky, S. Kurtze, L. Petko and S. Sonada 1982. Regulation of the heat-shock response in *Drosophila* and yeast. In *Heat shock from bacteria to man*, M. J. Schlesinger, M. Ashburner and A. Tissieres (eds), pp. 167–75. Cold Spring Harbor: Cold Spring Harbor Laboratory.

McGinnis, W., M. S. Levine, E. Hafen, A. Kuroiwa and W. J. Gehring 1984a. A conserved DNA sequence in homoeotic genes of the *Drosophila* Antennapedia and bithorax complexes. *Nature* **308**, 428–33.

McGinnis, W., R. L. Garber, J. Wirz, A. Kuroiwa and W. J. Gehring 1984b. A homologous protein-coding sequence in *Drosophila* homoeotic genes and its conservation in other metazoans. *Cell* **37**, 403–8.

Milner, R. J., F. E. Bloom, G. Lai, R. A. Lerner and J. G. Sutcliffe 1984. Brain-specific genes have identifier sequences in their introns. *Proc. Natl Acad. Sci. USA* **81**, 713–17.

Nusslein-Volhard, C. 1979. Maternal effect mutations that alter the spatial coordinates of the embryo of *Drosophila melanogaster*. In *Determinants of spatial organization*, S. Subtelny and I. R. Konigsberg pp. 185–211. New York: Academic Press.

Nusslein-Volhard, C. and E. Wieschaus 1980. Mutations affecting segment number and polarity in *Drosophila*. *Nature* **287**, 785–801.

Palmiter, R. D., R. L. Brinster, R. E. Hammer, M. E. Trumbauer, M. G. Rosenfeld, N. C. Birnberg and R. M. Evans 1982. Dramatic growth of mice that develop from eggs microinjected with metallothionein–growth hormone fusion genes. *Nature* **300**, 611–5.

Parker, C. S. and J. Topol 1984a. A *Drosophila* RNA polymerase II transcription factor contains a promoter-region-specific DNA-binding activity. *Cell* **36**, 357–69.

Parker, C. S. and J. Topol 1984b. A *Drosophila* RNA polymerase II transcription factor binds to the regulatory site of an *hsp* 70 gene. *Cell* **37**, 273–83.

Raff, R. A. and T. C. Kaufman 1983. *Embryos, genes and evolution*. New York: Macmillan.

Reudelberger, T. 1984. Upstream and downstream control of eukaryotic genes. *Nature* **312**, 700–1.

Scheller, R. H., J. F. Jackson, L. B. McAllister, J. H. Schwartz, E. R. Kandel and R. Axel 1982. A family of genes that codes for ELH, a neuropeptide eliciting a stereotyped pattern of behavior in *Aplysia*. *Cell* **28**, 707–19.

Scheller, R. H., J. F. Jackson, L. B. McAllister, B. S. Rothman, E. Mayeri and R. Axel 1983. A single gene encodes multiple neuropeptides mediating a stereotyped behavior. *Cell* **32**, 7–22.

Schlesinger, M. J., M. Ashburner and A. Tissieres 1982. *Heat shock from bacteria to man.* Cold Spring Harbor: Cold Spring Harbor Laboratory.

Scott, M. P. and A. J. Weiner 1984. Structural relationships among genes that control development: Sequence homology between the Antennapedia, Ultrabithorax and fushi tarazu loci of *Drosophila*. *Proc. Natl Acad. Sci. USA* **81**, 4115–19.

Struhl, G. 1984. A universal genetic key to body plan? *Nature* **310**, 10–11.

Strumwasser, F., L. K. Kaczmarek, A. Y. Chiu, E. Heller, K. R. Jennings and D. P. Viele 1980. Peptides controlling behavior in *Aplysia*. In *Peptides, integrators of cell and tissue function*, F. E. Bloom (ed.), pp. 197–218, Society of General Physiologists Series 35. New Haven: Raven.

Sturtevant, A. H. 1945. A gene in *Drosophila melanogaster* that transforms females into males. *Genetics* **30**, 297–9.

Waddington, C. H. 1962, *New patterns in genetics and development*, New York: Columbia.

Wahli, W., I. D. Dawid, G. U. Ryffel and R. Weber 1981. Vitellogenesis and the vitellogenin gene family. *Science* **212**, 293–304.

Wilson, A. C., L. R. Maxson and V. M. Sarich 1974. Two types of molecular evolution. Evidence from studies of interspecific hybridization. *Proc. Natl Acad. Sci. USA* **71**, 2843–7.

Wu, C. 1984a. Two protein-binding sites in chromatin implicated in the activation of heat-shock genes. *Nature* **309**, 229–34.

Wu, C. 1984b. Activating protein factor binds *in vitro* to upstream control sequences in heat-shock gene chromatin. *Nature* **311**, 81–4.

Zimmerman, J. L., W. Petri and M. Meselson 1983. Accumulation of a specific subset of *D. melanogaster* heat-shock mRNAs in normal development without heat shock. *Cell* **32**, 1161–70.

11

Population genetics, evolutionary rates and Neo-Darwinism

HAMPTON L. CARSON

ABSTRACT

Examination of the data of the fossil record has led some workers to espouse a theory ('punctuated equilibria') in which evolutionary change is hypothesised to be concentrated in relatively short periods of geological time. These metastable periods are then supposed to be followed by long periods without change. Such theories are based on limited data, both with regard to the lengths of time embraced and the biology of the characters involved. Punctuated equilibria cannot be directly applied to evolutionary theories that are based on the biology and genetics of contemporary populations. The accomplishments of studies on the genetics of natural populations in this century are reviewed. Empirically and theoretically, the approach to evolution through population biology (Neo-Darwinism) is highly robust. All evolutionary processes including varying rates and periods of stasis can be understood by recourse to studies of contemporary organisms. There is no reason to believe that any basic principle can be observed in the fossil record that it not also reflected in population biology.

INTRODUCTION

On punctuated equilibrium

The patterns . . . speak for themselves. They are there for all to see. We have not simply invented them to fit a set of preconceived notions. (Eldredge & Tattersall 1982)

I remain skeptical . . . but satisfied that the authors could make radioactive decay appear punctuated – all you have to do is to look closely enough. (White 1983)

The theory of punctuated equilibria was proposed more than a decade ago (Eldredge & Gould 1972). The idea turned out to be highly provocative, resulting in a flood of seminars, discussions and symposia. Articles either strongly pro or strongly con were numerous. This debate, in my opinion, has unfortunately been of poor quality and clarification has not been served by the considerable participation of reporters from various scientific journals, as well as from the non-technical press and proponents of creationism.

Provocation of one's peers, of course, is an important means of communication in the advancement of science, especially when the significance of new and little-understood data is apparently perceived by only a few persons. The history of science is replete with stories of the manner in which advocates of the 'conventional wisdom' have clung steadfastly to classical ideas that seemed at the time to be utterly unassailable. On the contrary, there is nothing more exciting for a scientist than the uncovering of data that appear to cause a restructuring of knowledge along lines that simultaneously clarify and simplify. Only a few workers, however, are destined by play this kind of rôle. The retrospective glamour of a Wegener arguing for continental drift or a McClintock announcing mobile genetic elements is a pervasive motivating influence for both peers and students who view the pioneer days from the comfort of current acceptance. Many eager scientists pressure their own thoughts to take theoretical directions that often outstrip the scope of their data. To some, like the great geneticist and evolutionist, Richard Goldschmidt, the overturn of what to him appeared as the orthodox complacency of the establishment became almost obsessive. It fuelled his drive as a scientist. Late in his career, he began one of his challenging articles with a quotation from a letter written by Orville Wright, six months before the first successful flight at Kitty Hawk:

'. . . but if we all worked on the assumption that what is accepted as true is really true, there would be little hope of advance.'

Whether they turn out to be right or wrong, such people are a necessary part of the social fabric of science; they require us to re-examine facts and entertain hypotheses that from some perspectives seem quite absurd. But today's 'absurdities' sometimes become tomorrow's laws.

A scientist who will not force himself to put his peers' theories at least to the mind's test condemns himself to be merely a pedestrian plodder. An annoying aspect is the necessity to discard most of these notions. Only the great among us are generous enough with their thought processes to remark, as Haldane once did after hearing one of Crodowaldo Pavan's ideas: ' . . . interesting – interesting – even if untrue . . .'

So, what rôle is the theory of punctuated equilibria to play? Must we dismiss Eldredge and Gould or must we restructure our basic thinking around their ideas?

PALAEONTOLOGICAL THEORY

In approaching these questions an important circumstance must be considered: the idea of punctuated equilibrium was proposed by palaeontologists from a consideration of the data of palaeontology. Were the debate over the idea to remain entirely within that data base, I believe the discussion of it would have been clearer and less acerbic. But the proposers chose to consider the idea as an alternative to what they labelled 'phyletic gradualism'. Although at first the gradualism idea was ascribed only to certain palaeontologists, it was later broadened and, thus, became the basis for a criticism of the whole structure of Neo-Darwinism. Thus the latter area, basically one of contemporary population biology, was also equated with undifferentiated gradualism. 'Punctuation' was proposed as the new and exciting alternative.

The trouble with this way of framing up the argument is that Neo-Darwinism is virtually equivalent, in moderns terms, to the sophisticated quantitative science of population genetics. The data of population genetics are wholly foreign to those of palaeontology. They must be considered separately and should have been considered separately by Eldredge and Gould.

In punctuated equilibrium, as I have pointed out earlier (Carson 1982), evolutionary change is supposed to be concentrated in a 'brief' metastatic period, the punctuation phase. Using the shortest punctuation that I could find in the examples cited (e.g. Gould 1980), the geological data suggest that it must have extended for about 55 000 years. The standard error on such values is not clear. Now, this may be a very short time from the palaeontological point of view, but from the perspective of modern population genetic biology which considers generations as the building blocks of change, it is substantial. The generation time in a relatively slow-breeding organism like man, for example, is generally taken as 30 years. This would provide 1833 generations for the operation of Neo-Darwinian forces such as mutation, recombination, selection, immigration and population size effects. *Drosophila melanogaster* is much faster; we may use the figure of 32 generations per year. In this case, the palaeontological punctuation under discussion would have extended over 1 760 000 generations. A great deal of genetic change can be incorporated into a population in ten or a hundred generations; and a thousand or a million is a sumptuous amount.

Dramatic changes in the phenotypic and genotypic characters in populations can be experimentally accomplished in as few as five generations of natural selection. A minor genetic alteration to an experimental population of *D. melanogaster* caused a tripling of population size and a drastic change in phenotype (Carson 1958). Under challenge from the insecticide DDT, resistance of an outbred population of *D. melanogaster* was increased by natural selection more than 2000 times in 3 years (Crow 1954). Many such instances could be cited. Some Neo-Darwinists (e.g. Grant 1977) refer to such episodes as times of 'quantum' change (a term originally employed by Simpson in 1944); they are part and parcel of the data of population genetics and should not be considered especially unusual or exceptional. Episodes of rapid change have also been associated with genetic revolutions (Mayr 1954), founder effects (see Carson & Templeton 1984) and hybridisations (Carson 1985). They acquire importance in the present discussion as they emphasise the existence of dramatic shifts in the *rate* of change, the subject of the symposium.

Much longer periods of geological time have been assigned to the equilibrium phase of the punctuated equilibrium idea. During these periods, it is alleged, the morphological attributes of the relevant fossils as seen in the fossil record do not undergo change. This is an aspect of the theory to which population genetics cannot contribute, since the vast periods of geological time involved cannot be tested or modelled experimentally from the point of view of contemporary population biology. As will be discussed below, however, theoretical population genetics may be able to make a contribution to the equilibrium concept. On the other hand, palaeontological data are far from robust on this point since the preservation of hard parts or casts do not provide much information on physiology, reproductive cycles or behaviour.

As the advocacy of punctuated equilibria has intensified, various questionable corollaries have been increasingly stressed. One suggests that some kind of macromutational change affecting a major developmental pathway is the key event that occurs during the punctuation phase. The original proponent of this idea, of course, was the genetic evolutionist, Goldschmidt (1940). Employment of this idea, however, especially as the key to a 'saltatory' change instrumental in the formation of species, goes back to Hugo de Vries and T. H. Huxley, if not further. Gould and Eldredge (1977) have extracted several delicious quotes from Huxley's writings, indicating that the geneticist William Bateson agreed with him on saltation but that Darwin had expressed 'disgust' at the idea.

Although downgraded by Simpson (1944 and later) the saltation idea has frequently been used by others, particularly by those to whom 'population thinking' is somewhat foreign. As a field naturalist, Darwin's approach to evolutionary change was definitely microevolutionary, that is, he invoked the accumulation of small increments of change in populations as the *modus operandi* of evolution. He did not, however, insist on gradualism and considered rapid rates of evolutionary change to be important (see Templeton and Giddings (1981) for quotes from Darwin on the subject); it was the abrupt change in phenotypes (saltations) that he felt were not important in natural populations.

Saltation, especially in species formation, appears to have few advocates among modern Neo-Darwinian population geneticists, either those who take an empirical or a theoretical approach. The feeling is widespread that homoeotic mutants, which all recognise as real, so disturb developmental pathways as to strongly decrease fitness in their carriers. This is not to say that a gene of moderately large effect may not increase in the population after the accumulation of modifier genes. The latter process, in which genetic recombination probably plays an important rôle, can be accomplished only over a number of generations.

The degree to which the punctuated equilibrium enthusiasts reject Neo-Darwinism is shown by the manner in which the origin of species is diagrammed. Thus, Stanley (1979) proposes that species formation be diagrammed by 'rectangular' phyletic branches, a procedure that appears to exclude the possibility of transitional stages.

POPULATION GENETICS AND EVOLUTION

Mention has been made earlier in this discussion of the tendency on the part of some palaeontologists to describe Neo-Darwinism in simplistic terms as a concept that reflects some basic preoccupation with phyletic gradualism. I believe the best way to counter such oversimplification is to state the nature of the Neo-Darwinian approach and to recount a few of the major 20th century discoveries of population genetics.

In contrast to those of palaeontology, the data of population genetics come from genetical analysis of individual organisms drawn directly from living populations. Unlike the palaeontologist, the geneticist has methods that can deal quantitatively with descent with change from generation to generation in both phenotype and genotype. The phenotypic attributes available for study include biochemical, developmental, behavioural, physiological and soft-part characters not available to the palaeontologist.

Biometrical or quantitative genetics involves methods borrowed by the evolutionist from the plant and animal breeder. Applied breeding is an elegant and sophisticated science that is able to partition environmental and genetic variance with great precision. This opens the way to an understanding of population alterations over a series of generations. Further, the science, dealing as it does with quantitative characters, finds itself clarifying the inheritance of those very characters that the systematist and the palaeontologist themselves stress as important in evolution. There are good and important reasons why these characters do not show simple single-factor Mendelian inheritance.

Although useful in theoretical constructs, evolutionary schemes based on simple allelic substitution (e.g. the industrial melanism case) should not serve as general examples of what is happening in most natural populations. Indeed, even the case of industrial melanism is far from being a single-locus effect: clear genome-wide modifier genes and associated heterotic effects caution us against the strict application of single-locus evolution as a model.

Increasingly, Neo-Darwinian evolutionists distinguish between the phyletic origin of adaptations and the origin of species. Both theoretically and in actual experiment, the origin of adaptations follows the Neo-Darwinian path of mutation, recombination and selection as it occurs within single populations of interbreeding organisms. Population genetics, however, is less able to deal with the origin of species, in which a sundering of a pre-existing gene pool must somehow be instituted. For this reason, theories as to the origin of species are less well grounded in data than is the intrapopulational origin of adaptations.

Over the past 50 years, a substantial literature on the genetics of natural populations of various species of *Drosophila* has accumulated. Although the subject has been reviewed repeatedly, the 'non-drosophilist' may have some difficulty seeing the woods for the trees. I now list, with brief comments, what appear to be the main results of these extensive studies of this genus as well as population studies of some other genera of fast breeding, cross-fertilising diploid organisms.

The genetic structure of natural populations

1 Natural populations contain abundant cryptic genetic variability sheltered in the heterozygous state. This variability has many components: there are oligogene visibles, chromosomal aberrations, protein-coding loci and specific regulatory and modifying genes. In the past few years, as DNA sequencing has proceeded, extensive heterozygosity has been observed at this, the ultimate source of genetic variation.

2 Most of this variability consists of genes of individually small phenotypic effect. The evidence for a substantial rôle for what Mather has called polygenes continues to be extended.

3 Genes of larger effect, if experimentally isolated, can be seen to have been strongly influenced in their expression by modifying or regulatory genes in the natural genetic environment.

4 Mutability in natural populations is not a species-wide constant but varies with the local population or stock, or even individual organism studied. In a wide variety of plants and animals much mutational activity can be ascribed to the effects of mobile genetic elements. Their effects are often suppressed in inbred populations but expressed in certain outcrossed populations. A major effort is now under way in genetics research to separate 'spontaneous' mutability from mutability mediated by mobile genetic elements.

5 Many naturally occurring variants are deleterious in the homozygous state yet tend to remain in the population rather than being eliminated by normalising selection. This widespread observation has led to the balance theory of the genetic structure of populations. Much of the variability is not fixed but exists in complex heterozygous balances capable of continual revision by meiotic recombination and renewed selection each generation. Included in this category are major chromosomal aberrations such as inversions and translocations. Although some species appear to lack them, others accumulate them as segregating genetic variants that are apparently also held in the population by balancing selection. Observations of isofemale lines (laboratory stocks started in the laboratory from a single wild female that was inseminated in nature) show that much of the inversion and soluble protein variation and indeed the DNA sequence variation as well, may be maintained over 50 generations or more even in the uniform laboratory environment. These balances appear to be virtually obligatory in some species; inbreeding and selection frequently lead to lowered fitness. As early as 1955, Dobzhansky wrote: 'The adaptive norm is an array of genotypes, heterozygous for more or less numerous gene alleles, gene complexes and chromosomal structures.'

This characterisation is even more true today than it was then, before modern electrophoretic and sequencing methods had been invented.

Hartl (1980) has stressed another aspect of our modern view of population structure. He calls attention to the interaction of the components of this variability:

> ...the genetic underpinning of fitness is extremely complex, involving alleles at many loci that influence fitness in two ways – one through the main affect of each locus (which typically may be an effect on some quantitative trait such as body size, growth rate, or developmental time) and the other, through pleiotropic effects on other traits related to fitness.

Over the years, Wright has emphasised gene interaction. Recently he wrote (Wright 1982):

> ...each gene ... (has) multiple pleiotropic effects because of interaction of its product with those of others ... the effects of these interactions are not in general additive. These are the inevitable consequences of the complex network of biochemical and developmental reactions that intervene between primary gene action and the ultimate effects subject to selection.

The implications of this situation for evolution are clear. Periods of intense selection, recombination, mutability and drift can greatly speed up evolutionary change, especially if they operate in small or attenuated subpopulations. Under these circumstances rapid ('quantum') evolutionary change can occur. These changes can result in the formation of new adaptive modes or species or both (see Wright 1982, Carson 1982, 1985).

The periods of quantum change, extending over perhaps 50 to 100 generations, may be followed by a return to a more stable state characterised by a balanced genetic variability.

However, not all evolutionary change appears to be concentrated in these bursts of evolutionary activity. As Wright emphasises, every character important to the life of the organism appears to require maintenance by the all-pervading action of the principal Darwinian force, that is, natural selection. Thus, if selection is removed, the pertinent character will tend to decay back to the level at which selection again operates. The active 'decay' in the genetic material is caused by the inexorable process of random mutational change in the DNA. Without natural selection as a force leading to conservation, the character will start to become unstable. For an example, see Carson et al. (1982).

All of these concepts have grown out of modern Neo-Darwinism and are available for confirmation and further direct study in living contemporary populations. They may be applied to the intrapopulational origin of adaptive traits as well as to the mode of origin of new isolated populations and species.

The facts that have led some palaeontologists to espouse punctuated equilibria are unfortunately not robust enough to warrant meaningful comparisons to the principles that have been established by the Neo-Darwinian evolutionist and population geneticist. Accordingly, only rough, imprecise analogies of questionable significance can be drawn between 'punctuation' and 'quantum' change and between 'equilibrium' and 'balanced genetic polymorphism'. Despite some claims to the contrary, the major edifice of Neo-Darwinism is unaffected by these arguments and is stronger than ever in 1985.

ACKNOWLEDGEMENT

The work of the author is supported by NSF grand DEB 79–26692.

REFERENCES

Carson, H. L. 1958. Increase in fitness in experimental populations resulting from heterosis. *Proc. Natl Acad. Sci. USA* 44, 1136–41.

Carson, H. L. 1982. Speciation as a major reorganization of polygenic balances. In *Mechanisms of speciation*, C. Barigozzi (ed.) pp. 411–33. New York: Alan R. Liss.

Carson, H. L. 1985. Unification of speciation theory in plants and animals. *Syst. Bot.* 10, 380–90.

Carson, H. L. and A. R. Templeton 1984. Genetic revolutions in relation to speciation phenomena: the founding of new populations. *Ann. Rev. Ecol. Syst.* 15, 97–131.

Carson, H. L., L. S. Chang and T. W. Lyttle 1982. Decay of female sexual behavior under parthenogenesis. *Science* 218, 68–70.

Crow, J. F. 1954. Analysis of a DDT-resistant strain of *Drosophila*. *J. Econ. Entomol.* 47, 393–8.

Dobzhansky, Th. 1955. A review of some fundamental concepts and problems of population genetics. *Cold Spring Harb. Symp. Quant. Biol.* 20, 1–15.

Eldredge, N. and S. J. Gould 1972. Punctuated equilibria: an alternative to phyletic gradualism. In *Models in paleobiology*, T. J. M. Schopf (ed.), pp. 82–115. San Fransisco, Cooper.

Eldredge, N. and I. Tattersall 1982. *The myths of human evolution*. New York: Columbia University Press.

Goldschmidt, R. 1940. *The material basis of evolution*. New Haven, Conn.: Yale University Press.

Gould, S. J. 1980. Is a new and general theory of evolution emerging? *Paleobiology* 6, 119–30.

Gould, S. J. and N. Eldredge 1977. Punctuated equilibria: the tempo and mode of evolution reconsidered. *Paleobiology* 3, 115–51.

Grant, V. 1977. *Organismic evolution*. New York: W. H. Freeman.

Hartl, D. E. 1980. *Principles of population genetics*. Sunderland, Mass.: Sinauer Press.

Mayr, E. 1954. Change of genetic environment and evolution. In *Evolution as a process*, J. Huxley, A. C. Hardy and E. B. Ford (eds), pp. 157–80. London: Allen & Unwin.

Simpson, G. G. 1944. *Tempo and mode in evolution*. New York: Columbia University Press.

Stanley, S. M. 1979. *Macroevolution: pattern and process*. New York: W. H. Freeman.

Templeton, A. R. and L. V. Giddings 1981. Macroevolution conference. *Science* 211, 770–1.

White, T. 1983. Talking patterns. *Nature* 303, 90.

Wright, S. 1982. Character change, speciation and the higher taxa. *Evolution* 36, 427–43.

12

Genetic systems and evolutionary rates

ALAN R. TEMPLETON

ABSTRACT

A genetic system is defined by its basic mode of inheritance (bisexual versus unisexual, diploid versus haploid) and, when referring to a specific phenotype, by its genetic architecture (number and relative importance of loci, the pattern of dominance, epistasis and pleiotropy, linkage relationships, etc.). Both influence the rate of evolution, but this chapter will focus only on basic mode of inheritance.

The interaction of cladogenesis and mode of inheritance is illustrated by examining the impact of founder events in *Drosophila*. Basic theory predicts that X chromosomes (a haplo-diploid genetic system) should be more sensitive to founder effects than autosomes (a bisexual diploid system). An examination of the genetic basis of evolved differences in both natural and experimental founder populations reveals a strong preferential involvement of X-linked loci over autosomal loci. Further evidence for the increased sensitivity of X chromosomes to founder events is provided by the fact that X chromosomal inversions are preferentially fixed over autosomal inversions in Hawaiian *Drosophila*, a group for which founder events are important, but there is no preferential fixation in continental *Drosophila*. Basic theory also predicts that mitochondrial DNA (mtDNA) should be more sensitive to founder events than nuclear DNA. This prediction is tested by contrasting two lineages of Hawaiian *Drosophila*, one for which founder events have apparently been important in speciation and the second for which founder events were not important. The pattern of nuclear versus mtDNA evolution is discordant between these two lineages, and the difference is shown to be due to a greatly accelerated rate of mtDNA evolution in the lineage subjected to inter-island founder events.

The importance of anagenesis in determining rates of evolution is illustrated by contrasting rates of evolution in mtDNA and nuclear DNA in the hominoid primates. The human lineage has had a large increase in generation time and a simultaneous expansion of population size over at least the last two million years. Theory predicts that these trends should cause an evolutionary slowdown in all genetic systems. This prediction is supported by a statistical analysis that indicates that mtDNA, nuclear DNA and karyotypes all have slower rates of evolution in humans than in African apes.

In summary, the rate of evolution of genetic systems is sensitive to both cladogenesis and anagenesis, but the degree of sensitivity may differ among the genetic systems. The heterogeneity in evolutionary rates does not mean that molecular data cannot be used to estimate divergence times; rather, it is shown that more accurate estimates are possible when rate heterogeneity is taken into account. Moreover, patterns of differences and uniformity among different genetic systems can therefore be used to make inferences concerning the events that have influenced the evolution of the group under study.

INTRODUCTION

A genetic system is defined by its basic mode of inheritance (bisexual versus unisexual, diploid versus haploid) and, when referring to a specific phenotype, by its genetic architecture (number and relative importance of loci, the pattern of dominance, epistasis and pleiotropy, linkage relationships, etc.). Templeton (1982a,b) has shown that the type of genetic architecture strongly interacts with the mode of speciation and the type of anagenetic evolution in determining rates of evolution. Because genetic architecture has been discussed elsewhere, this chapter will focus only on basic mode of inheritance and its relationship to evolutionary rates.

Previous work on the impact of basic mode of inheritance on evolutionary rates has traditionally dealt with the problem of the evolution of sex. Thus, there is an extensive literature comparing rates of evolution in sexually reproducing, diploid species versus asexual haploids or parthenogenetic diploids (Templeton 1982c). However, intraspecific comparisons are also possible because most organisms have several different genetic systems rather than just one. For example, in *Drosophila* and mammals, four distinct genetic systems are readily identified. First, there is the autosomal system that is inherited as a bisexual diploid. Secondly, the X chromosome is bisexually inherited, but is diploid in females and haploid in males. Thirdly, the Y chromosome is a paternally inherited haploid and finally, the mitochondrial DNA (mtDNA) is inherited as a maternal haploid. Unlike comparisons across species, the genetic systems imbedded within the same species obviously share many common evolutionary constraints; yet, their differing modes of inheritance allow these contraints to display a variety of potential impacts. The purpose of this chapter is to illustrate the impact of both cladogenetic and anagenetic evolutionary constraints on the rates of evolution of diverse genetic systems, showing that this impact across genetic systems imbedded within the same species can be homogeneous in some cases but heterogeneous in others.

CLADOGENESIS AND EVOLUTIONARY RATES

Founder events and genetic systems

The impact of cladogenesis on the evolutionary rates of diverse genetic systems will be illustrated by speciation events induced by founder effects.

Carson and Templeton (1984) have recently reviewed the evidence and theory for founder-induced speciation, and they have concluded that speciation events are likely only when the founder event is followed by a rapid and large increase in population size.

Nei *et al.* (1975) examined the genetic impact of such founder-flush events. They showed that founder events of size 10 have very little impact on overall levels of genetic variation and fixation when followed by a rapid increase in population size. Even the most extreme founder event possible, a population of two individuals, carried over a minimum of 65% of its genetic variation. The rôle of the founding number was investigated in more detail by Senner (1980). Restricting attention only to the case when the founding event was followed by a large increase in population size, Senner found an inflection point at about four individuals. Above four, most genetic variation is preserved and little fixation occurs. Below four, the amount of fixation is greatly influenced by the actual number of founding individuals, with decreasing numbers of founders resulting in increased fixation.

It is important to realise that both the Nei *et al.* (1975) and Senner (1980) models refer to bisexual, diploid genetic sytems. Hence, the inflection point of about four individuals actually translates into an inflection point of about eight haploid genomes. A founder event consisting of N individuals will consist of diverse numbers of haploid genomes for the various genetic systems imbedded within those individuals. For example, much of the speciation in the Hawaiian *Drosophila* is most likely due to founder events involving a single gravid female (Carson & Templeton 1984). Suppose the female had been inseminated by M males, so that the founder size in terms of number of individuals is $N = M + 1$. Then, the number of autosomal genomes present in the founder population is $2M + 2$, the number of X chromosomes is $M + 2$, the number of Y chromosomes is M, and the number of mitochondrial genomes is 1. Obviously, the various genetic systems can show extremely different sensitivities to the same founding event. For example, if $M = 3$, the autosomal system should be relatively insensitive to the founding event, but the X, Y and mtDNA will show progressively increasing sensitivies in terms of loss of genetic variation and increased fixation rates.

X chromosomal versus autosomal genetic systems

Experimental founder events   For several years, I have been investigating the capacity for parthenogenetic reproduction in a natural, and normally bisexually reproducing, population of *Drosophila mercatorum* (Templeton *et al.* 1976, Templeton 1983a). As argued elsewhere (Templeton 1979), these screenings for parthenogenetic capacity constitute an experimental investigation of founder events that result in the most extreme inbreeding possible. The parthenogenetic lines were established from the virgin female offspring of single wild-caught females. Hence, the sources of genetic variation available for selection in the establishment of any given parthenogenetic strain is the genetic variation brought into the laboratory by a single gravid female. Moreover, most wild-caught females are multiply inseminated, with the probability of two offspring of the same female having different fathers being 0.32 (A. R. Templeton, unpublished observations). Hence, these parthenogenetic strains are established from isofemale founding lines for which the founder effect will be much greater for the X chromosome than for any autosome.

Once a parthenogenetic strain is established, a sexual analogue can be bred (see Templeton 1983a, for details). Hence, one can focus directly on the genetic impact of these experimental and extreme founder events rather than parthenogenesis *per se*. The most extensive work to date has been done on one of the first parthenogenetic lines established, K28-0-Im (Templeton *et al.* 1976). Several isolating barriers evolved in this line, so it is an ideal experimental model of speciation.

First, this line is characterised by extremely strong premating isolation from certain other strains of *D. mercatorum* (Carson *et al.* 1977). Chromosomal contrasts have been performed that demonstrate that the sexual isolation depends upon a strong epistatic interaction between two chromosomes: the X chromosome and the acrocentric II autosome (Templeton 1986a). No other chromosomes are involved.

Given that mating occurs, other isolating barriers are unmasked. Many of the F_2 and backcross males lack motile sperm and are sterile (Templeton 1986a). The genetic basis of sterility is similar to that of premating isolation; sterile males are produced by simultaneous hemizygosity of the K28-0-Im X chromosome and homozygosity for the K28-0-Im acrocentric II autosome. Although the same chromosome pair is involved in both premating isolation and male sterility, these traits are separable genetically.

Finally, there is an F_2 breakdown in viability associated with the phenotypic syndrome known as abnormal abdomen (*aa*) (Templeton 1979). Extensive genetic and molecular studies have been performed upon the *aa* syndrome. Genetically, this syndrome is associated with two necessary genetic components which are found on the X chromosome less than 1 map unit apart (Templeton *et al.* 1985). Molecularly, the syndrome also depends upon two necessary components (DeSalle & Templeton 1986). First, a majority of the 28S ribosomal genes must have a 5 kb insertion into their coding region. The ribosomal genes are found in a tandem cluster of about 200 copies on the X chromosome. Second, during the formation of polytene tissue (such as fat bodies), there must be no preferential replication of non-inserted over inserted 28S genes. Many X chromosomes have the ability to replicate preferentially the non-inserted sequences both intra- and interchromosomally. If such a preferential replication occurs, *aa* is suppressed. In addition to these major and necessary X-linked elements, the expression of *aa* can be modified by genes located on the autosomes. Although the major genetic elements for *aa* are on the X chromosome, this syndrome is also separable through genetic recombination from the sexual isolation and male sterility X-linked elements (Templeton 1986a).

Notice that all three isolating barriers have one or more major genes on the X chromosome, but most autosomes are not involved or have only minor modifiers. Given that the X chromosome constitutes only about 20% of the *mercatorum* genome, this preferential involvement of X-linked genetic systems is indicative of the increased sensitivity of the X chromosome to founder events over that of the autosomes. An alternative explanation is that genes involved with isolating mechanisms are clustered upon the X chromosome. This later explanation seems unlikely, since a broader review of the genetic basis of isolating mechanisms in the genus *Drosophila* shows extensive autosomal involvement (Templeton 1981, and references therein).

Natural speciation events   There is evidence for this same pattern of preferential involvement of X-linked genetic systems in natural speciation events associated with founder effects. Founder-induced speciation is probably the dominant mode of speciation among the Hawaiian *Drosophila* (Carson & Templeton 1984). Two of these Hawaiian species are *D. silvestris* and *D. heteroneura*. These species are morphologically distinct and are reproductively isolated by premating barriers that can be broken down in the laboratory. Since no postmating barriers are present, it is possible to perform genetic analyses of the traits differentiating these two species. One of the more dramatic differences is in head shape. *Drosophila heteroneura* has a hammer head, whereas *D. silvestris* has a round head. Head shape is controlled by a major gene or genes on the X chromosome epistatically interacting with several minor autosomal modifiers (Templeton 1977). The species also differ in carina and groove (face) colour, with *D. heteroneura* being yellow and *D. silvestris* black. Both aspects of face colour are controlled by an X-linked locus interacting with one autosomal gene (Val 1977). Two different aspects of wing pattern are each controlled by interactions between an X-linked and autosomal locus. Differences in the pigmentation of the mesopleurae are due to an autosomal locus (Val 1977). Hence, five of the six characters analysed involve major genes on the X chromosome. Thus, both experimental and natural founder populations display a preferential involvement of X-linked loci for their evolved differences.

Further evidence for the preferential sensitivity of X-linked genetic systems to founder events can be obtained by looking at the pattern of fixation of chromosomal inversions among the Hawaiian picture-winged *Drosophila*. Of 127 inversions fixed in 103 species, 59 (46%) are on the X chromosome despite the fact that there are four autosomes of about equal size (Carson 1983). One might explain this by assuming that the X chromosome has a greater rate of inversion mutation, but this seems unlikely. Table 12.1 presents the data from Carson (1983) on the number of fixed and polymorphic inversions found on each of the major chromosomes of 103 species of Hawaiian *Drosophila*. If fixation rate is merely a consequence of mutation rate, the proportion of fixed to polymorphic inversions should be the same across all chromosomes. The contingency chi-square test of this hypothesis is 23.19 with four degrees of freedom, a result that is significant at the 0.00001 level of significance. Excluding the X chromosome from the analysis, the resulting contingency chi-square is 0.66 with three degrees of freedom. Hence, the autosomal inversions are being fixed at a rate that is proportional to their polymorphism rates, but the X chromosomal inversions are being fixed

Table 12.1 Number of fixed and polymorphic inversions among 103 Hawaiian species of *Drosophila* (from Carson 1983) and among the *repleta* group of *Drosophila* (from Wasserman 1982).

Chromosome:	X	2	3	4	5
Hawaiian					
fixed	59	11	14	32	11
polymorphic	13	9	18	35	11
Repleta					
fixed	12	79	16	2	8
polymorphic	7	86	15	5	5

at a rate that is far in excess of their polymorphic rates. More specifically, the data from Table 12.1 indicate that X chromosomal inversions are being fixed at a rate 4.9 times that of autosomes. As a further check on the inference that these differences can be attributed to founder events rather than some peculiar property of the X chromosome, Table 12.1 also presents the comparable data set on inversion evolution in the *repleta* group of *Drosophila* as summarised in Wasserman (1982). This is a speciose continental group for which there is no evidence for extensive founder-induced speciation. In great contrast to the Hawaiian *Drosophila*, only 10% of the fixed inversions are on the X chromosome, and the contingency chi-square between fixed versus polymorphic inversions as a function of chromosomal location is 3.62 with four degrees of freedom. Hence, in the *repleta* group, the rate of inversion fixation is proportional to rate of inversion polymorphism, and there is no discrepancy between X versus autosomal inversions. The contrast between Hawaiian versus *repleta* group *Drosophila* provides strong support for the theoretical prediction that X-linked genetic systems are far more sensitive to founder events than autosomal genetic systems.

Mitochondria DNA versus nuclear genetic systems

The theory outlined above predicts that mtDNA should be more sensitive to single-female founder events than nuclear systems. To test this prediction, DeSalle (1984) performed an extensive study of mtDNA restriction site evolution in two lineages of Hawaiian *Drosophila*; the alpha and beta lineages of the *planitibia* subgroup. To understand why these two lineages where chosen, it is first necessary to review some pertinent facts concerning the geological history of the Hawaiian Islands. Most of the major islands of the Hawaiian chain are separated by deep ocean channels. The fluctuating sea levels that occurred during the Pleistocene could never have connected these major islands. Hence, the traditional arguments for single female founder events (Carson & Templeton 1984) are applicable to instances of speciation involving transfers between these major islands. Most speciation in the beta lineage falls into this category. However, Maui,

Molokai and Lanai (known collectively as Maui Nui) are islands that were connected to one another at various times in the past as a function of fluctuating sea levels. Hence, the *Drosophila* on these islands, which includes the alpha lineage, were much more likely to be separated by subdivision of large populations rather than by founder events.

Behavioural studies provide further evidence that founder events were not important in speciation processes occurring on Maui Nui. Kaneshiro (1976) discovered a pattern of mating asymmetry between ancestor and descendant species that exists among many Hawaiian *Drosophila*. Basically, derived females will mate with ancestral males, but ancestral females will not readily mate with derived males. This pattern has also been observed in non-Hawaiian *Drosophila* for which biogeographical evidence indicates founder-induced speciation (Giddings & Templeton 1983). In addition, this mating asymmetry evolves in experimental founder events involving both Hawaiian and non-Hawaiian *Drosophila* (Giddings & Templeton 1983). This pattern is *not* observed in contrasts between continental *Drosophila* species for which the biogeographical evidence indicates founder events to be unlikely and is *not* present in *Drosophila* speciation experiments that do not involve founder effects (Giddings & Templeton 1983). Consequently, this pattern of mating asymmetry is one that appears to be specifically associated with founder-induced speciation in *Drosophila* (see Giddings & Templeton (1983) for the theoretical reasons why this specificity is expected). Most Hawaiian *Drosophila*, including the beta lineage, display this asymmetry. The only exceptions discovered so far involve species from the Maui Nui complex (Giddings & Templeton 1983), thereby implying that speciation on Maui Nui does not involve founder effects.

To see if mtDNA is more sensitive to founder events than nuclear genetic systems, comparisons were made between DeSalle's genetic distances on mtDNA using Ewens' (1983) 'p' statistic with Roger's genetic distances calculated by Johnson *et al.* (1975) on isozyme data (Table 12.2). The Ewens' and Roger's distances are not directly comparable, so the absolute values of the distances are not important, but rather the relative pattern within and between lineages.

The pattern of distances within and between lineages can be summarised by taking the appropriate averages. For the isozyme data, the average distances within lineages are 0.10 and 0.27 for alpha and beta respectively, and the average intralineage distance is 0.62. Hence, in terms of the nuclear genetic distances, the interlineage distance is considerably larger than intralineage distances. Moreover, as Carson (1976) has shown by overlaying these genetic distances on the known geological time scale of these islands, the distances in flies from both lineages can be explained by the same rate of evolution. In contrast, the average intralineage p-distances for the mtDNA are 0.028 and 0.053 for alpha and beta respectively, with the interlineage distance being 0.059. This pattern differs dramatically from that observed for the isozymes. Note that the interlineage distance is only slightly larger than the within beta distances. This pattern implies that the beta lineage has a much faster rate of mtDNA evolution than the alpha.

Table 12.2   The genetic distances between various species of Hawaiian *Drosophila* from the alpha and beta lineages of the *planitibia* subgroup. The distances above the diagonal are the Ewens' p values for mtDNA given in DeSalle (1984), and the distances below the diagonal are Roger's distances from the isozyme data of Johnson *et al.* (1975). The alpha lineage species abbreviations are 'cyr' (*cyrtoloma*), 'mel' (*melanocephala*) and 'npk' (*neoperkinsi*). The beta lineage species are 'sil' (*silvestris*), 'dif' (*differens*), 'plb' (*planitibia*) and 'hem' (*hemipeza*).

Drosophila spp.	Lineages						
	cyr	mel	npk	sil	dif	plb	hem
cyrtoloma	—	0.014	0.034	0.065	0.052	0.061	0.060
melanocephala	0.04	—	0.035	0.063	0.050	0.057	0.059
neoperkinsi	0.12	0.13	—	0.063	0.054	0.060	0.063
silvestris	0.68	0.67	0.77	—	0.061	0.040	0.059
differens	0.48	0.46	0.59	0.29	—	0.048	0.049
planitibia	0.60	0.60	0.70	0.26	0.15	—	0.058
hemipeza	0.60	0.60	0.70	0.44	0.26	0.22	—

The hypothesis that the beta lineage is evolving more rapidly than the alpha lineage for mtDNA can be tested directly using the statistical procedure outlined by Templeton (1983b, 1986b). This procedure uses the character-state data rather than distances. A great advantage of character-state data over distance data is that it is often possible to reconstruct the ancestral intermediates between the taxa being compared. This property has long been used in constructing intermediate states in the inversion phylogenies of *Drosophila*, and the logic is identical in reconstructing intermediate states of restriction sites (Templeton 1983b). This is an important property because it allows one to estimate the number of mutations accumulated along any branch segment with an ancestral-state estimation procedure that does not assume the molecular clock *a priori*.

A second desirable property of this test is that it is non-parametric. Hence, no assumptions are made about the underlying sampling model induced by the evolutionary process. This is important because we simply do not know at this time what the appropriate sampling model ought to be; yet, the conclusions based on parametric tests of the molecular clock are very sensitive to the sampling model assumed (e.g. Langley & Fitch 1974 versus Gillespie & Langley 1979). However, recent work indicates that the molecular clock does not hold (Hudson 1983, Gillespie 1984).

The details of this test are given in Templeton (1983b, 1986b), so they will not be repeated here. Nevertheless, some of the limitations of this approach should be mentioned. First, the test requires a knowlege of the topology of the phylogenetic tree. Fortunately, Templeton (1983b, 1986b) gives statistical critera for judging how well a topology has been estimated. If more than one topology cannot be statistically distinguished, it is best to test the molecular clock under each topology to check for the robustness of the conclusions with respect to the ambiguities regarding the true topology. DeSalle (1984) used these statistics on the Hawaiian species being

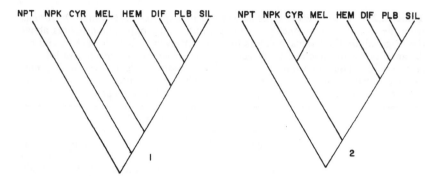

Figure 12.1 Alternative phylogenies for the Hawaiian *Drosophila* species under consideration. The species abbreviations are the same as those given in Table 12.2.

considered, and concluded that there were two statistically indistinguishable phylogenies. These are shown in Figure 12.1, and the tests of the clock are repeated for each topology.

Second, given a topology, the ancestral states themselves may not have been estimated in an unambiguous fashion. The reason for this is that some alternative ways of allocating mutational events or even the number of events are basically equally probable (Templeton 1986b). To prevent this estimation ambiguity from eroding the power of the test, it is important to only test data sets for which convergences of any sort are rare. Fortunately, DeSalle's data set falls into this category.

Table 12.3 presents the results of DeSalle's (1984) statistical tests. The pattern is very clear cut; the beta lineage is evolving at a greatly accelerated rate compared to the alpha lineage, Hence, founder events do seem to accelerate the rate of mtDNA evolution over that of nuclear genetic systems. When these results are coupled with the earlier results concerning X-linked genetic systems, it is clear that the mode of cladogenesis interacts with the nature of the genetic system to determine rate of evolution.

ANAGENESIS AND GENETIC SYSTEMS

The debate concerning punctuated equilibrium has focused much attention on the mode of speciation as a primary determinant of rate of morphological evolution. In contrast, most models of molecular evolution have ignored the details of speciation mechanisms and explained rates of molecular evolution solely in terms of continuously operating anagenetic mechanisms, primarily genetic drift. As shown by the contrast of the alpha and beta lineages given above, this disregard for speciation mechanism by molecular evolutionists is clearly not justified. Templeton *et al.* (1981) give additional examples of the importance of speciation in determining rates of molecular evolution, and Gillespie (1984) has recently performed a statistical analysis that indicates an episodic pattern of molecular evolution.

Table 12.3 Wilcoxon matched pairs signed rank tests of the molecular-clock hypothesis as given in Templeton (1983b). The test was applied to both phylogenies 1 and 2 of Figure 12.1. The taxa are designated by the first letter of their species name.

Comparison	Smallest signed rank sum (Phylogeny 1/2)	Non-zero ranks	Probability ($P<$)
Within β			
S × P	40/35	15/14	0.30/0.30
S × D	46.5/37	17/16	0.15/0.125
S × H	50/35.5	17/16	0.25/0.10
P × D	41/43.5	16/15	0.20/0.40
P × H	48.5/43	16/15	0.30/0.40
D × H	51.5/41	14/13	0.50/0.80
Between α and β			
S × C	15.5/9	19/20	0.001/0.0001
S × M	15/0	18/20	0.001/0.0001
S × N	14/10.5	20/19	0.001/0.001
P × C	3.5/0	15/16	0.001/0.0001
P × M	11/0	15/16	0.005/0.0001
P × N	27/16.5	18/18	0.01/0.005
D × C	22.5/13.5	15/16	0.03/0.001
D × M	20/15	15/17	0.025/0.001
D × N	22.5/22	15/17	0.04/0.01
H × C	18.5/13	16/17	0.01/0.001
H × M	17/9	15/17	0.015/0.001
H × N	4/3.5	13/15	0.005/0.001
Within α			
C × M	2.5/1.5	4/3	0.65/0.125
C × N	14/19.5	8/9	0.70/0.80
M × N	15.5/12.5	8/9	0.80/0.3027

However, anagenesis is still important in determining evolutionary rates, as will now be shown.

The importance of anagenetic trends as a determinant of rates of molecular evolution will be illustrated by examining humans versus African apes. Given the reasonable assumption that genetic drift is the primary determinant of rate of fixation of most molecular variants, there are at least two anagenetic trends that should cause rate heterogeneity between humans and African apes. First, one of the important features of human evolution has been a dramatic decline in the rate of development and an attendant increase in generation time over the past three million years (Cutler 1975, Lovejoy 1981). The rate of evolution through genetic drift is proportional to generation time unless one makes the assumption that mutation rates are proportional to absolute time. This assumption is not consistent with the empirical evidence among eukaryotes. For example, estimated mutation rates in *Drosophila*, mice, corn and humans are generally within an order of magnitude of one another (Strickberger 1976, pp 540–1) even though the generation lengths of these organisms differ by three orders of magnitude. Although mutation rates are hard to estimate accurately, it is difficult to reconcile such grossly discrepant results with the idea that mutation rates are constant in absolute time in higher eukaryotes. Indeed, Kimura (1979), long one of the leading advocates of mutation rate constancy in absolute time, has abandoned that position and now models his version of the neutral theory in terms of mutation rates that are constant on a per generation basis. Hence, the straightforward prediction of the anagenetic trend to longer generation times in humans is that the rate of molecular evolution in humans should be slower than that of African apes. Moreover, this slowdown should affect all genetic systems.

Another trend in the human lineage, but not in the African apes, has been a consistently expanding population size starting at the very least with *Homo erectus* which had a fossil distribution covering the entire Old World and not just Africa (Lovejoy 1981). Nei (1976) has shown that expanding population size decelerates the rate of molecular evolution under absolute neutrality, and Kimura (1979) has shown that the deceleration is even more severe when one includes mutations with very small selection coefficients. Once again, this slowdown should affect all genetic systems in which molecular evolution proceeds by random fixation of neutral or nearly neutral mutations.

These predictions can be tested by comparing the rates of molecular evolution in humans versus African apes for a number of genetic systems. First, Templeton (1983b, 1986b) has already tested the rate of mtDNA evolution in these primates and has shown that there is a statistically significant slowdown in the rate of mtDNA evolution in humans relative to African apes. This slowdown also appears to exist for nuclear genetic systems as well. For example, Marks (1982) concluded that humans have a slower rate of karyotypic evolution than the African apes. More recently, Sibley and Ahlquist (1984) performed extensive studies on nuclear DNA–DNA hybridisation between these species. Testing the molecular-clock hypothesis is far more difficult and less clean-cut with distance data than with character-state data and, unfortunately, the DNA–DNA hybridisation data is inherently a distance measure. Nevertheless, inferences can be made concerning the clock, but they must be made carefully rather than from a casual inspection of the data.

For example, Sibley and Ahlquist (1984) argued that their data is consistent with the clock because it satisfied the triangular inequality. They then quoted Farris (1981) to the effect that this metricity implies 'clocklike' behaviour. However, what Farris (1981) really said was that metricity is a necessary, but not sufficient, condition for clocklike behaviour. Commenting on Sibley and Ahlquist's quotation of Farris, Farris and Kluge (personal communication) corrected this error and pointed out that 'a distance may be metric, yet show strong variation in rates of evolution'.

A better way of testing the molecular-clock hypothesis is to contrast the distances between two closely related species with a third, more distantly related, outgroup species. This is known as the relative rate test (Sarich & Wilson 1967). The relative rate test, however, is biased in favour of the clock unless one can correct for the covariances of the distances or estimate the ancestral distance in a manner that is not biased towards the clock. Unfortunately, the covariances for DNA–DNA hybridisation distances cannot be calculated unless one invokes a parametric model of unknown validity. The second course of action, estimating the ancestral distance states, is flawed by the fact that distance estimation algorithms are biased in favour of, or explicity assume, the clock. For example, Nei *et al.* (1985) recently gave formulae for estimating the position and variance of branch points for various types of distance data. However, their equations are only valid if the molecular clock is exactly true, so their estimation procedure cannot be used to test the clock.

Although it is difficult to test the clock hypothesis with the Sibley and Ahlquist data set, it can easily be shown that these data contain internal inconsistencies which cannot be reconciled with the hypothesis of rate constancy (Templeton 1985). For example, the distance between humans and chimps is significantly smaller (at the 0.1% level) than that between humans and gorillas, but in contrast the genetic distance between gorillas and chimps is significantly smaller (at the 1% level) than that between humans and gorillas. These conclusions are based on *t*-tests that were uncorrected for the covariance between distances, and hence these tests are strongly biased in favour of the clock. Despite this bias, these significant *t*-test results cannot be reconciled with the clock. If the clock were true, the first significant test would imply that humans and chimpanzees form a clade, but the second test implies that gorillas and chimpanzees form a clade. Since there can be only one phylogenetic topology, the molecular clock obviously cannot hold for all these species. In a survey of the available molecular data, Templeton (1985) showed that there is one and only one branching order that is statistically compatible with all the molecular data; namely, the tree that clusters the African apes together as a clade. As further discussed by Templeton (1985), the pattern of internal inconsistencies found in the data of Sibley and Ahlquist (1984) can be readily explained by a slowdown in the human lineage under this topology because, although humans are the most distantly related in a topological sense, the human slowdown in rate of molecular evolution could yield very similar expected distances among all three species.

Thus, both the nuclear and mtDNA genetic systems appear to have been slowed down in humans relative to the African apes. This pattern supports the predictions discussed above based on the well-documented anagenetic trends in the human lineage of increasing population size and generation times.

DISCUSSION

The examples given in this chapter illustrate that rates of molecular evolution are not constant, but vary as a function of both cladogenesis and anagenesis. This variation in rates of molecular evolution does *not* mean that molecular data cannot be used for making inferences about times of divergence. The molecular data can still yield much information concerning divergence times, and indeed, the identification of specific heterogeneities can allow times of divergence to be estimated in an even more precise manner (Templeton 1983b). For example, consider estimating the times of the divergence nodes given in Figure 12.1 for phylogeny 2. As mentioned above, DeSalle (1984) estimated the number of mutational events along all branches in order to perform the tests given in Table 12.3. These estimated numbers of mutations will be used as a distance measure. This distance can be calibrated in absolute time by noting that the divergence between *D. planitibia* and *D. silvestris* involved a founder event on the Hualalai Volcano on the Island of Hawaii, which is about 400 000 years old. Assuming this founder event occurred shortly after the formation of Hualalai, the *silvestris/planitibia* divergence is set equal to 400 000 years. Another potential calibration point is the divergence between the alpha and beta lineages. The chromosomal and biogeographical data indicate that these lineages diverged on West Maui, which is 1.3 million years old. However, for now, only the *planitibia/silvestris* calibration will be used.

Using this calibration with no correction for the rate heterogeneity indicated by Table 12.3, one obtains the divergence times shown in part A of Figure 12.2. The divergence times of *differens* and *hemipeza* are consistent with the available biogeographical data; a result that is not too surprising given the fact that the clock holds within the lineages (Table 12.3). However, the divergence between the alpha and beta sublineages is estimated as 835 000 years, a figure that is much too small. Moreover, the estimated times of divergence within the alpha lineage are low (74 000 years for *crytoloma* and *melanocephala*, and 185 000 years for *neoperkensi*).

Because highly significant rate heterogeneity was detected between the alpha and beta lineages, but not within them (Table 12.3), the divergence times were next estimated by using the original calibration on the *silvestris/planitibia* node only within the beta sublineage. The rate of mtDNA evolution within the alpha lineage was regarded as being 3.23 times slower than that within the beta. This figure was obtained by dividing the average number of mutations on the branches leading to beta species from the alpha–beta node by the average number of mutations on the branches leading to alpha species from this node. Using the calibration corrected for rate heterogenity produces the estimated divergence times shown in part B of Figure 12.2. These estimated times make much more sense in the light of the known biogeographical data. For example, the alpha–beta split is now estimated to have occurred at 1.27 million years; a figure that is in excellent agreement with the 1.3 million year age of West Maui. Recall also that the behavioural and biogeographical evidence mentioned above implies that the alpha-lineage flies did not speciate through interisland founder events, but rather through the more traditional erection of geographical barriers between Maui Nui subpopulations. Since the *neoperkinsi* strain currently lives on Molakai (1.5 to 1.8 million years) and the *cyrtoloma* and *melanocephala* strains on East Maui, the speciation events separating these

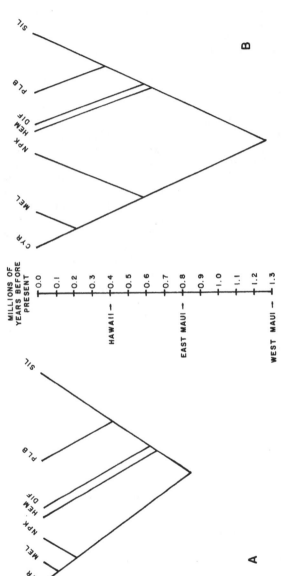

Figure 12.2 The estimated times of divergence for the Hawaiian *Drosophila* species under consideration. The species abbreviations are the same as those given in Table 12.2. An absolute time scale showing the times of formation of the relevant volcanoes is in the middle. In part A, the times of divergence are estimated using the *planitibia/silvestris* calibration node of 400 000 years with no correction for rate heterogeneity between the alpha and beta lineages. In part B, the times are corrected for the acceleration of mitochondrial DNA evolution within the beta lineage.

groups could only have occurred after the formation of East Maui if this interpretation of their speciation pattern is correct. Note that the estimated divergence times in part B of Figure 12.2 are consistent with this interpretation, since it indicates that the first split in these alpha species occurred shortly after the formation of East Maui.

The results reported in this chapter not only reveal that cladogenesis and anagenesis can affect the rate of molecular evolution, but in addition the results show that sometimes the impact on evolutionary rates is uniform on all genetic systems within a species (as shown by the human example), but in other cases the impact is quite heterogeneous (the *Drosophila* examples). Hence, different genetic systems imbedded within a single species are differentially sensitive to some of the historical contraints that shaped or even created that species. By simultaneously studying several different genetic systems, patterns emerge that can be used to make inferences about the history of the species being examined that could not be made just by studying each genetic system in isolation. For example, Charlesworth *et al.* (1982) argued that founder events were unlikely in Hawaiian *Drosophila* because they have maintained high levels of isozyme heterozygosity. However, Charlesworth *et al.* (1982) noted that this observation does not exclude founder events followed by a population flush because a depletion of isozyme heterozygosity is only expected if the population 'bottleneck' is 'prolonged over several generations'. Consequently, the isozyme data, considered in isolation, are consistent with two hypotheses: one, no founder or bottleneck effects occurred; or two, founder events occurred followed by large increases in population size. This ambiguity in interpretation is clearly resolved when the mtDNA data is overlaid on the isozyme data, for now the no-founder hypothesis can be rejected on the basis of the dramatic rate of increase of mtDNA. Consequently, a simultaneous examination of evolutionary rates among various genetic systems offers a potentially powerful tool for making inferences about the cladogenetic and anagenetic events that shaped the evolution of the species we observe today.

ACKNOWLEDGEMENT

This work was supported by NIH grant R01 GM31571.

REFERENCES

Carson, H. L. 1976. Inference of the time of origin of some *Drosophila* species. *Nature* **259**, 395–6.

Carson, H. L. 1983. Chromosomal sequences and interisland colonizations in Hawaiian *Drosophila*. *Genetics* **103**, 465–82.

Carson, H. L. and A. R. Templeton 1984. Genetic revolutions in relation to speciation phenomena: The founding of new populations. *Ann. Rev. Ecol. Syst.* **15**, 97–131.

Carson, H. L., L. T. Teramoto and A. R. Templeton 1977. Behavioral differences between isogenic strains of *Drosophila mercatorum*. *Behav. Gen.* 7, 189–97.

Charlesworth, B., R. Lande and M. Slatkin 1982. A Neo-Darwinian commentary on macroevolution. *Evolution* 36, 474–98.

Cutler, R. G. 1975. Evolution of human longevity and genetic complexity governing aging rate. *Proc. Natl Acad. Sci. USA* 72, 4664–8.

DeSalle, R. 1984. *Evolutionary genetics of Hawaiian Drosophila*. PhD thesis. St Louis, Missouri: Washington University.

DeSalle, R. and A. R. Templeton 1986. The molecular through ecological genetics of abnormal abdomen in *Drosophila mercatorum*. III. Tissue specific differential replication of ribosomal genes modulates the abnormal phenotype. *Genetics* 112, 877–86.

Ewens, W. J. 1983. The rôle of models in the analysis of molecular genetic data, with particular reference to restriction fragment data. In *Statistical analysis of DNA sequence data*, D. R. Weir (ed.), pp. 45–73. New York: Marcel Dekker.

Farris, J. S. 1981. Distance data in phylogenetic analysis. In *Advances in cladistics*, V. A. Funk and D. R. Brooks (eds), pp. 3–23. New York: New York Botanical Garden.

Giddings, L. V. and A. R. Templeton 1983. Behavioral phylogenies and the direction of evolution. *Science* 220, 372–8.

Gillespie, J. H. 1984. The molecular clock may be an episodic clock. *Proc. Natl Acad. Sci. USA* 81, 8009–13.

Gillespie, J. H. and C. H. Langley 1979. Are evolutionary rates really variable? *J. Mol. Evol.* 13, 27–34.

Hudson, R. R. 1983. Testing the constant-rate neutral allele model with protein sequence data. *Evolution* 37, 203–17.

Johnson, W. E., H. L. Carson, K. Y. Kaneshiro, W. W. M. Steiner and M. M. Cooper 1975. Genetic variation in Hawaiian *Drosophila*. II. Allozymic differentiation in the *D. planitibia* subgroup. In *Isozymes, vol. 4, Genetics and evolution*, C. L. Market (ed.), pp. 563–84. New York: Academic Press.

Kaneshiro, K. Y. 1976. Ethological isolation and phylogeny in the *planitibia* subgroup of Hawaiian *Drosophila*. *Evolution* 30, 740–5.

Kimura, M. 1979. Model of effectively neutral mutations in which selective constraint is incorporated. *Proc. Natl Acad. Sci. USA* 76, 3440–4.

Langley, C. H. and W. M. Fitch 1974. An examination of the constancy of the rate of molecular evolution. *J. Mol. Evol.* 3, 161–77.

Lovejoy, C. O. 1981. The origin of man. *Science* 211, 341–50.

Marks, J. 1982. Evolutionary tempo and phylogenetic inference based on primate karyotypes. *Cytogenet. Cell Genet.* 34, 261–4.

Nei, M. 1976. Mathematical models of speciation and genetic distance. In *Population genetics and ecology*, S. Karlin and E. Nevo (eds), pp. 723–65. New York: Academic Press.

Nei, M., T. Maruyama and R. Chakraborty 1975. The bottleneck effect and genetic variability in populations. *Evolution* 29, 1–10.

Nei, M., J. C. Stephens and N. Saitou 1985. Methods for computing the standard errors of branching points in an evolutionary tree and their application to molecular data from humans and apes. *Mol. Biol. Evol.* 2, 66–85.

Sarich, V. M. and A. C. Wilson 1967. Immunological time scale for hominid evolution. *Science* **158**, 1200–4.

Senner, J. W. 1980. Inbreeding depression and the survival of zoo populations. In *Conservation biology*, M. E. Soule and B. A. Wilcox (eds), pp. 209–24. Sunderland, Mass.: Sinauer.

Sibley, C.G. and J. E. Ahlquist 1984. The phylogeny of the hominoid primates, as indicated by DNA–DNA hybridization. *J. Mol. Evol.* **20**, 2–15.

Strickberger, M. W. 1976. *Genetics*, 2nd edn. New York: Macmillan.

Templeton, A. R. 1977. Analysis of head shape differences between two interfertile species of Hawaiian *Drosophila*. *Evolution* **31**, 630–42.

Templeton, A. R. 1979. The unit of selection in *Drosophila mercatorum*. II. Genetic revolutions and the origin of coadapted genomes in parthenogenetic strains. *Genetics* **92**, 1283–93.

Templeton, A. R. 1981. Mechanisms of speciation – a population genetic approach. *Ann. Rev. Ecol. Syst.* **12**, 32–48.

Templeton, A. R. 1982a. Genetic architectures of speciation. In *Mechanisms of speciation*, C. Barigozzi (ed.), pp. 105–21. New York: Alan R. Liss.

Templeton, A. R. 1982b. Adaptation and the integration of evolutionary forces. In *Perspectives on evolution*, R. Milkman (ed.), pp. 15–31. Sunderland, Mass.: Sinauer.

Templeton, A. R. 1982c. The prophecies of parthenogenesis. In *Evolution and genetics of life histories*, H. Dingle and J. P. Hegmann (eds), pp 75–101. New York: Springer-Verlag.

Templeton, A. R. 1983a. Natural and experimental parthenogenesis. In *The genetics and biology of Drosophila*, M. Ashburner, H. L. Carson and J. N. Thompson (eds), Vol. 3C, pp. 343–98. London: Academic Press.

Templeton, A. R. 1983b. Phylogenetic inference from restriction endonuclease cleavage site maps with particlar reference to the evolution of humans and the apes. *Evolution* **37**, 221–44.

Templeton, A. R. 1985. The phylogeny of the hominoid primates: a statistical analysis of the DNA–DNA hybridization data. *Mol. Biol. Evol.* **2**, 420–33.

Templeton, A. R. 1986a. Founder events and the evolution of reproductive isolation. In *Genetics, speciation, and the founder principle*, L. V. Giddings, W. W. Anderson and K. Y. Kaneshiro (eds). Oxford: Oxford University Press, in press.

Templeton, A. R. 1986b. Relation of humans to African apes: a statistical appraisal of diverse types of data. In *Evolutionary processes and theory*, E. Nevo and S. Karlin (eds). New York Academic Press, in press.

Templeton, A. R., H. L. Carson and C. F. Sing 1976. The population genetics of parthenogenetic strains of *Drosophila mercatorum*. II. The capacity for parthenogenesis in a natural, bisexual population. *Genetics* **82**, 527–42.

Templeton, A. R., T. J. Crease and F. Shah 1985. The molecular through ecological genetics of abnormal abdomen in *Drosophila mercatorum*. I. Basic genetics *Genetics* **111**, 805–18.

Templeton, A. R., R. DeSalle and V. Walbot 1981. Speciation and inferences on rates of molecular evolution from genetic distances. *Heredity* **47**, 439–42.

Val, F. C. 1977. Genetic analysis of the morphological differences between two interfertile species of Hawaiian *Drosophila*. *Evolution* **31**, 611–29.

Wasserman, M. 1982. Evolution of the *repleta* group. In *The genetics and biology of Drosophila*, M. Ashburner, H. L. Carson and J. N. Thompson (eds), Vol. 3B, pp. 61–139. London: Academic Press.

13

The origin, nature and significance of genetic variation in prokaryotes and eukaryotes

D. C. REANNEY

ABSTRACT

This chapter is based on the premise that natural selection acted to minimise the deleterious effects of (expressed) genetic errors during the early phases of evolution, when many familiar features of the genetic apparatus were 'fixed'. According to this thesis many key features of the information-transmitting system of cells may be viewed as adaptive responses to the heavy error load placed on early genomes by their inability to correct errors made during copying or introduced from the environment. These features include:

(1) diploidy and reciprocal recombination in eukaryotes;
(2) genome segmentation among RNA viruses and;
(3) the processing system which excises introns from RNA precursors.

INTRODUCTION

The title of this chapter is something of a misnomer. I am less concerned with the origin and nature of genetic variation than with its significance for evolution. Let me start by showing how sloppy use of language un-consciously shapes our ideas, often to the detriment of the facts. Compare the words 'variation' and 'error'. In the context of genetics, both usually refer to mutations in DNA (or RNA), that is, they refer to the same thing. But the mental images they evoke are quite different: variation somehow implies 'useful change' – it links with the notion that mutation is a 'good' thing because it fuels evolution. By contrast error implies 'malfunction' –

it links with the idea that mutation is a 'bad' thing because it causes information to decay. This confusion of meanings must be avoided. The hard fact is that although mutations may cause genomes to 'vary' they almost always do so in a *destructive* sense. To use an overworked analogy, when a typist copies a sheet of instructions almost any accidental mistake results in a deterioration of the quality of the information. The chances that a misspelled word or a transposed sentence will result in a more meaningful set of instructions are remote.

Part of the problem is that for two generations most textbooks of biology have enshrined the idea that sexual reproduction and reciprocal recombination have been positively selected because they generate diversity. The implication is that evolution needs more diversity than stochastic processes provide. Although most of us now consciously reject these arguments as 'group selectionist' their influence on our thinking remains deep-rooted and pervasive.

These qualitative statements can be put on a quantitative basis. The error rates in most ancient genes can be estimated by measuring the mutation frequency of polynucleotide copying in the absence of enzymes. The documented values are in the range of 10^{-1} to 10^{-3} substitutions per base per generation (Eigen & Schuster, 1977, Inoue & Orgel 1983). This is to be contrasted with the estimated mutation rate of 10^{-9} to 10^{-11} in modern DNA replication (Drake 1974, Loeb & Kunkel 1982). Simple comparisons like these show that, during the course of evolution, the level of noise in the gene copier mechanism has been slashed by a factor of 100 000 000 or so (see Reanney 1984a).

This drop in error levels demonstrates that, at least during much of evolution, natural selection has worked powerfully to decrease the frequency of mutation or to minimise its harmful effects. The best known mechanism of error reduction is correction. That is, secondary channels exist which allow a variety of error-detecting and error-correcting systems to repair premutational lesions in genetic systems. Such corrective processes operate not only during DNA synthesis but afterwards, in the resting state (postreplicative mismatch repair).

THE PROTECTIVE EFFECT OF GENETIC REDUNDANCY

The present concern is not with error correction but error compensation (see Reanney *et al.* 1983). The most widespread form of compensation is *redundancy*. The selective advantage, K, of redundancy can be modelled mathematically as:

$$K = [1 - (1 - q^L)^n] / q^L$$

where q is the base copying fidelity, n is the number of copies and L is the length of the genetic molecule (see Reanney *et al.* 1983). The best known example of this protective effect is seen in the structure of the genetic code where the identity of important amino acids is preserved by representing their cognate codons many times over (Goldberg & Wittes 1966, Ycas

1969). Fourfold redundancy in the glycine codons, for example, means that 33% of substitution errors in codons will have no effect whatsoever on the phenotype of the encoded polypeptide (assuming that base positions within codons do not affect mutational frequency).

The protective effects of redundancy can be modelled non-mathematically in terms of Figure 13.1. Here we see four DNA molecules encoding two information modules A and B. Provided that there is a critical number of whole genomes, a level of error sufficient to score one 'hit" per molecule per generation should not eliminate intact copies of total information. Thus a reiterated genetic system can survive a degree of genetic scrambling which would consign a single-copy DNA to its doom in short order.

A striking example of protection due to redundancy may be offered by the unusual prokaryote *Dienococcus radiodurans*. This bacterium is extraordinarily resistant to ionising radiation. This radio-resistance may be linked to the presence of 4—10 'genome equivalents' in each cell of this organism (Tigari & Moseley 1980). These DNAs appear to be functionally separate and capable of independent segregation. As we have pointed out elsewhere (Reanney *et al.* 1983), in this situation chance works for the preservation of genetic identity following exposure of *D. radiodurans* to ionising radiation. This is because the *random* nature of mutation ensures that the chances of a given *single* gene being 'hit' in *all* reiterated copies at *equivalent* positions are remote. The presence of multiple copies of essentially the same information in one 'nucleus' permits recombination between them and this also improves the survival potential of the information since large numbers of multiple crossovers may regenerate some wild-type genomes from damaged ones. I will return to this key point shortly.

Dienococcus radiodurans is adapted to a level of mutation that few other living systems could tolerate. Beyond doubt the most familiar examples of genetic systems protected from error damage by genome redundancy are the *diploid* genomes of eukaryotic cells. It is selectively useful for systems that encode relatively 'large' amounts of genetic information to present that information to a 'noisy' environment in duplicated form so that mutations in key alleles do not immediately eliminate the organism. The fact that such

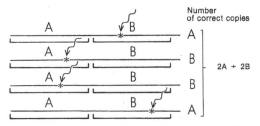

Figure 13.1 Diagrammatic representation of the protective effect of genetic redundancy. Four DNA molecules encoding two information modules A and B are shown where, even with a high level of errors, a sufficient number of correct copies are produced.

reiterated systems can store errors as recessive mutations is an 'incidental' consequence of this primary protective strategy. This is not to say that diploidy in modern cells is *maintained* by selection for its protective value. As is often the case in evolution, something that originates in response to one selective pressure may then be subverted to serve quite a separate function. Since diploidy is usually linked to the capacity for sexual reproduction, one can speculate that diploidy arose because it allowed higher cells to alternate between protected (diploid) and unprotected (haploid) phases (Reanney & Pressing 1984). Such reversible phases have the capacity to cull out unacceptably damaged genes during cool parts of the seasonal cycle (when the level of heat-induced error is low) and to protect cells during warm intervals (when the level of heat-induced error is high). According to this theory (Reanney & Pressing 1984), the familiar alternation of *generations* between haploid and diploid phases had its origins in the equally familiar alternation of *seasons* between cool and hot periods, with the consequent alternation of error rates.

THE PROTECTIVE EFFECT OF GENETIC SUBDIVISION

Redundancy protects information by *increasing* the mass number of nucleotides per system, i.e. the genetic 'bulk' of the system rises in proportion to the level of protection offered. There is an escalating energy cost to this kind of protection and it is proper to ask if there are alternative mechanisms which allow information to be maintained without incurring such penalties. The answer is 'yes'. It is possible to improve the efficiency of data transfer not by repeating the information but by *subdividing* the information into shorter modules which present a smaller target size to the various error-promoting agents. The principle here is simple: the *shorter* a module of information the *greater* its chances of passing *undamaged* through a noisy channel (Reanney 1982,1984c).

Such divided genomes are today found almost exclusively among the RNA viruses and it can hardly be a coincidence that these tiny genetic objects retain the high error level (3×10^{-4}) characteristic of unrepaired polynucleotide synthesis (Domingo *et al.* 1978).

The protective effect offered by genome subdivision can be readily visualised in terms of ultraviolet target theory. Consider an idealised situation in which ultraviolet dosage is calibrated so that it introduces one lethal 'hit' per molecule in a large population of RNAs all of length X. If each RNA is divided into two equal-sized modules, A and B, of length X/2 then the same ultraviolet radiation dosage is unlikely to inactivate the whole population. This is because a 'hit' in module A need not affect the integrity of the information in module B, and vice versa (Fig. 13.2). The logic is similar when mutations introduced by an error-prone polymerase are not a result of outside factors (Reanney 1982,1984a, Pressing & Reanney 1984).

However, a fundamental distinction must be made between what happens when the modular RNAs are packaged in one capsid (monocompartment viruses) and what happens when each RNA is separately accom-

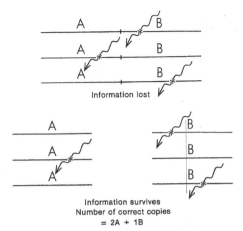

Information lost

Information survives
Number of correct copies
= 2A + 1B

Figure 13.2 Diagrammatic representation of the protective effect of genetic sub-division. Each RNA molecule is divided into two equal-sized modules A and B: an error in module A need not affect the integrity of the information in module B and vice versa. For further details, see the text.

modated in a discrete particle (multicompartment viruses). Consider the above example again, and assume that the error rate of the replicase that reproduces the two equal-length modules A and B is such that 50% of pro-geny carry lethal errors. The combinations of modules shown in Figure 13.3 are then possible, where letters with asterisks indicate lethally damaged modules, and letters without asterisks indicate viable modules. In the first case, each pair of modular RNAs can be accommodated in one capsid so a single particle can initiate infection. However, this strategy limits the number of particles carrying acceptable copies of the genetic information to

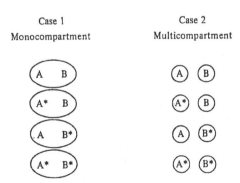

Case 1   Case 2

Monocompartment   Multicompartment

Figure 13.3 Diagrammatic representation of the protective effect of genetic sub-division in monocompartment and multicompartment viruses. Letters with asterisks indicate lethally damaged modules, and letters without asterisks indicate viable modules. For further details, see the text.

25%. In this instance, unless there is some preferential association of non-damaged modules (see below), a segmented genome has no advantage over a non-segmented one. In the second case, 50% of the population of progeny particles carry viable copies of units of the genetic information. This gives the multicompartment strategy a distinct survival advantage, so long as the transmission of the various encapsidated RNAs remains random and in-dependent. This requirement is met in nature.

The monocompartment viruses appear to have followed an independent evolutionary route. Where multiple RNAs are encapsidated in one coat, sequence-specific interactions must occur if there is to be provision for any kind of discrimination against miscopied information. Does such discrimination exist? Consider the reoviruses. During reovirus maturation *each* particle accumulates the *correct* combination of 10 to 12 *different* modular RNAs. The molecular basis of this specificity is believed to be a set of RNA–RNA or RNA-protein interactions that operate while the RNAs are single stranded (Silverstein & Acs 1976, Joklik 1981).

Lane (1979) has proposed a co-operative process in which the binding for one RNA molecule during assembly alters the nucleation complex to create a binding site for a second RNA and so on. If the binding of a given RNA is faulty, the subsequent RNAs have a lesser chance of entering the nascent particle. Since miscopied RNAs are less likely than faithfully copied RNAs to recognise sequence-specific elements in complementary RNA or RNA binding sites in proteins, their self-amplifying interactions constitute a crude form of molecular proof reading.

Since sequence-specific elements like this are likely to be quite short, one can ask why such interactions – however specific – would select out those molecules whose overall information has been well copied. In answering this question, it is crucial to remember the fundamental difference between a single-stranded and a double-stranded polynucleotide. In double-stranded DNA, the base-pairing capacity of one strand is fully satisfied by its partner, whereas with single-stranded RNA base-pairing capacity is satisfied inter-nally. This means that with single-stranded RNAs *shape* depends on *se-quence*. Or, to put it another way, with single-stranded RNA, selection can directly monitor changes in the coding *sequence* through their effects on the three-dimensional *topology* of the molecule. This means that a sequence required to be unpaired (so as to recognise some matching element) must be maintained in its correct configuration by multiple secondary or 'long-range' tertiary bonds distributed throughout the rest of the molecule. Only if this underlying three-dimensional scaffold is correct can the target sequences adopt the optimal conformation for a faithful interaction either with a complementary sequence on a different RNA or with a stereo-specific RNA binding site in a protein. The target sequence therefore acts as a form of 'quality control' because its availability, i.e. its ability to recognise a mat-ching element, is a sensitive function of the correctness of information elsewhere in the genome.

The reoviruses therefore also conform to my chief thesis in this chapter that genome subdivision is an adaptive response (at least in part) to the high error burden placed on the transmission of information encoded in RNA.

Significantly, one can show mathematically that the kind of molecular 'proof reading' envisaged in this model should allow the upper size limit on RNA genomes to be exceeded. It may be no accident that reoviruses, which have the most highly divided genomes of all segmental RNA viruses (10–12 RNA modules), also have the largest genome sizes $(12–20 \times 10^6$ nucleotides) (see Pressing & Reanney 1984).

'SPLIT' GENES

Sequence-specific interactions among reoviral RNAs provide an intriguing insight into a much wider process. These selective interactions may act as error-screening mechanisms because the reoviral genome is 'split'. But 'split' genes are common in nature. Indeed in the chomosomes of plant and animal cells they are closer to the rule than the exception.

Split genomes and split genes: is there any connexion? I believe there is. Moreover, I believe this connexion may allow us to explain how the exon–intron structure of split genes arose in the first place. The reasoning is as follows.

First, there is now a fairly widespread consensus among biologists that the split gene structure pre-dated the continuous (for supporting data and arguments see Doolittle 1978 and Darnell 1978). Perhaps the most compelling evidence for this view is the fact that exons often correspond to modular peptide substructures, which may have discrete functions or rôles (for a recent review see Blake 1985).

Secondly, a large number of biologists now appear to accept Crick's view (Crick 1968) that genetic information was originally encoded in single-stranded RNA not double-stranded DNA (for supporting data and arguments see Eigen *et al.* (1981) and Reanney (1979)).

Thirdly, as we have seen, the error rate in the 'first' genes was very high. If we take RNA virus replication as indicative of unrepaired nucleic acid synthesis, we can infer that the error frequency in early genomes was of the order of 10^{-3} to 10^{-4} errors per base per generation (see Reanney 1984a). The early copier mechanism was extremely error prone. These postulates together indicate:

(a) early genetic data was encoded in RNA;
(b) such early RNA genomes were 'split';
(c) early replicative processes were very 'noisy'.

This last point is critical. Eigen and Schuster in 1977 worked out a relationship which inversely relates the level of copy error in a self-reproducing system to the maximum size of the information that can be stably transmitted by that system, without experiencing an 'error catastrophe': *the higher* the error rate, the *shorter* the amount of reproducible information (Eigen & Schuster 1977). This means that a genetic system without repair capabilities is locked into an 'information freeze' – it cannot evolve towards higher information contents and hence, in an important sense, it cannot evolve at all.

How could genetic systems escape this freeze? Obviously one cannot invoke repair processes such as we see today because these universally use the information specified by the undamaged strand of a duplex molecule to guide restorative processes on its damaged complement. If primitive genetic systems were made of single-stranded RNA it is hard to see how a corrective mechanism that depends on a duplex structure could have evolved or worked.

What options which do not require the *de novo* creation of complex repair mechanisms would be open to such early systems? The most plausible mechanism is genetic redundancy, as this protects the information for the reasons already given, and requires nothing more complicated than the collection of multiple copies of the same information in one 'cell'. Such RNA 'polyploidy' has interesting implications: it immediately allows for recombination between the RNA genomes. Elementary recombination like this need not require protein catalysis in its start-up phases. A model for non-enzymic RNA recombination exists in the self-splicing ribosomal RNA of the ciliate *Tetrahymena* (Cech 1985). This transesterification reaction requires strand breakage, strand switching and strand reunion as does DNA recombination (Cech 1985), and it occurs in the total absence of proteins (Zaug *et al.* 1983). Significantly, the splicing of RNA precursors in the nuclei of higher cells passes through a series of intermediate phases in which branched 'lariat' structures are produced (Padgett *et al.* 1984). There are some tempting similarities between the formation of these branched intermediates and the formation of the postulated intermediates in the auto-excision of RNA from the precursor RNA in *Tetrahymena*.

As Padgett *et al.* (1984) note of their model for RNA splicing in higher cells:

> ...this reaction... is consistent with the ribosomal RNA self-catalysed splicing process seen in *Tetrahymena*. In this latter process, the 3' hydroxyl group from the 5' exon attacks the phosphate group at the 3' splice site in a transesterification reaction to produce the spliced exons and the excised intervening sequence. This reaction is similar to the second step in Fig. 6.

I believe then it is entirely reasonable to propose that RNA recombination arose early in evolution for the same reason as DNA recombination did later on (Reanney 1984b). It provided a means of eliminating errors from the pool of mutationally damaged genomes, by regenerating undamaged copies of the genetic information and hence allowing some upward mobility of information content. The selective advantages of recombination between single-stranded RNA molecules are in fact identical to those proposed by Maynard-Smith (1978) for double-stranded DNA recombination. Both function essentially as forms of repair! (Reanney 1984b).

If RNA splicing had its origins as a mode of *inter*molecular recombination, what relationship (if any) does it have to the *intra*molecular splicing which occurs today in the nuclei of higher cells? In assessing this question, it is important to bear in mind two points: firstly, the synthesis of RNA from DNA templates retains the high error rate of unrepaired nucleic synthesis; and secondly incorrect splicing has a multiplier or magnifier effect on cell physiology because of its ability to generate out-of-phase messages. In the light of these considerations, it seems likely that the capacity of the modern RNA processing mechanism to discriminate against 'incorrect' sequences is considerable. For example, if errors accumulate in the splice signals or in the sequences that determine topology, then an RNA with multiple exons is unlikely to proceed safely through all the steps needed to remove its introns. This is because RNAs blocked at particular processing steps by mutations in DNA may be degraded in the nucleus. The same is obviously true of errors introduced at equivalent positions during transcriptions. All the processing machinery 'sees' is what is there, whether it came from DNA/DNA or DNA/RNA copying is irrelevant. It follows therefore that the RNA processing machine functions potentially as a kind of molecular 'sieve' or 'filter' ensuring firstly that the average quality of the information reaching the decoding system as message is, in an important sense, *better* than the population of RNA precursors from which it was derived, and secondly that the cell does not waste its energy making nonsense (i.e. out-of-phase) proteins, some of which could damage its internal physiology. The only proviso is that for this system to work, cells must have some mechanism for *separating* primary RNA transcripts and messenger RNAs in a *selective* fashion. It is tempting to ask whether the *nucleus* that characterises cells with a splicing process was not originally selected for just this reason, as a form of biochemical 'quarantine' to keep in the unproofed primary RNA copies of key genes and let out only those message molecules that had passed the fitness test posed by the multiple-step processing machinery.

According to this thesis, RNA splicing has been retained in the nuclei of cells that encode 'large' amounts of information because trancriptional copying never evolved efficient repair mechanisms. In DNA, however, the development of a formidable battery of error-detecting and error-correcting processes has seen the error rate fall from about 10^{-3} to 10^{-11} per base per generation. In modern genomes therefore the contribution of genetic noise to the measured rate(s) of mutation is very small. Yet the genomes of higher plants and animals are extraordinarily plastic. Mammalian DNA in particular is something of a genetic museum littered with the 'fossil footprints' of past events due to the insertion and spread of 'selfish' DNA molecules.

DISCUSSION

I believe these two processes – the evolution of DNA editing mechanisms and the development of 'selfish' modular sequences – go hand in hand. Error-correcting mechanisms like the 'cut and patch' process carried out by the *Escherichia coli* DNA polymerase 1 combine the features of recombination (strand cleavage and ligation) and replication (fill-in synthesis). These kinds of processes could lead rather easily to the kind of 'replicative recombination' that allows semi-autonomous sequences like transposons to multiply in the genome (for a recent review of mechanisms see Shapiro 1985).

This speculation can be put on a more secure footing by going back to Eigen and Schuster's inverse relationship between error rate and genome size. A genetic system operating near the upper limit of its information capacity (as determined by its error frequency) could not tolerate the disruption nor carry the energy burden that the insertion and transposition of large numbers of self-amplifying foreign sequences would bring about. As the error rate falls with improved repair, the maximum information-carrying capacity rises. As genome sizes expand, the cell's energy 'bill' for DNA synthesis also increases. However, beyond a certain point the cell may not, as it were, 'notice' the energy cost of replicating redundant sequences which do not serve any selectively useful function. Such sequences may thus be freed from the selective constraints that dominated previous evolution.

According to this scenario, mutation in modern cells is due less to copy error than to the existence of a multiplicity of mobile genetic elements which have the ability to integrate promiscuously into chromosomal DNA. Evolution in higher plant and animals cells has thus been powered not so much by mutational rewriting of the genetic script as by changes in the 'ecology' of key genes. This idea is consistent with the views that evolution in higher cells is chiefly due to the re-wiring of genetic regulatory circuits and the frequent rearrangements of executive and housekeeping genes (Britten & Davidson 1971).

SUMMARY

This chapter has explored the origin and evolution of genetic variation in the light of the premise that selection has consistently acted to minimise rather than promote errors in the information-transmitting systems of living things since most genetic changes are harmful. The strength of historical selection pressures to *reduce* error can be quantified: the error rate of non-enzymic polynucleotide synthesis is about 10^{-1} to 10^{-2} substitutions per base per generation. This gives an indication of the mutation rate in the most ancient genes. By contrast, the mutation rate of modern eukaryotic DNA is often about 10^{-9} to 10^{-11}. Thus, during the course of evolution, the level of 'noise' in the gene copier mechanism has been reduced by about six to eight orders of magnitude.

I believe it is reasonable to view a number of familiar genetic mechanisms as adaptive responses to the need to minimise the deleterious effects of genetic error.

These mechanisms are as follows:

(1) *Reciprocal recombination*: As Maynard-Smith and others have pointed out, recombination between mutated chromosomes can regenerate the wild-type sequence; hence recombination probably arose as a form of 'repair'.

(2) *Diploidy in eukaryotes*: In terms of information theory one of the best-

known ways to transmit information through a noisy channel is to send it more than once. Genetic redundancy at the level of both gene and genome protects information because it is statistically unlikely that a lesion introduced at one point in a sequence will be duplicated at the corresponding point of its allele.

(3) *Genome segmentation*: RNA viruses are the only surviving genetic objects that retain the high error rate about 10^{-3} to 10^{-4} characteristic of unrepaired nucleic acid synthesis. These viruses have exploited a second mechanism for passing information safely through a noisy channel, namely division of the information into smaller modules which present a lower target size to error promoting agents.

(4) *RNA splicing*: It is now widely accepted that the first genes were made of RNA not DNA and these earliest genes were 'split'. I have suggested elsewhere that RNA splicing arose as a means of preserving the stability of the genetic information in the face of high level of noise in the RNA copier process (cf.today's RNA viruses). RNA–RNA 'splicing' is really a mode of RNA recombination and, as such, has the same selective advantages as reciprocal recombinations between DNA molecules. Splicing still may serve as an error-screening function in modern cells since the level of transcriptional error in the nucleus (about 10^{-4}) is roughly the same as that in RNA virus replication and there is evidence suggesting that unacceptably 'mutated' nuclear RNA precursors are less likely to pass through the hierarchy of RNA processing steps than faithful copies of chromosomal genes.

As the level of genetic error fell during evolution, with the development of effective repair processes, mechanisms such as sequence duplication, which may originally have been favoured for their protective value, could have been released from the normal constraints of natural selection. Such reiterative mechanisms may have been the progenitors of 'selfish' RNA or DNA modules. During later phases of evolution it is likely that fewer 'mutations' were due to copy error and more to the existence of these many mobile 'parasitic' elements which possessed, or gained, the ability to integrate promiscuously into chromosomal DNA.

REFERENCES

Blake, C. C. F. 1985. Exons and the evolution of proteins. In *Genome evolution in prokaryotes and eukaryotes*, D. C. Reanney and P. Chambon (eds), pp. 149–87. New York: Academic Press.

Britten, R. J. and E. H. Davidson 1971. Repetitive and non-repetitive DNA sequences and a speculation on the origins of evolutionary novelty. *Q. Rev. Biol.* **46**, 111–38.

Cech, T. 1985. Self-splicing RNA; implications for evolution. In *Genome evolution in prokaryotes and eukaryotes*, D. C. Reanney and P. Chambon (eds), pp. 3–22. New York: Academic Press.

Chambon, P. 1981. Split genes. *Scient. Am.* **244**(5), 48–59.

Crick, F. H. C. 1968. The origin of the genetic code. *J. Mol. Biol.* **38**, 367–79.

Darnell, J. E., Jr. 1978. Implications of RNA-RNA splicing in evolution of eukaryotic cells. *Science* **202**, 1257–60.

Domingo, E., D. Sabo, T. Taniguchi and C. Weissman 1978. Nucleotide sequence heterogeneity of an RNA Phage population. *Cell* **13**, 735–44.

Doolittle, W. F. 1978. Genes in pieces; were they ever together? *Nature* **272**, 581–2.

Drake, J. W. 1974. The role of mutation in microbial evolution. *Symp. Soc. Gen. Microbiol.* **24**, 41–58.

Eigen, M. and P. Schuster 1977. The hypercycle. *Naturwissenschaften* **64**, 541–65.

Eigen, M., W. Gardiner, P. Schuster and R. Winkler-Oswatitisch 1981. The origin of genetic information. *Scien. Am.* **244**, 78–94.

Goldberg, A. L. and R. E. Wittes 1966. Genetic code: aspects of organization. *Science* **153**, 420–4.

Inoue, T. and L. E. Orgel 1983. A non-enzymatic RNA polymerase model. *Science* **219**, 859–62.

Joklik, W. 1981. Structure and function of the reovirus genome. *Microbiol. Rev.* **45**, 438–501.

Lane, L. C. 1979. The RNA's of multipartite and satellite viruses of plants. In *Nucleic acids in plants*, T. C. Davies and J. W. Davies (eds), Vol. 2, pp. 65–110. Boca Raton: CRC Press.

Loeb, A. A. and T. A. Kunkel 1982. Fidelity of DNA synthesis. *Ann. Rev. Biochem.* **51**, 429–57.

Maynard-Smith, J. 1978. *The evolution of sex.* Cambridge: Cambridge University Press.

Padgett, R. A., M. M. Konarska, P. J. Grabowski, S. F. Hardy and P. Sharp 1984. Lariat RNA's as intermediates and products in the splicing of messenger RNA precursors. *Science* **255**, 898–903.

Pressing, J. and D. C. Reanney 1984. Divided genomes and intrinsic noise. *J. Mol. Evol.* **20**, 135–46.

Reanney, D. C. 1982. The evolution of RNA viruses. *Ann. Rev. Microbiol.* **36**, 47–73.

Reanney, D. C. 1984a. Genetic noise in evolution? *Nature* **307**, 318–19.

Reanney, D. C. 1984b. RNA splicing as an error-screening mechanism. *J. Theor. Biol.* **110**, 315–21.

Reanney, D. C. 1984c. Molecular evolution in viruses. *Symp. Soc. Gen. Microbiol.* **36**, 175–96.

Reanney, D. C. and J. Pressing 1984. Temperature as a determinative factor in the evolution of genetic systems. *J. Mol. Evol.* **21**, 72–5.

Reanney, D. C., D. G. MacPhee and J. Pressing 1983. Intrinsic noise and the design of the genetic machinery. *Aust. J. Biol. Sci.* **36**, 77–91.

Shapiro, J. 1985. Mechanisms of DNA reorganisation in bacteria. In *Genome evolution in prokaryotes and eukaryotes*, D. C. Reanney and P. Chambon (eds), pp. 25–56. New York: Academic Press.

Silverstein, S. C. and J. K. Acs 1976. The reovirus replicative cycle. *Ann. Rev. Biochem.* **45**, 375–408.

Tigari, S. and B. E. B. Moseley 1980. Transformation in *Micrococcus radiodurans*; measurement of various parameters and evidence for multiple, independently segregating genomes per cell. *J. Gen. Microbiol.* **119**, 287–96.

Ycas, M. 1969. *The biological code*. Amsterdam: North Holland.

Zaug, A. J., P. J. Grabowski and T. R. Cech 1983. Autocatalytic cyclization of an excised intervening sequence RNA is a cleavage-ligation reaction. *Nature* **301**, 578–83.

14

Old and new theories of evolution

J. LANGRIDGE

ABSTRACT

Various theories of evolution which have been advanced in opposition to Neo-Darwinism, as well as certain currently controversial aspects of Neo-Darwinism itself, have been reconsidered from the viewpoint of modern genetics. With advances in knowledge, the vitalist, finalist and creationist arguments are now even less tenable in the scientific view than they ever were. Geneticists are divided over the relative evolutionary importance of allelic states of loci versus restructurings of the genome, and evolutionists disagree about whether evolutionary events are smooth extrapolations from populational variation or occur as quite abrupt transitions. Since in no case is the origin and nature of variation of evolutionary significance known, any generalisation must await further information. As well as those theories questioning the materials of evolution, others are designed to minimise the random element in current evolutionary theory. In contrast to the vitalistic theories, which usually provide both a course and an aim of evolution, these 'directional' theories emphasise that any evolutionary advance determines to some degree the nature of future change. Evolution appears to progress by multiple mechanisms, varying in importance according to the organism's genetic complexity or evolutionary status. With this in mind, it seems that some previously discarded theories should again be taken into account in developing modern explanations of the evolutionary process.

INTRODUCTION

Although the theory of evolution is said to be the great unifying principle of biology, it has been a source of much disunity among those who study it. It is not the concept of evolution itself that is usually in dispute, but rather the mechanics of its processes. The main cause of disagreement has been a lack of knowledge of the genotypic changes that lead to adaptive evolution; very little is known of them even today.

The nature of evolutionary processes at the transspecific level in particular has provided the ground for continuous criticism of the 'synthetic', i.e. populational, theory of organic evolution. It was, and is, generally agreed that the differential effects of natural selection on random genetic variation are sufficient to diversify a population and even to produce new species. There has always been much less belief in the assertion that these same processes, if continued for sufficiently long time, by themselves produce the morphological and physiological differences characterising families, orders and phyla. In consequence, various theories alternative to a part or the whole of that of Darwin and his successors have arisen. Generally they fall into four groups: one of them is in opposition to the concept of evolution itself; another, while accepting natural selection, is in dispute over the nature of adaptive variation; yet another questions the rôle of natural selection; and a fourth group imposes some form of order in the evolutionary process. They are considered below under the titles: vitalism, mutation, selection and direction.

VITALISM

It may be asked why one should consider metaphysical or supernatural evolutionary theories based on concepts that are unscientific, in the sense that they cannot be subjected to scientific analysis and are not accepted by most scientists. But disbelief alone is not a counter argument in the absence of rational alternatives, because it is merely the opposite face of the belief that lies behind the vitalistic theories.

The problem is whether matter and life are different aspects of the same thing or whether the universe is composed of two things or principles. The more generally held view among philosophers and some scientists is that life and matter are distinct principles in the sense that neither can be derived solely from the other (e.g. Joad 1948). Life is considered to begin when some non-material principle or force forms an association with matter at a particular stage of its development. Matter of a certain type is then used to maintain and multiply the life principle. This principle has been variously thought of as a specially created force or as an inner purpose which directs the course of evolution of life forms. The opposing (scientific) view regards life merely as matter at a certain rare and complex level of organisation.

A discussion of the special arguments of vitalism (Bergson 1907), finalism (du Noüy 1947, de Chardin 1959) or 'scientific creationism' (Morris 1978) to name the more widespread theories, will not contribute to a resolution of the problem outlined above. These theories can be countered best by showing that their invocation is unnecessary, that the characteristics of life are merely realisations of properties that were latent in the purely material beginnings and that evolution is the agent of their realisation.

However, before attempting to do this, one must take account of the argument that there may be two distinct subjects involved, the origin of life and its continued evolution. Thus Simon (1971) writes that it is not possible to consider the origin of life in terms of the explanations (mutation and natural selection) of biology. Similarly, Mora (1965) maintains that the principle of selection cannot be used to account for the building-up of the more improbable and complex from the more probable and less complex. At this point, the doctrine of uniformitarianism, on which all evolutionary study is based, can be invoked. It holds that the same physical and chemical laws as operate today would apply to perturbations at the atomic and molecular level in the epoch of origin; perturbations of structure, which in non-living matter lead to disintegration but which in living (i.e. replicating) matter provide the variation necessary for evolution. With regard to the argument about selection and complexity, it has been shown that, under the influence of hydrated electrons produced by the irradiation of water, single nucleotides decrease in reactivity by one to two orders of magnitude when polymerised into polynucleotides (Scott 1975). Evidently there can be conditions of selection in chemical systems comparable with biological ones in which increased molecular stability, and thus survival, is correlated with increased complexity. Evolution, as driven by mutation and selection, is not solely a property of living systems, but of any replicative one.

The evidence, in refutation of vitalistic theories of evolution, that the 'emergence' of the characteristics of living things can be related to organisation inherent in atomic and molecular structure, has been advanced before (Langridge 1982). Such fundamental properties of the living state as self-multiplication, catalysis of reactions, shape and form, movement and environmental response, seem merely to be amplifications of qualities latent in abiological matter from the beginning. In brief the nucleic acids have a structure, not possessed by other polymers, with appropriately situated amino and keto groups for complementary hydrogen bonding, as required for template replication. The combination of a very stable resonating structure with potentially reactive groups attached, provides both the stability required in a transmissible compound and the spontaneous variation necessary for evolution. In some artificially synthesised polypeptides, the position and nature of amino acid side chains is such that they can catalyse a number of bond rearrangements and in some cases appear to be regenerated after reaction. Other polypeptides of this sort appear to contain various amounts of α-helical structure and, since there is evidence that muscular contraction rests on a transition between the helical and coil configuration of α-helices (Harrington 1979), a molecular basis for movement is evident. The presence in the polypeptides of patches of hydrophobic amino acids leads them to aggregate in water, a property that has been modified by evolution to provide microtubules and other determinants of morphogenesis. Finally, when unspecific protein is mixed with lipid, semi-permeable membrane-like sheets result. In the presence of ions, an electrical potential is generated across the membrane and, when this 'resting potential' is disturbed, a flow of ions occurs. With the evolution of cells designed to detect such changes in current, the basis of a nervous system is laid.

These examples, even in bare outline, may be sufficient to indicate that there is no necessary reason to suppose that evolution relies on anything other than the known laws of physics and chemistry which generate change (mutation) and differential stability (selection). Selection, of course, can only operate when the molecule is capable of self-multiplication. When that is so, a form of matter (self-duplicating) that can multiply itself at the expense of inanimate (non-duplicating) matter must increase in mass with time.

MUTATION

Some of the topics in this section will be discussed in detail by other contributors to this book, so only a very general view of changing attitudes to evolutionary variation will be given here.

In the *Origin of Species*, Darwin (1859) wrote that evolution advanced by small steps, quoting the canon 'natura non facit saltum'. This was soon contested by the Swiss physiologist, von Kölliker (1864) who considered that evolution is brought about by leaps rather than by the slow accumulation of minute advantageous variations. According to his view and the later conclusions of Goldschmidt (1940) and Schindewolf (1950), evolution is periodic or saltatory. One hundred and twenty years after Darwin, this is still an open question.

On the origin of hereditary variation, Darwin wrote only that 'the conditions of life may be said to cause variation either directly or indirectly'. The actual nature of such variation was thought to be shown by the experiments of de Vries (1901) on *Oenothera* where large inherited changes (mutations) occurred in a single step. This finding led to theories of mutationism involving preadaptation in which evolutionary changes are said to arise spontaneously in organisms and are selected or rejected without much regard to adaptive value. The mutationist theories are generally answered by pointing out that, although adaptation is clearly non-random and increases the fit of the organism to its environment, mutational preadaptation would be entirely random. De Vries's mutations were soon shown to be due to recombination between chromosomal rings and thus not the usual stuff of evolution. The real stuff of evolution was taken to be the spontaneous mutants found by Morgan (1911) in *Drosophila*. However, it now appears that most spontaneous mutations in *Drosophila* are due to the insertion of transposable elements into genes. In general then, those considering evolutionary mechanisms relied for their source of genetic variation on 'mendelising gene differences' (within species of course), chromosomal rearrangements including position effects, X-ray-induced mutations (mostly deletions) and spontaneous mutations (mostly tranpositions) in *Drosophila*. On such a basis of random, small and continuously occurring mutation, the genetic theory of natural selection and the synthetic theory of evolution was built by Haldane, Fisher, Wright and Chetverikov using the mathematics of chance.

Both measured rates of spontaneous mutation, and the genetic variation found in populations, indicated that evolutionary change would be a slow process, assuming that these were its raw materials. This opinion changed when the techniques of gel electrophoresis were used in 1966 to detect differences in protein charge and demonstrated that very many genes in wild populations are represented by more than one allele. One view of this discovery is that populations in their natural state must contain a very large amount of genetic variation for evolution to draw on because protein charge variants are only symptomatic of more general sequence changes and represent a mere third of all amino acid substitutions. Indeed, Ayala (1984) has calculated, taking into account sequential electrophoresis, heat denaturation and peptide mapping, that the protein-coding genes of invertebrates are, on average, 28% heterozygous, vertebrates 22% and plants 27%. An alternative view of the data is that the mutations reflected in protein differences are merely an unselected or weakly selected residue of the mutations occurring naturally, and that the majority of other classes of mutation have been removed by negative selection.

The idea that mutations of any sort, even those to synonomous codons, could be unaffected by natural selection was not well received, because it cast doubt on the efficacy of selection to detect genetic change and on the goodness of fit of an organism and its parts to the environment. However, some clear examples of mutant neutrality in the genetically best known eukaryotic protein, haemoglobin, have recently been given by Perutz (1983). He reviewed the structure and function of haemoglobin in several animal phyla at the molecular and stereochemical level. Most of the amino acid replacements between species (about 400) were found to be neutral or nearly so, caused by the random drift of selectively equivalent mutant genes. There were a few amino acid replacements which had been selected because they advantageously increased certain properties of haemoglobin. In the evolution from cartilaginous to bony fish, for example, a shift and flattening of the oxyhaemoglobin dissociation curve (which allows oxygen to be transferred to the swim bladder) has been brought about by a single substitution of alanine by serine in the haemoglobin.

Estimates of the relative numbers of neutral and positively selected changes in amino acids in haemoglobin may be derived from nucleic acid sequencing. A comparison of nucleotide sequences of parts of the messenger RNA for haemoglobin from man and rabbit shows that the rate of incorporation of 'silent' mutations (ones that do not alter the amino acid sequence) has been tenfold higher than the rate of amino acid substitution (Fitch 1976). Yet it can be calculated from the genetic code that, in the absence of selection, only 25.5% of nucleotide replacements should cause no change in amino acid sequence. This 40-fold discrepancy between the expected and observed frequency of 'silent' mutations means that, in the haemoglobin molecule, many amino acid substitutions are selected against and that 'neutral' mutations, other than synonomous ones, can form only a small fraction of the range of spontaneous mutations.

Unfortunately, these data from haemoglobin cannot be extrapolated to protein-coding genes generally, because proteins differ so much in their susceptibility to mutation. In soluble metabolic proteins such as those studied with respect to electrophoretic and thermal variation, the majority of amino acid changes have little effect on catalytic activity. Thus, in the β-

galactosidase of *Escherichia coli*, of 733 amino acid substitutions examined, only 1.5% reduced enzyme activity below 50% of the wild-type level (Langridge 1974). Such inertness to alterations in amino acids is due to the buffering properties of the genetic code, the protein structure and the mode of enzyme synthesis. But in the galactoside permease, which has to fit into the structural organisation of the membrane, one-third of amino acid replacements inactivate the protein, and nearly all such replacements inactivate the repressor which normally binds to DNA.

Evidently, the proportion of amino acid substitutions that are neutral, or that have little effect on protein activity, varies according to the function of the protein and may not be large. It is also apparent that some alterations in the amino acid sequences confer benefits in adapting proteins to altered intracellular conditions and are positively selected. However, the adaptive changes observed in haemoglobin were ones that occurred over many millions of years and so are unlikely to be representative of the protein changes seen in the much shorter periods involved in species formation.

Although species frequently show marked genetic differences in protein-coding genes from related species, in that at least one-third of the loci are allelically distinct (Avise 1977), there are reasons to believe that these differences are not associated with the evolution of reproductive isolation, but accumulate following its completion. Consequently, most protein differences observed between species will not be ones that have occurred during the formation of the species increasing adaptation to their environments. The evidence for this assertion comes from comparisons of changes in amino acids in groups which, although reproductively separate, are morphologically similar, and from ones with markedly different evolutionary rates. Thus reproductive isolation can occur with relatively little overall genetic divergence (e.g. brown trout – Ryman *et al.* 1979), and morphologically identical species can sometimes represent groups that have been genetically distinct for long periods of evolutionary time (e.g. bonefish – Sharklee & Tamaru 1977). Similarly, placental mammals, which have undergone recent rapid evolution, show a degree of protein divergence comparable with that of frogs with relatively static morphology (Wilson 1976). There appears, at least in these instances, to be no correlation between either reproductive isolation in speciation or in rates of morphological and physiological evolution and degree or rate of change in protein sequence. This conclusion accords with the proposition that the degree of amino acid difference is mainly a function of the time that the organisms containing the coding genes have been separated.

Since changes in structural genes, as shown by alterations in protein sequence, do not seem to play a major rôle in evolution, attention has turned to 'regulatory' genes, ones that influence the timing, placement and degree of structural gene action. The rôle of regulatory genes in adaptation is well known in bacteria, especially in relation to the utilisation of new carbon and nitrogen sources. They are mostly variations in gene expression due to mutations affecting the transcription or translation of genes. Another type of 'regulatory' variation is that of genes which affect the expression or timing of the products of other genes. Both of these types of regulatory gene variation are frequently found as part of the natural variation of eukaryotic populations (McDonald 1983). However, in addition to the arguments presented above, the relative unimportance of enzyme-coding genes in evolution is provided by the observation of Britten and Davidson (1971) that higher and lower organisms appear to have perhaps 90% of their enzymes in common. A third type of proposed regulatory mutation is that which alters gene position, and thereby function, by chromosomal rearrangements. Chromosomal mutations are demonstrably important in species formation where they act as fertility barriers, but any rôle they may have in changing gene regulation is still unclear. A correlation has been found between frequency of chromosomal change and degree of evolutionary divergence (Wilson *et al.* 1975), but these chromosomal observations may be a consequence of divergence and not a cause of it. Experiments in *Drosophila* show that a gene may be placed in a large number of locations in euchromatic regions of the genome without altering that gene's normal regulation (Spradling *et al.*1983).

Mutations of the genes that control the developmental processes in unknown ways remain as a major source of variation to adapt organisms to their surroundings. The genes controlling such processes may be acting in the conventional manner by coding for enzymes, but there is no evidence for this. In fact, the few developmental genes whose products have been identified, make low concentrations of rather small proteins that are confined to the nucleus and probably bind to DNA (North 1984).

An indication of some of the processes that can be genetically altered to produce new features of differentiation and morphogenesis, such as are required in evolution, comes from the analysis of certain mutants. In summary, altered organisation in plants can occur by changes in meristem activity with respect to both rates of cell division and auxin production, local alterations in planes of cell division, shifts in the position and extent of hormone induction in vascular tissue, and so on. In the animals, small changes in tissue induction such as in amount and potency of mesoderm, variation in response to inducing tissues, shifts in interacting zones and changes in maintenance factors appear to be particularly important in developmental change. In both plants and animals, more general evolutionary phenomena such as change in adult size, allometric growth parameters, the self-limiting capacity of organ growth or the onset of neoteny, can materially alter the phenotype.

Information on the nature of genetic variation in such developmental processes, although meagre, is now beginning to accumulate at an increasing rate. It is made possible by applying the techniques of recombinant DNA to certain well-known mutants in *Drosophila*. The mutants are of 'homoeotic' genes which, in segmented organisms such as insects, act to control the particular form and cell pattern characteristic of each segment. Their mutation can cause such changes in development as the formation of a pair of legs in place of a pair of antennae. The homoeotic genes occur as complexes, tightly linked clusters of genes with closely related functions, and some of them have been isolated, cloned and sequenced. A notable finding from the DNA sequences is that a short sequence of 180 nucleotides is common to these developmental genes of *Drosophila* (McGinnis *et al.* 1984). It is believed to be associated with selector genes responsible for anterior–posterior specification and periodicity genes concerned with the segmentation pattern, and has been found in a wide range of organisms from beetles to man. There thus appear to be DNA sequences that control the pattern of development and, although their known mutants may be evolutionary dead ends, their study should lead to general principles governing development and, on earlier reasoning, evolution.

With molecular analysis narrowing the distinction between phenotype and genotype, certain aspects of DNA structure and turnover have become implicated in the speciation process. Some of the DNA of the eukaryote nucleus, a little more than one per cent in man, is in individually unique sequences that code for proteins. Another class is repeated to an intermediate degree, although it may be repeated up to 10^5 times, and is either dispersed along the chromosomes or in clusters. Some of this DNA is transcribed into ribosomal or transfer RNA and some is translated into proteins such as histones and immunoglobulins, but the function of most of it is unknown. A third class, satellite DNA, has short simple sequences repeated more than 10^6 times; it appears to be neither transcribed nor translated. However, this fraction of the DNA, and to a lesser extent the intermediate repeated sequences, seem to be characteristic of a species, the satellite DNA of one species differing markedly from that of even a closely related species.

There is great uniformity of sequence in the repeats of a family of sequences, indicating that some mechanism must be operating to counteract their tendency to diverge by mutation. More importantly from the evolutionary point of view is the conclusion that individuals, at least in small populations, change in unison, a process called 'molecular drive' (Dover 1982). If, as has been proposed (Fry & Salser 1977), an abrupt but uniform change in satellite sequence disturbed chromosome pairing between individuals possessing the new sequences and the rest of the population, it would ensure reproductive isolation and incipient species formation. Even if such a process facilitates the origin of species, there is no evidence yet that the sequence changes that provide 'genome resetting' have any morphological effect. However, it would fit in with palaeontological data which indicate that evolution on the larger scale involves long periods during which individual species remain virtually unchanged, followed by much shorter periods during which a new species arises from the ancestral population. This mode of evolution has been called 'punctuated equilibrium' by Eldredge and Gould (1972). It should also be remembered that, if the normal course of macroevolution is one of punctuated equilibria, there are theories other than the saltatory origins of new satellite sequences to account for it; e.g. a partial loss of developmental homoeostasis (Carson 1975) or a breakdown in stabilising selection (Stebbins & Ayala 1981).

One of the main issues in evolution, that of the nature and source of the genetic changes that allow organisms to change in an adaptive fashion, seems to be rather more obscure now that analysis has progressed to the molecular level, than it was formerly. In the past, evolution was discussed in more abstract terms such as polygenic control, super genes, coadapted gene complexes, increased fitness etc. and these were generally considered satisfactory despite their unavoidable vagueness. Now we have to choose among phyletic gradualism, punctuated equilibria, even punctuated gradualism to designate the course of evolution, and genome resetting, molecular drive and 'mutations in homoeotic boxes' to explain its mechanism. The newer terms and their implications cannot be simply substituted for the older ones because selection experiments within populations indicate that practically all characters that are determined by many genes can readily be selected for increased or decreased expression. Numerous examples of changes in morphological, physiological and behavioural characters brought about by artificial selection are given by Lewontin (1974). The type of variation selected and the number of genes involved are not known, but this genetic variation does not appear to be of the type usually measured in populations.

SELECTION

The reality of natural selection as proposed by Darwin seems intuitively obvious. Since some of the differences between individuals in natural populations are hereditary, and since some of these affect the number of progeny produced, it follows that certain genotypes will appear in changed proportion in the next generation. Put in terms of population genetics, natural selection ensures that adaptively advantageous genes are spread to all individuals in a population and thus it determines evolutionary change. The only additional circumstance recognised is that of genetic drift when one or a few individuals become reproductively isolated from the parental population and contain only a small fraction of its total genetic variation. The effect of environmental selective agents is thus expected to be the gradual elimination of the less fit genotypes, ones resulting in fewer progeny with reference to the average of the population, and the propagation of the more fit, those resulting in more progeny. Dissent from this proposition has been directed, not to the former aspect of negative selection, but to the latter which is held to be primarily responsible for the course of evolution.

Some of these dissenting views have already been mentioned. Thus de Vries (1901) contended that mutants important for evolution were preadapted in the organism, thus minimising the rôle of positive selection. Similarly, Whyte (1965) writes that 'internal selection would here, in a certain degree, directly *determine* the variations resulting in phylogeny (provided they pass the Darwinian test), rather than merely *select* from given biologically arbitrary variations'. The opinion of Berg (1926) was that evolution proceeded by natural law, not by natural selection, and that 'not infrequently evolution proceeds, as it were, in the face of the environment, in a direction leading the organism to destruction'.

As discussed in the next section on direction in evolution, these objections seem to be based on the necessity for newly arisen mutations to be compatible with the results of evolution that has occurred earlier. As put by Whyte

(1965) the crucial issue is 'are the mutated genotypes whose consequences are submitted to external selection random or not in relation to the subsequent direction of evolutionary change'. Clearly they are not random.

Of a quite different sort is a recently proposed mechanism, molecular drive (Dover 1982), which would relegate natural selection from a primary rôle in certain aspects of evolution to a subsidiary one. When a new species is formed, there is apparent rapid 'turnover' of the repeated sequence fraction of DNA, possibly due to gene conversion, unequal crossing over or transposition. It results in uniformity of repeats within a given family of sequences which are, however, dissimilar from the repeated sequences in a family from another species. There can be, in a sexually reproducing population 'a progressive concerted change in phenotype of a group of individuals effected by internal processes of genomic turnover between chromosomes'. Natural selection is involved only to the degree that if the phenotypic change is deleterious, it will be selected against. It appears to be a random process and as such subject to the same reservations as those that apply to the preadaptation model for mutation; that is, that the direction of evolution is not in accord with any random mechanism of fixation.

Finally, recent studies of environmental influences on inherited characters reopens the question of whether the environment may not only select adaptive genetic changes but also induce certain of them. Campbell (1984) has described and discussed many reproducible phenomena of this sort. In flax (*Linum*) and some other plants, conditions during seedling growth can cause both expansions and contractions of the moderately repeated fraction of nuclear DNA sequences. These DNA changes, which may become stabilised and inherited indefinitely, depend on the presence of a particular genetic mechanism for their occurrence. The resultant phenotypic effects include changes in size, in isozyme constitution and in various morphological features of the plant. Examples of environmentally induced genetic change in animals include the pronounced inherited effects that may result from environmental interference with the numerous neuroendocrine hormones. The changes, which affect a range of metabolic and physical characters, are specific to the regulatory hormone involved and appear to be in the DNA since they are transmitted through the male parent. The advantages of certain hereditary changes being inducible by the appropriate environment are, if they are specific in phenotypic effect and restricted in the characters modified, of such great potential value to the genetic adaptation of an organism that it would not be too surprising if a mechanism has evolved to accommodate them. If it has, it seems likely that it developed using existing pathways; that is, interactions evolved for other purposes which relate certain conditions of the body of the organism to its nucleic acids. The polypeptide and steroid hormone systems are clear candidates.

DIRECTION

One of the principal criticisms of the concept of evolution has always been the apparent inability of random events, even with the sieving effects of natural selection, to generate the extreme levels of organisational complexity attained by living matter. This difficulty has been partly or wholly the reason for the initiation of three groups of theories in addition to that of Neo-Darwinism – the vitalist theories discussed above, Lamarckism as already mentioned, and the proposals of Berg (1926) and Whyte (1965). All these theories are concerned to introduce a further element of order into the simple mutation–selection concept of evolutionary change.

There is general agreement that the directions taken by evolution have not been random; they give the appearance of being oriented. Even before biological evolution began, matter underwent changes of state as the environment altered, the last state being the living one.

A glance at the evolutionary record shows that later evolved organisms are generally more complex than the ones from which they sprang. By 'more complex' is meant that the later organism has a greater number of facets to its organisation and that these facets are more diverse. Such organisms are said to be more advanced or to show biological progress, meaning that they have developed improvements in structure and function which are not mere specialisations, but ones that set the stage for still further improvement. This trend towards increasing complexity gives an obvious direction to the course of evolution.

There are also other trends, on a smaller scale, which indicate that some evolution is in particular directions and not in others that would seem equally possible. Notable among them are the sustained trends that may occur in the development of a particular organ or feature, the progressive reduction with time of certain other characters, the tendencies for evolution to involve the same structure in unrelated lines of descent and generally the occurrence of parallel and convergent evolution.

Such considerations as these led Berg (1926) to develop his theory of nomogenesis, or evolution determined by law. He wrote that 'characters occur which owe their development to inner causes, inherent in the very nature of the organism, which we call autonomic, independently of any effects from the environment'. These inner causes he believed were connected with the stereochemical properties of cytoplasmic proteins which limited the development of organisms to predetermined directions. Berg's views were not supported by the Neo-Darwinists who maintained that successive changes over a prolonged time in particular structures are produced by the environmental selection of favourable mutations. They further concluded that the major, if not the only non-random factor in evolution is adaptation. The reasoning that leaves adaptation as the sole directive influence is based on the realisation that all evolutionary change is ultimately to be referred to gene mutations, the belief that spontaneous mutation is a completely random process, and the implication that phenotypic characters will consequently vary in random ways.

There are, however, several reasons why gene-controlled characters will not vary in a random fashion, even if a random mechanism produced the mutations affecting them. At least for developmental genes, it is evident that many of them are part of interacting complexes arranged in various levels of order and complexity. Such co-ordinated genetic subassemblies are apparently necessary for directing the formation of complex organs, but they also restrict the range of evolutionary change that can occur. So, the greater the degree of organisation, the lower the probability of randomness entering the system by mutation because of structural and functional limitations to the types of modifications that can be made. Despite the fact that mutation is not now seen as the degenerative process it was once thought to be, only mutations concerned with peripheral features, which are relatively unimportant, or terminally developed characters, which are in the line of the most recent evolution, are likely to spread and contribute to adaptive change.

In part, also, the canalisation of development, already referred to, narrows down the evolutionary potential of mutations. The concept of canalisation refers to the observation that pathways in ontogeny leading to specific shapes, organs etc. appear to be under very close control. This control is expressed as an inbuilt tendency for the particular pathway to be followed in the face of potentially disturbing genetic and environmental influences. Canalisation is presumably a property conferred on the organism by the selection of systems of genes that interact in a self-stabilising fashion. It is much more pronounced in animal than in plant development, probably because the more extensive and intricate pathways in animal ontogeny leave little latitude for variation.

It is perhaps these phenomena of somatic organisation and canalisation which lie behind the hypothesis of Whyte (1965) on internal factors in evolution. He distinguished internal selective forces which determine whether a given mutation is transferred to the organism's progeny, and Darwinian external selection which determines the eventual fate of mutant individuals.

As mentioned above, the chance that a newly arisen mutation affecting development will be selected for, depends on whether its effects are compatible with the maintenance of the orderly ontogenetic progression. Those mutations that act very early in ontogeny, and whose expression thus becomes amplified as growth continues, and those that affect multiple processes are likely to be selected against during embryogeny. In addition, mutations that grossly disturb the canalisation of the structures or processes they act upon are equally unsuitable for positive selection. It is frequently the case that a mutation not only alters the morphogenic pathways but also disturbs the buffering system to a degree that would probably render the new phenotype unacceptable in all environments.

For the above reasons, the spectra of phenotypically expressed mutations in individuals of related groups such as species of a genus or genera of a family would be expected to be similar, and provide changes in certain phenotypes more readily than in others. Hence the frequent occurrence of parallel evolution. Of all the genes forming the genome, only a small fraction can be altered and still leave a viable organism. As developmental complexity increases, this fraction must become even smaller and the avenues for new evolutionary change narrower and fewer. This is the general situation, but there are aspects of development such as sex-organ formation in animals and leaf and flower production in plants, which are not closely integrated into the whole process of ontogeny, and these are capable of independent variation.

CONCLUSION

Although this chapter was not meant to serve as an apologia for ancient and long-rejected theories of the evolutionary process, a re-examination in the light of modern genetics suggests that many of them may have been discarded rather too quickly. The error of their originators, and indeed also of the adherents of Neo-Darwinism, was to take a partial mechanism of evolution for the whole process. Evolution, as it is 'understood' today includes, to varying degrees, the mutations of small effect of Fisher and Wright, the saltations of Goldschmidt and Schindewolf, the directional elements of Berg and Whyte, and perhaps even the inheritance of acquired characters of Lamarck; vitalism alone is not under serious reconsideration. The main reasons for this expanded view are insights into evolutionary matters given by molecular examination, the much refined analysis of palaeontological data and cracks in the apparently solid body of theory that comprised Neo-Darwinism. This is not to say that evolution is better understood today than it has been, but rather that the focus of scrutiny has shifted to what appears to be the more significant aspects of the mechanisms of evolution.

REFERENCES

Avise, J. C. 1977. Is evolution gradual or rectangular? Evidence from living fishes. *Proc. Natl Acad. Sci. USA* **74**, 5083–7.

Ayala, F. J. 1984. Molecular polymorphism: how much is there and why is there so much? *Devel. Genet.* **4**, 379–91.

Berg, L. S. 1926. *Nomogenesis*. London: Constable.

Bergson, H. 1907. *L'evolution créatrice*. Paris: Alcan et Guillaumin.

Botstein, D. 1980. A theory of modular evolution for bacteriophages. *Ann. NY Acad. Sci.* **354**, 484–91.

Britten, R. J. and E. H. Davidson 1971. Repetitive and non-repetitive DNA sequences and a speculation on the origins of evolutionary novelty. *Q. Rev. Biol.* **46**, 111–13.

Campbell, J. H. 1984. A biological interpretation of evolution. In *Biology and new philosophy of science*, B. H. Weber and D. J. Depew. (eds).

Carson, H. L. 1975. The genetics of speciation at the diploid level. *Am. Nat.* **109**, 83–92.

Chardin, T. P. de 1959. *The phenomenon of man*. New York: Harper.

Darwin, C. 1859. *The origin of species*. London: Murray.

Dover, G. 1982. Molecular drive: a cohesive mode of evolution. *Nature* **299**, 111–7.

Eldredge, N. and S. J. Gould, 1972. Punctuated equilibria: an alternative to phyletic gradualism. In *Models in paleobiology*, T. J. M. Schopf, (ed.) pp. 82–115. New York: W. H. Freeman.

Fitch, W. M. 1976. Molecular evolutionary clocks. In *Molecular evolution*, F. J. Ayala (ed.) pp. 160–78. Sunderland Mass: Sinauer.

Fry, K. and W. Salser, 1977. Nucleotide sequences of H S-α satellite DNA from kangaroo rat *Dipodmys ordii* and characterization of similar sequences in other rodents. *Cell* 12, 1069–84.

Goldschmidt, R. 1940. *The material basis of evolution*. New Haven: Yale.

Harrington, W. F. 1979. On the origin of the contractile force in skeletal muscle. *Proc. Natl Acad. Sci. USA* 76, 5066–70.

Joad, C. E. M. 1948. *Guide to modern thought*. London: Pan.

Kölliker, A. von 1864. *Über die Darwinische Schöpfungstheorie; ein Vortrag*. Leipzig.

Langridge, J. 1974. Mutation spectra and the neutrality of mutations. *Aust. J. Biol. Sci.* 27, 309–19.

Langridge, J. 1982. Precambrian evolutionary genetics. In *Mineral deposits and the evolution of the biosphere*, H. Holland and M. Schidlowski (eds), pp. 83–101. Berlin: Dahlem Konferenzen.

Lewontin, R. C. 1974. *The genetic basis of evolutionary change*. New York: Columbia.

McDonald, J. F. 1983. The molecular basis of adaptation. A critical review of relevant ideas and observations. *Ann. Rev. Ecol. Syst.* 14, 77–102.

McGinnis, W., R. L. Garber, J. Wirz, A. Kuroiwa and W. J. Gehring, 1984. A homologous protein coding sequence in *Drosophila* homeotic genes and its conservation in other metazoans. *Cell* 37, 403–8.

Mora, P. T. 1965. The folly of probability. In *The origins of prebiological systems*, S. W. Fox (ed.), pp. 39–52. New York: Academic Press.

Morgan, T. H. 1911. An attempt to analyze the constitution of the chromosomes on the basis of sex-limited inheritance in *Drosophila*. *J. Exp. Zool.* 11, 365–412.

Morris, H. M. 1978. *The twilight of evolution*. Michigan: Baker.

North, G. 1984. How to make a fruitfly. *Nature* 311, 214–6.

Noüy, L. du 1947. *Human destiny*. New York: Mentor.

Perutz, M. F. 1983. Species adaptation in a protein molecule. *Mol. Biol. Evol.* 1, 1–28.

Ryman, N., F. W. Allendorf and G. Stahl 1979. Reproductive isolation with little genetic divergence in sympatric populations of brown trout (*Salmo trutta*). *Genetics* 92, 247–62.

Schindewolf, O. H. 1950. *Grundfragen der Paläontologie*. Stuttgart: Schweizerbartse.

Scott, J. E. 1975. Composition and structure of the pericellular environment. *Phil. Trans. R. Soc. Lond.* B271, 235–42.

Sharklee, J. B. and C. S. Tamaru 1977. Biochemical and morphological evidence of sibling species of bonefish, *Abula vulpes*. *Am. Zool.* 17, 973.

Simon, M. A. 1971. *The matter of life*. New Haven: Yale.

Spradling, A., B. Wakimoto, S. Parks, J. Levine, L. Kalfayan and D. de Cicco 1983. Genetic manipulation of *Drosophila* with transposable P elements. *International*

Conference on Genetic Manipulation, 179–86. Cambridge University Press: Cambridge.

Stebbins, G. L. and F. J. Ayala, 1981. Is a new evolutionary synthesis neccessary? *Science* 213, 967–71.

Vries, H. de 1901–1903. *Die Mutationstheorie*. Leipzig: Veit.

Whyte, L. 1965. *Internal factors in evolution*. London: Tavistock.

Wilson, A. C. 1976. Gene regulation in evolution. In *Biochemical evolution*, F. J. Ayala (ed.), pp. 225–34. Massachusetts: Sinauer.

Wilson, A. C., G. L. Bush, S. M. Case and M. C. King 1975. Social structuring of mammalian populations and rate of chromosomal evolution. *Proc. Natl Acad. Sci. USA* 72, 5061–5.

15

From genome to phenotype

GEORGE L. GABOR MIKLOS and
BERNARD JOHN

ABSTRACT

The classical genetic approach to the analysis of evolution was based on a mathematics that reduced the entire process to the study of changes in gene frequencies within populations. In effect, the developmental interactions responsible for producing evolutionary changes were ignored. In Neo-Darwinian theory, selection, coupled sometimes with neutral drift, was regarded as the only mechanism responsible for the spread and establishment of biological novelty. In treating the genome simply as a collection of individual genes, however, this theory fails to come to terms with the implications of the turnover of multigene families implicit in the concept of molecular drive. It is simply not adequate to graft such recent molecular findings onto conventional evolutionary theory. Thus two decades of analysing the polymorphisms that are evident in single genes has given no insight into the origin of morphological change. The signal failure of this approach has, in turn, led to recourse to 'explanations' in terms of 'regulatory' genes and their rôle in generating evolutionary change.

Molecular analyses of genome structure, by contrast, have highlighted the rapid flux that characterises the genomes of eukaryotes. In addition it has drawn attention to the fact that most of these genomes are choked with DNA sequences that represent nothing more than evolutionary debris. These sequences are the natural outcome of processes involving the replication, recombination, amplification, insertion, excision and conversion of the DNA molecules within a genome. Many of these sequences are mobile, and most of them make no significant contribution to developmental programmes and so do not impinge on phenotype.

Molecular analyses of the variation in multigene and repetitive DNA families, both within and between species, as well as the experimental addition of specific cloned DNA sequences to a genome, emphasise that many DNA sequences are naturally apostolic and missionary oriented since they continuously convert each other. These findings have provided the basis for the theory of molecular drive which has added a new dimension to the spread of variation through a population and has directed attention to the need to examine the developmental rôle of multigene families.

Molecular analyses are also now focusing attention on executive genes, as opposed to these which service and maintain universal metabolic back-up systems. One class of these executive genes, typified by the Bithorax complex of *Drosophila*, has been shown to have homologies in all major classes of vertebrates. The structure and mode of expression of such executive genes draws attention to the concept of developmental circuits and the ways in which alterations in circuitry lead to phenotypic changes of evolutionary significance. Many genes affecting early embryogenesis in *Drosophila* have also now been cloned, and the localisation of their transcripts and protein products has revealed unexpected distributions.

Finally, the genetic analyses of neural cell lineages in the roundworm *Caenorhabditis elegans*, the monoclonal antibody studies on neuronal pathfinding in the grasshopper, and the recombinant DNA analyses of RNA populations in the vertebrate brain, have revealed both unexpected simplicities and complexities of neuronal developmental lineages.

These developmental insights, stemming from reductionist molecular approaches to the genome, offer for the first time to provide an objective basis for defining and analysing the modes of morphological change that play a major rôle in evolution.

INTRODUCTION

I pass with relief from the tossing sea of cause and theory to the firm ground of result and fact. Winston Spencer Churchill

Recombinant DNA technologies and their spin-offs have revealed that eukaryote genomes are remarkably variable in DNA amounts, chromosomal architecture, gene number and distribution. Enormous variances in amounts of DNA can persist in the face of almost identical morphologies as in frogs, for example, whereas radically different morphologies may stem from what initially appears as a near constancy of genome size as in mammals.

We argue that most of the DNA of eukaryote genomes is irrelevant both to basic metabolism and developmental programmes and hence also to phenotype. It is effectively evolutionary flotsam and jetsam. Most organisms are simply unable to avoid the production and turnover of DNA sequences that have little or no influence on phenotype. Such sequences are by-products of the enzymic processes which replicate and reorganise DNA sequences (Ohno 1982).

That part of the genome that contributes significantly to developmental decisions, and hence to morphogenesis, consists of a small and exclusive set of genes (Garcia-Bellido *et al.* 1979). These genes are now rapidly being cloned in *Drosophila* and the first of these, Bithorax, has been shown to have homology with genes in amphibia, chicken, mice and men. Such homologies provide us with the first real evidence of any molecular commonality between the development of vertebrates and invertebrates.

We contrast this explosion of knowledge and the new evolutionary vistas it has revealed with the lack of meaningful progress that has been made under the largely mathematical aegis of Neo-Darwinism.

A GENOMIC OVERVIEW

DNA amounts

Examination of the total amount of nuclear DNA in fish, for example, reveals a staggering range which is certainly not paralleled by an equivalent range of morphological complexity (Table 15.1). When one compares related species within a single amphibian genus, the morphological differences between species are slight, yet in these there is a threefold difference in genome size (Table 15.1). On a grander scale, *Plethodontid* salamanders which again show little morphological differentiation, vary from 19 000 to 64 000 million base pairs (mbp). The chromosome numbers in both these examples are constant so that it has been wholesale amplification, deletion or movement of DNA sequences which has led to the present day disparity in genome sizes. Whatever the genomic rearrangements have involved, the excess DNA of the larger genomes has clearly made no contribution to morphological variation.

In contrast to these examples of morphological monotony in the face of cataclysmic genome reorganisation, mammals illustrate quite the reverse. Here a relatively constant genome size of approximately 3300 mbp, has

Table 15.1 Nuclear genome sizes in different vertebrates (from B. John & G. L. G. Miklos 1987), *n* denotes haploid chromosome number.

Vertebrate	Genome size (million base pairs)	*n*
Fish		
puffer (*Tetraodon fluviatilis*)	360	
salmon (*Oncorhynchus kisutch*)	2700	
shark (*Squalus acanthias*)	6 600	
lungfish (*Protopterus aethiopicus*)	130 000	
Frogs		
Limnodynastes ornatus	900	11
Limnodynastes terraereginae	1900	11
Limnodynastes dumerilii	3100	11
Plethodontid salamanders		
Plethodon c. cinereus	19 000	14
Plethodon vehiculum	34 000	14
Plethodon vandykei	64 000	14

been no impediment to the evolution of such morphological extremes as bats, whales and humans.

Transcriptive activity

The gross transcriptive activity of various vertebrate genomes is instructive in this regard. When the total RNA complexity is measured for different tissues or different developmental stages, it is effectively constant irrespective of genome size (Table 15.2). Thus, although the two amphibians *Xenopus laevis* and *Triturus cristatus* differ at least sevenfold in amounts of their nuclear DNA, their RNA complexities are quite comparable. Similarly, although the chicken genome is only a third that of a mammal, its somatic RNA complexities (excluding brain tissues) are equivalent. When egg RNA complexities are examined in both vertebrate and invertebrate genomes, whose sizes vary from almost 100 mbp to 10 000 mbp, they show only about twofold variation (Hough-Evans *et al.* 1980). These examples reveal that only a small proportion of the total genome is used in a transcriptive sense. The remainder is essentially debris from a phenotypic viewpoint.

Gross genomic organisation

Major karyotypic changes have often been considered as being of paramount evolutionary significance in terms of generating novel morphologies. Examination of related species, however, reveals that gross genomic organisation can be remarkably fluid in the absence of any meaningful phenotypic change (Table 15.3). In related species of muntjak deer the haploid genome can be organised as 3 or 23 chromosomes. Thus a genome that has been truly 'blitzed' in terms of rearrangements remains largely unperturbed in terms of its phenotypic expression. The zebra and the horse, which can hardly be described as anatomically or physiologically very different, have 16 and 32 chromosomes respectively. The muntjak parallel

Table 15.2 RNA sequence complexities in different vertebrates (from B. John & G. L. G. Miklos, 1987).

Species	Genome size (million base pair)	Tissue/stage	polyA$^+$ mRNA (million nucleotides)
Xenopus laevis	3 000	egg	35
Triturus cristatus	23 000		40
chicken	1 200	liver	20
		oviduct	30
		myofibril	32
mouse	3300	embryo	20
		kidney	20
		liver	21

Table 15.3 Genomic rearrangements and morphological similarities in related eukaryotes (from B. John & G. L. G. Miklos, 1987).

Animal	Haploid chromosome number
Deer	
Muntiacus muntjak	3
Muntiacus reevesi	23
Equids	
Equus zebra	16
Equus equus	32
Yeasts	
Schizosaccharomyces pombe	3
Saccharomyces cerevisiae	17
Ciliates	
Tetrahymena pyriformis	5
Oxytrichia similis	(thousands of gene-sized pieces 2.2 kb)

in fungi is provided by two related yeasts which have 3 and 17 chromosomes respectively. Finally, in ciliate protozoans, while *Tetrahymena* has five conventional chromosomes in its macronucleus, *Oxytrichia* has its macronuclear genome shattered into thousands of gene-sized pieces averaging 2200 bp.

The conclusions that flow from these and similar examples are that the genome can be mercilessly modified by gross karyotypic rearrangements and still produce equivalent phenotypes. Indeed, in no case has it been demonstrated that an evolutionarily significant morphological change has been the direct consequence of any chromosomal rearrangement, as opposed to a point mutation in a gene or its controlling sequences, or gene inactivation stemming from the mobile sequences that often abound in a genome. The requirement, then, is to differentiate between chromosomal rearrangements as causes of evolutionary novelty, as opposed to other genomic changes that have gone on in parallel.

GENOMIC DISSECTION BY RECOMBINANT DNA TECHNOLOGIES

Drosophila melanogaster

Although an invertebrate, this organism is proving *the* most useful starting point for the analysis of eukaryote genome organisation and developmental biology simply because it is the best known eukaryote in terms of genetic and molecular organisation. The information it provides offers an objective basis for an evaluation of similar data from vertebrates. The 165 million base pairs of haploid nuclear DNA in the fly consists roughly of, first, 50

mbp of highly repetitive sequences localised predominantly around cen-
tromeres (Brutlag 1980, Miklos 1985), secondly, 20 to 25 mbp of nomadic
or mobile DNA sequences together with some dispersed repetitive sequences
whose status as nomadic or sedentary has as yet not been finalised
(Spradling & Rubin 1981, Rubin 1983), and thirdly, 90 mbp of sequences
that are classically defined as unique or single-copy DNA.

The localised highly repetitive sequences are certainly dispensable from
both the metabolic and the developmental arena since large deficiencies and
duplications of this material produce individuals which, in a somatic sense,
are phenotypically indistinguishable from controls and whose early embryogenesis is likewise unperturbed (Miklos 1982). These sequences in both
vertebrates and invertebrates suffer almost every conceivable enzymic insult
from point mutation, amplifications, deletions, conversions, insertions and
rearrangements without causing overt phenotypic change. Furthermore.
some sequences are totally absent from one species, yet present in its closely
related relatives (Brutlag 1980, Singer 1982, Strachan *et al*. 1982, Lam &
Carroll 1983, Brown & Dover 1980, Maresca *et al*. 1984, Miklos 1985).
None of this is really surprising since these repetitive sequences are not
transcribed into RNA except by default when readthrough occurs from
adjacent promoters. They are thus oblique to discussions on phenotypic
change during evolution.

The nomadic sequences of *D. melanogaster* consist of about 50 families
which are structurally heterogeneous and are evolutionary transients (Rubin
1983) and some are intimately related to vertebrate retroviruses (Saigo *et
al*. 1984). The nomadics differ not only in frequency between different
strains of *D. melanogaster* but also between species of *Drosophila*. When
nomadic elements are isloated fom *D. melanogaster* and tested against the
sibling species *D. simulans* (Dowsett & Young 1982), it is evident that *D.
simulans* has only about one-eighth the frequency of *D. melanogaster*
nomadics and has not supplemented them in any substantial way with different families. Thus these sibling species differ by about 20 million base
pairs of mobile DNA and yet are morphologically nearly identical. These
mobile elements can act as internal mutagens since they insert and excise
from genic landscapes and cause predictable lesions. They can also mobilise
chromosomal segments and move them to other genomic landscapes
(McGinnis *et al*. 1984, Collins & Rubin 1984).

All eukaryotes contain families of nomadic, or formerly nomadic,
elements which are heterogeneous in structure and in family membership
(Singer 1982, Singer *et al*. 1983, Ullu & Tschudi 1984). The prominent
human nomadic family, termed the *Alu* family, exists in roughly 300 000
copies per genome and is found inserted in a variety of genomic locations,
genic and non-genic (Lee *et al*. 1984) as are some retroviral sequences
(Harbers *et al*. 1984). As in the case of highly repetitive DNAs, there is no
evidence that the middle repetitive fraction, which often consists
predominantly of mobile sequences, plays any significant rôle *per se* in
morphogenesis. The situation was appreciated by Doolittle and Sapienza,
as early as 1980 in their critique of the phenotype paradigm. They argued
convincingly that no phenotypic or evolutionary rôle was necessary for

middle repetitive sequences and that the obsessive search for functional explanations 'may prove, if not intellectually sterile, ultimately futile'.

DEVELOPMENT, GENES AND PHENOTYPE

We now turn our attention to the genic compartments of the genome and consider whether any subsection exerts a greater influence than any other in development. Morphological complexity stems ultimately from the correct developmental expression of integrated genic circuits. These genic cascades invariably mean that it is difficult to classify eukaryote genes as 'regulatory' or 'structural' because of the numerous interactions that occur between genes, their products and the intra- and intercellular transport of these products. When, for example, polymorphisms for structural genes have been studied and found to be near identical between chimpanzees and man, it has been a convention to resort to 'regulatory' genes as the prime movers in phenotypic differences. This just defers the issue by semantics. Unlike prokaryotes, where the distinction is relatively straightforward (Glass 1982), the concept of distinct 'regulatory' and 'structural' genes is too ill defined to be useful for eukaryotes. It is more pragamatic to classify the available genes in terms of their relative importance as developmental switches (Garcia-Bellido *et al.* 1979, Botas *et al.* 1982).

In terms of 'gene regulation' *per se*, current fashion in recombinant DNA studies focuses heavily on the upstream and downstream landscapes surrounding genes in an effort to elucidate the important sequences used in the control of gene transcription. The existence of various chromatin configurations, enhancers, hypersensitive sites (Elgin 1984, Fritton *et al.* 1984, North 1984b), differential processing at the RNA and protein levels (Tamkum *et al.* 1984) and mRNA stability (Raj & Pitha 1983) is relevant only to when, where, and how rapidly, certain proteins are made, modified and dispensed with. They tell us little about developmental regulation, much of which involves interaction between active sites or structural components of proteins and cell–cell interactions. Such interactions are only now being examined by the newer technologies. We focus on them because we believe that it is only by considering cellular phenomena that we will be able to decide which gene pathways are initially worthwhile pursuing in a developmental context.

Cellular interactions

At gastrulation, entire *sheets* of cells undergo movement resulting in the particular disposition of the germ layers which allows correct organogenesis to subsequently proceed. What gene programmes underpin these cell determinations and cell movements? If we had some inkling of the genetic pathways involved, we could rationalise how changes in these processes at the cell and tissue level lead to morphological variation. We could then estimate which parts of the pathways were more subject to genetic and environmental stress and hence discover what perturbations are likely to cause significant phenotypic changes during evolution.

Oster and Alberch (1982) provide a striking illustration of a developmental system, epithelial morphogenesis, which brings home the importance of cell–cell interactions and their effects on morphology. The sequence of events that culminates in the formation of feathers, scales, teeth and hair depends initially on cell shape changes involving the epidermis. Here a basal lamina separates an epidermal from a dermal cell layer. The epidermal layer thickens to yield epidermal placodes with cell shape changing from cuboidal to columnar, and cells within the dermal layer then condense beneath the placodes. A binary decision is now made; if *envagination* of the epidermal layer occurs, then either scales (reptiles) or feathers (birds) are formed; if *invagination* is the order of the day, hair or skin glands results (mammals). The critical bifurcation in this system is the inward or outward folding of the epithelial layer. Some relevant questions to be asked in dissecting such a system mechanistically are thus; how do we determine the biochemical events that underlie this process? Is it possible to isolate key genes or key proteins which, in the first instance, allow the determinative events to occur?

So shockingly rudimentary is our knowledge of the control of developmental processes that we are forced to start at any chink in the biological armour. Without any doubt, the most promising avenue is the molecular analysis of development in *Drosophila*. Here the application of recombinant DNA and monoclonal technologies, classical genetics, embryology and gene transfer techniques has already revealed that a number of developmentally significant executive genes have a common 61 amino-acid protein domain, termed the homoeo box, which is also found in most vertebrates. It appears that the most profitable approach is to continue to isolate, characterise and perturb *Drosophila* systems and then, by raiding vertebrate genomic libraries, to evaluate if the genes and developmental processes are similar or divergent.

Developmental genes in Drosophila

Despite 30 years of research, the molecular bases of pattern formation are still obscure. We still do not know how the position-dependent activity of molecules is laid down and interpreted. The situation is now rapidly changing owing to the isolation of three major classes of genes which control spatial organisation in the *Drosophila* embryo. There are: (1) maternal effect genes that control spatial co-ordinates; (2) segmentation genes that determine body segment number and polarity by interpreting the positional information supplied by the maternal effect genes; and (3) homoeotic genes that specify segment identity. Not only are these genes being cloned and characterised, but in the case of both the sequentation genes and the homoeotics their sites of transcription within the early embryo are being defined by the use of suitable molecular probes.

Nusslein-Volhard and Wieschaus (1980) undertook a saturation coverage of the genome to search for embryonic lethals which altered the segmentation pattern of the embryo (Fig. 15.1). They uncovered more than a dozen loci which were implicated in various stages of segmentation but

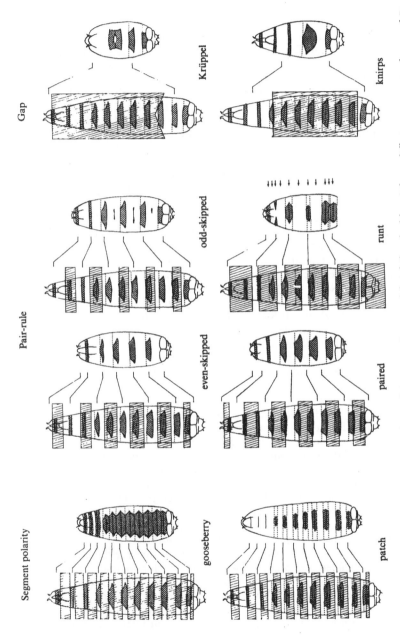

Figure 15.1 Diagrammatic representations of the areas that are deleted (hatched bars) form different mutant embryos of *D. melanogaster*. The underlying normal pattern is on the left-hand side in each case (from Nusslein-Volhard & Wieschaus 1980).

none of which affected the overall polarity of the embryo. There were genes that effected pattern duplication in each segment, pattern deletion in alternating segments and deletion of adjacent segments.

Other genes have been isolated in less complete searches. One gene, *fushi tarazu* (lack of enough segments), is of particular interest. In homozygous mutant *fushi tarazu* embryos alternate body segments are lacking and the embryos are lethal. The gene has been cloned and characterised and the distribution of its 1.9 kb poly A^+ RNA transcripts in the wild-type embryo determined (Kuroiwa *et al.* 1984, Hafen *et al.* 1984, Weiner *et al.* 1984). The RNA is expressed only from early blastoderm to the gastrula stage and is first detected during nuclear cleavage. At that time, the transcripts are equally localised between 15 and 65% of the egg length with zero being at the posterior pole. The transcripts then gradually become restricted so that by the blastoderm stage they form seven evenly spaced bands and give the embryo a zebra pattern which anticipates its eventual segmentation.

The currently most fashionable developmental genes are the homoeotics, which specify segment identity. These are genes which, when mutated effectively, transform one cell lineage into another. They are thus switch genes. The reason for the current excitement is that a number of them have been found to contain a conserved 61 amino-acid protein domain which has the molecular characteristics of a DNA binding region. These domains have now been found in developmentally regulated vertebrate genes and the accumulated genetic knowledge of *Drosophila* is now being tested there (Slack 1984, North 1984a).

The 61 amino-acid protein domain of the various *Drosophila* homoeotic genes (Scott & Weiner 1984) is more than 90% conserved between flies and humans (Levine *et al.* 1984), although most interestingly in humans it is embedded in a totally non-homologous sequence to that in *Drosophila*. It is also highly conserved between flies and mice (McGinnis *et al.* 1984) and between flies and amphibians (Carrasco *et al.* 1984, Muller *et al.* 1984). In *X. leavis* one homoeo-box containing gene is abundantly transcribed during oogenesis, whereas another is expressed at late gastrulation. All this lends credence to the hypothesis that these protein domains may represent some form of genetic regulatory element that may be involved in the early determination of embryonic cell types in vertebrates.

In *Drosophila*, a sequence from the Bithorax locus has been expressed as a fusion protein in bacteria and then used to raise monoclonal antibodies (White & Wilcox 1984). The antibodies reveal that the Bithorax protein is localised exclusively in nuclei, and it may well be a DNA binding protein whose specificity of binding regulates other genes. Alternatively, it may be far less spectacular and represent a homing device which instructs the protein to return to the nucleus (Raff & Raff 1985).

The general significance of all these results is that for the first time genes have been defined and analysed which are developmentally important and which have been isolated and characterised on the basis of impeccable developmental pedigrees. Whether other genes or domains which regulate these various maternal effects, segment identity, and homoeotic classes will also be found to be present in vertebrates, is an intriguing question.

Finally, the way in which developmental circuits have altered over evolutionary time should at last be amenable to analysis. The finding of such high degrees of amino acid conservation means that the 'active sites' of the homoeo-box domain are not overly prone to accepting alterations. Since the homoeo-box domain has been so well conserved between organisms such as yeast (which is not segmented) and man, and indeed has good homology to DNA binding proteins in prokaryotes, its evolutionary significance in morphogenetic terms is difficult to evaluate. Since these organisms differ considerably in phenotype, there must be modes of escaping from this strict conservatism, if indeed it is the homoeo-box domain which is so important. Alternatively, it may well be that other domains within the same protein in which the homoeo-box is embedded are of developmental significance. Whatever the outcome, the homoeo box has labelled some eukaryote genes which, in the amphibia at least, are expressed in oogenesis and gastrulation, two critical periods in development.

NEURAL CELL LINEAGES

The central nervous system has, for obvious reasons, been a much neglected component of evolutionary considerations. However, it is the very first system to be defined developmentally and, as Gans and Northcutt (1983) have pointed out, vertebrates have probably evolved from protochordates principally by expansion of the epidermal nerve plexus. Thus many of the sensory and integrative functions are derived embryologically from the neural crests and the epidermal placodes. In effect, the vertebrate head is an enlarged addition to a protochordate body. Moreoever, more than half the transcribed portion of a vertebrate genome is expressed in the brain so, with the majority of gene expression involved in its embryonic architecture and functioning, it is an appropriate place to examine genes and phenotype.

Conventionally, evolution is thought of in terms of new structures with new functions. Thus the changes that have given rise to the lungs of vertebrates, the limbs of tetrapods, the feathers of birds and the inner ear of mammals all enable the organism to perform a new function. The evolution of the nervous system operates on a quite different principle. Here, new functions arise by using the same neuronal components in different and more complex combinations. The basic building blocks of the nervous system, the neurons, have been available since the origin of multicellularity and it has been new and altered neuronal *connectivity* that has been largely responsible for evolutionary changes in the nervous system. It is all well and good to evolve sight, sound, taste, smell and touch and a neuromuscular system as well as mechanisms for monitoring and co-ordinating internal activities, but these systems need to be integrated in multimodal convergence to give rise to sensible behaviour, and such integration is a function of neurons. Since each human neuron synapses on average with 1000 others, the total connections approach 10^{15}. How are these connections laid down faithfully in any nervous system? What genes or gene families control these processes and how are they altered during evolution to yield novel behaviours?

Once again it will be easier to glean rudimentary principles from invertebrates, but should the reader find this a somewhat presumptuous exercise, we emphasise that nearly half of the monoclonal antibodies directed against *Drosophila* nervous tissue cross-hybridise to the human brain (Miller & Benzer 1983).

Studies using the simple roundworm *Caenorhabditis elegans* make it clear that the development of the nervous system operates through binary decision-making processes (Horvitz *et al*. 1983). Consider, for example, the fate of six ectoblast cells of the lateral ectoderm of this organism (Fig. 15.2). Five of them undergo a common developmental lineage giving rise to hypodermal cells of the cell surface. One cell, V5, produces both hypodermal and neural cells. This lineage also includes programmed cell death as well as producing a lateral sensory structure known as a postdeirid. This structure consists of a dopaminergic neuron, a non-dopaminergic neuron, a sheath cell and a socket cell.

The homoeotic mutation *lin-22* disrupts this lineage in a specific way (Horvitz *et al*. 1983). Cells V1, 2, 3 and 4 now express the exact characteristics of V5, so the organism finds itself with superfluous sensory structures and reciprocally lacks hypodermal cells (Fig. 15.2).

There is also a gene, *ced-3*, the mutant form of which rescues cells that normally are destined to undergo programmed cell death. In the V5 lineage described above, a non-dopaminergic neuron and cell death was the normal

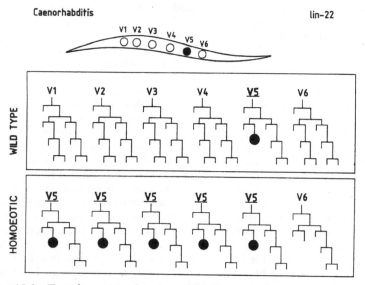

Figure 15.2   Transformation of the lateral ectoblasts of *Caenorhabditis elegans* V1 to V4 by the homoeotic mutation *lin-22* (from Horvitz *et al*. 1983).

outcome. In the presence of *ced-3*, however, death does not occur and instead a supernumerary dopaminergic neuron is generated.

Caenorhabditis is also in the vanguard for the analysis of heterochronic genes. Changes in the temporal pattern of gene expression, and their concomitant morphological novelties has always ranked high as prime movers of evolution. In this organism, two major classes of mutants can be summarised as follows: in *delayed* (or retarded) cases juvenile characteristics are expressed in the adult, whereas in *early* (or precocious) cases adult characteristics appear in immature individuals. Ambros and Horvitz (1984) have isolated a number of such mutants. In *lin-14*, *-28* and *-29*, for example, certain developmental events occur either earlier or later than normal. Particular *lin-14* alleles can lead to reiterations where moulting and larval cuticle events are repeated. In other alleles, a hypodermal blast cell does not produce its normal progeny during the L2 stage but generates a lineage produced a larval generation earlier by its grandparental precursor cell.

The important point to note from this simple organism is that genes can be identified which cause heterochronic changes. Their relevance in an evolutionary context will be discernable when they are cloned and their protein products analysed, and any cross homologies to other organisms evaluated.

Neural pathfinding genes and phenotype

When one examines the setting up of a nerve network, the problems, the genes and the final phenotype, are far from simple. Once again the initial inroads stem from molecular approaches in invertebrates since here the developing neurons consist of a segmental chain which arises from the metameric segmentation of the arthropod body. The studies of Thomas *et al.* (1984) and Goodman *et al.* (1984) reveal that in the grasshopper the early growth cones of nerve cells send out filopodia and use cell surface cues to build an initial scaffold along which later developing neurons find their way. Thus a developing G neuron, for example, makes a choice when it encounters the A/P nerve bundle. Which of the four neurons within this bundle will it follow? It makes its choice on the basis of cell-surface molecules on axons and their selective adhesion properties.

Rather surprisingly, when other arthropods are examined, such as a crayfish (*Procambarus*) and *Drosophila*, astounding similarities are seen in the location of identified neurons, in the location of cell bodies, in the axonal scaffolds and in the bifurcations that are made by the respective growth cones. For example, the G neuron in *Drosophila* has near it at least 25 nerve bundles and yet it invariably selects the A/P bundle and within it the P, but not the A, axons, just as does the grasshopper.

Thus, in examining the descriptive biology of how the embryonic nervous system develops, a remarkable early conservatism is seen in arthropods as different as crayfish, grasshopper and *Drosophila*. The important future steps will be first, to describe the network and discover the embryonic rules to see how they may be perturbed; secondly, to determine what causes different neurons to express specific surface labels; thirdly, to determine how growth cones are guided to their final destinations on the basis of selective substrate affinities and fourthly, to determine the molecular basis of cell recognition during neuronal development.

In order to answer some of these questions, one needs to isolate the molecules involved or the genes that code for them. The use of monoclonal antibodies has already revealed a specificity that is not apparent on morphological inspection. In the grasshopper, the monoclonal antibody *Mes-2* recognises an antigen that is present on only four neurons out of 1000 in a hemisegment. *Mes-3* and *Mes-4*, on the other hand, distinguish the MP1–dMP2 nerve bundle from the other 25 longitudinal fascicles that occur at this developmental stage (Goodman *et al.* 1984).

These findings mean that we are on the threshold of unravelling simple circuits in terms of cell-surface molecules. What is still unknown, is whether these phenomena are controlled by single genes or multigene families and whether there are many different proteins, or whether the proteins are few but their modification, by glycosylation, leads to a diverse array of cell-surface types (Edelman 1983).

Since the wiring diagram of the central nervous systems of both vertebrates and invertebrates is so poorly described, it is difficult to take the alternative genetic approach of mutagenising the genome and looking for specific connectivity dislocations in anything but giant fibres or anatomically distinct cell types. This approach, however, has also begun in *Drosophila* (Wyman & Thomas 1983).

In the vertebrate brain the molecular situation is almost palaeoneurobiological. Even so, the surprises are already with us. One will recall that, relative to other tissues, the poly A^+ RNA complexity of the vertebrate brain is enormous, near 120 million nucleotides, whereas that of liver and kidney, for example, is of the order of 30 million nucleotides. Milner and Sutcliffe (1983) and Sutcliffe *et al.* (1983) have isolated polyA$^+$ RNA from adult rat brain, converted it to cDNA and have found that brain specific RNAs are larger and rarer than most metabolic transcripts. They estimate that there are at least 30 000 polyA$^+$ RNAs in the brain, most of which are brain specific. This is a larger number of transcripts than all other somatic tissues combined. Furthermore, brain-specific messages contain an 82 bp core sequence, termed an identifier (*ID*) (Milner *et al.* 1984). There are about 100 000 of these sequences per genome and they are generally found in the introns of brain genes where they have been hypothesised to be important in brain-specific gene control.

Sutcliffe *et al.* (1983) also determined the nucleotide sequence of one of their clones (IB236), identified the sense strand and its long open reading frame and translated it into the amino acid sequence. They then chemically synthesised short peptides, raised antisera against such molecules, and then probed thin brain sections in immunocytochemical reactions. These experiments revealed a neuronal specificity which was not apparent from neuronal morphology alone. Clone IB236 is synthesised in certain cell bodies and transported via certain axons to particular areas of the cerebellum, hippocampus and cortex.

These experiments, although only in the pioneering stage, lift the lid off the problems, complexities and evolutionary hurdles that need to be faced in the central nervous systems of vertebrates and invertebrates. Some of the structural components will clearly be held in common, as we saw from the *Drosophila* monoclonal cross-hybridisation to human brain, but others will clearly be distinctive. Furthermore, if brain-specific transcripts, or the protein products, really are larger than conventional metabolic transcripts, then the developmental pathways so involved may have a greater complexity. It is already known that several neuropeptide precursors such as propiomelanocortin are processed to yield several proteins.

All these results, both anatomical and molecular, are still too patchy to give us a firm idea of the underlying principles involved in setting up the architecture of the brain. However, the pioneering approaches we have illustrated show that the gap between genes and phenotype is rapidly narrowing.

GENES, PHENOTYPE AND EVOLUTIONARY STUDIES

When one examines the controversies that still rage in population genetics, namely selectionists versus neutralists, gradualists versus punctualists, and regulatory versus structural genes as prime movers of evolutionary novelty, one is struck by how far removed they are from the real problems of evolution. In the case of the structural versus regulatory gene argument, it has quite correctly been pointed out by MacIntyre (1982) that no single compelling observation was responsible for this spotlighting of the putatively significant evolutionary rôle of regulatory genes. In fact '. . . in the absence of detailed molecular and developmental studies, any interpretation of interspecific differences involving gene regulation are strictly hypothetical' (Charlesworth *et al.* 1982).

We turn finally from the reductionist molecular approach, which is now pouring out information on genomic architecture, and ask how much of it impinges on evolutionary studies. Here, much of the data are being used to examine rates of change in nucleotide and protein sequences, and to refine phylogenies by the estimations of evolutionary distances. Very little of it addresses itself to the underlying mechanisms of phenotypic change that have given rise to morphological novelties. Molecular evolution is currently still in a DNA data collecting phase, and does not address itself to what many consider to be the main tasks of population genetics, namely '. . . the description and prediction of changes in the genetic and phenotypic composition of populations responding to the forces of mutation, recombination, selection, migration and systems of mating' (Charlesworth *et al.* 1982).

Multigene families and molecular drive

In general, the mechanisms by which new genes and new phenotypes arise is largely based on speculation. Multigene families (Hunkapiller *et al.* 1983) were, until recently, not an integral part of modern theories, and as yet they still sit uneasily with Neo-Darwinists.

The concept of molecular drive as developed by Dover (1982), and its implication for evolutionary change as well as phenotypic novelty (Dover 1984, Dover & Flavell 1984) has exposed the fragility of some aspects of Neo-Darwinism. This concept is based on a feature of many multigene and non-coding families (ribosomal DNA, globins, immunoglobulins, histones and repetitious 'satellite' sequences) in which a continuous homogenisation process produces a greater within-species homogeneity than between-species homogeneity.

In utilising turnover in multigene families, drive focuses on the spread and establishment of novelties within populations by internal processes of flux such as unequal exchange, transposition and gene conversion. This literally leads to the transformation of a population under its own steam without the involvement of any conventional selective forces. The example of the bat-eared fox used by Dover (1984), illustrates the situation, though it could be applied to any presumed adaptation.

Bat-eared foxes have abnormally large bat-like ears and the conventional 'explanation' for such extremities is that they have evolved under the aegis of natural selection for increased efficiency in tracking prey. They would conventionally be considered as evidence of an adaptation that increases the survival of the animals possessing them, as most adaptations are so interpreted (Lewontin 1978). There are alternative explanations, however, particularly if multigene families are involved, and we consider them below.

What Dover's theory encapsulates is that once the nexus is broken between chromosome behaviour and gene behaviour, as occurs by internal turnover mechanisms in the genome, then there can be unexpected transformations of a population with the changes occurring in a cohesive manner. Different individuals in a population undergoing molecular drive will be expected to show little variance between members of the particular multigene family in question. Individual members of a multigene family are not constrained by a chromosome, since chromosomes no longer provide refugia from the genomic turnover processes, which can be both inter- and intrachromosomal. Thus when the rate of spread of a variant between chromosomes differs from the rate of gametic mixing, a multigene family can be homogenised in a synchronous fashion in all individuals. This means that a *population* is slowly changing from one DNA composition to another *without* a large variance between individuals. Thus *if* the bat-like ears of the fox are largely determined by one or a number of multigene families (and this is as yet unknown), then the change to large ears may well be a consequence of genomic turnover processes. This increase in ear size affects *all* members of the population because it is a *cohesive* molecular change. The change in phenotype can be gradual or sudden, and there are no biological incompatibilities between individuals.

This last point circumvents the problem of how a *single* mutant individual arising in a population can lead to subsequent fixation of a new phenotype. Under molecular drive, the DNA of all individuals can be converted slowly and synchronously. Thus the change of ear size in foxes could simply be a fortuitous by-product of changes in multigene families that impose no real problems of differential survival. Nor do such changes need to be adaptive.

We see from examples such as these that much of the fabric of population genetics theory is under pressure to accommodate the molecular data bases. Furthermore, it is particularly striking that the field not only shows strong resistance to any challenge of orthodox beliefs, but is itself preoccupied with what Levin (1984) has referred to as the 'two camp syndrome':

> Some hypotheses are more often championed than tested. Positions (almost two for any given issue) are fiercely defended even when they are not mutually exclusive. New ideas are liable to attack for no reasons other than their real, or even apparent, violation of orthodoxy.

It is clear to us that this preoccupation with Darwinian orthodoxy enshrouds the field to little benefit. Dover's concept of molecular drive is a bold new incursion into the orthodox realm. It is based on *known* mechanisms of genomic turnover processes, whereas selection has always been, and still is, more a logical construct than a well-documented process. Neo-Darwinism as an 'explanation' for evolutionary change is in fact in need of a serious new outlook. It needs to focus back to the developmental gene circuits that are critical for phenotypic change and hence for evolutionary change.

REFERENCES

Ambros, V. and H. R. Horvitz 1984. Heterochronic mutants of the nematode *Caenorhabditis elegans*. *Science* **226**, 409–16.

Botas, J., J. M. Del Prado and A. Garcia-Bellido 1982. Gene-dose titration analysis in the search of transregulatory genes in *Drosophila*. *EMBO J.* 1, 307–10.

Brown, S. D. M. and G. A. Dover 1980. Conservation of segmental variants of satellite DNA of *Mus musculus* in a related species: *Mus spretus*. *Nature* 285, 47–9.

Brutlag, D. L. 1980. Molecular arrangement and evolution of heterochromatic DNA. *Ann. Rev. Genet.* 14, 121–44.

Carrasco, A. E., W. McGinnis, W. J. Gehring and E. M. De Robertis 1984. Cloning of an *X. laevis* gene expressed during early embryogenesis coding for a peptide region homologous to *Drosophila* homeotic genes. *Cell* 37, 409–14.

Charlesworth, B., R. Lande and M. Slatkin 1982. A Neo-Darwinian commentary on macroevolution. *Evolution* 36, 474–98.

Collins, M. and G. M. Rubin 1984. Structure of chromosomal rearrangements induced by the FB transposable element in *Drosophila*. *Nature* 308, 323–7.

Doolittle, W. F. and C. Sapienza 1980. Selfish genes, the phenotype paradigm and genome evolution. *Nature* 284, 601–3.

Dover, G. 1982. Molecular drive: a cohesive mode of species evolution. *Nature* 299, 111–17.

Dover, G. A. 1984. Forces of evolution. *Kos* 5, 129–36.

Dover, G. A. and R. B. Flavell 1984. Molecular co-evolution: rDNA divergence and the maintenance of function. *Cell* **38**, 622–3.

Dowsett, A. P. and M. W. Young 1982. Differing levels of dispersed repetitive DNA among closely related species of *Drosophila*. *Proc. Natl Acad Sci. USA* **79**, 4570–4.

Edelman, G. M. 1983. Cell adhesion molecules. *Science* **219**, 450–7.

Elgin, S. C. R. 1984. Anatomy of hypersensitive sites. *Nature* **309**, 213–4.

Fritton, H. P., T. Igo-Kemenes, J. Nowock, U. Strech-Jurk, M. Theisen and A. E. Sippel 1984. Alternative sets of DNase 1 – hypersensitive sites characterize the various functional states of the chicken lysozyme gene. *Nature* **311**, 163–5.

Gans, C. and R. G. Northcutt 1983. Neural crest and the origin of vertebrates: a new head. *Science* **220**, 268–74.

Garcia-Bellido, A., P. A. Lawrence and G. Morata 1979. Compartments in animal development. *Scient. Am.* **241**, 102–10.

Glass, R. E. 1982. *Gene function*. London: Croom Helm.

Goodman, C. S., M. J. Bastiani, C. Q. Doe, S. Du Lac, S. L. Helfand, J. Y. Kuwada and J. B. Thomas 1984. Cell recognition during neuronal development. *Science* **225**, 1271–9.

Hafen, E., A. Kuroiwa and W. J. Gehring 1984. Spatial distribution of transcripts from the segmentation gene *fushi tarazu* during *Drosophila* embryonic development. *Cell* **37**, 833–41.

Harbers, K., M. Kuehn, H. Delius and R. Jaenisch 1984. Insertion of retrovirus into the first intron of 1(1) collagen gene leads to embryonic lethal mutation in mice. *Proc. Natl Acad. Sci. USA* **81**, 1504–8.

Horvitz, H. R., P. W. Sternberg, I. S. Greenwald, W. Fixsen and H. M. Ellis 1983. Mutations that affect neural cell lineages and cell fates during the development of the nematode *Caenorhabditis elegans*. *Cold Spring Harbour Symp. Quant. Biol.* **48**, 453–63.

Hough-Evans, B. R., M. Jacobs-Lorena, M. R. Cummings, R. J. Britten and E. H. Davidson 1980. Complexity of RNA in eggs of *Drosophila melanogaster* and *Musca domestica*. *Genetics* **95**, 81–94.

Hunkapiller, T., H. Haung, L. Hood and J. H. Cambell 1983. The impact of modern genetics on evolutionary theory. In *Perspectives on evolution*, R. Milkman (ed.), pp. 164–89. Sunderland Mass.: Sinauer Associates.

John, B. and G. L. G. Miklos 1987. The eukaryote genome in development and evolution. George Allen and Unwin, London.

Kuroiwa, A., E. Hafen and W. J. Gehring 1984. Cloning and transcriptional analysis of the segmentation gene *Fushi tarazu* of *Drosophila*. *Cell* **37**, 825–31.

Lam, B. S. and D. Carroll 1983. Tandemly repeated DNA sequences from *Xenopus laevis*. I. Studies on sequence organization and variation in satellite I DNA (741 base pair repeat). *J. Mol. Biol.* **165**, 567–85.

Lee, M. G., C. Loomis and N. J. Cowan 1984. Sequence of an expressed human β-tubulin gene containing ten *Alu* family members. *Nucl. Acids Res.* **12**, 5823–36.

Levin, B. R. 1984. Science as a way of knowing – molecular evolution. *Am. Zool.* **24**, 451–64.

Levine, M., G. M. Rubin and R. Tjian 1984. Human DNA sequences homologous to a protein coding region conserved between homeotic genes of *Drosophila*. *Cell* **38**, 667–73.

Lewontin, R. C. 1978. Adaptation. *Sci. Am.* **239**, 213–30.

MacIntyre, R. J. 1982. Regulatory genes and adaptation. In *Evolutionary biology*, M. K. Heckt, B. Wallace and G. T. Prance (eds), Vol. 15, pp. 247–85. New York: Plenum Press.

Maresca, A., M. F. Singer and T. N. H. Lee, 1984. Continuous reorganization leads to extensive polymorphism in a monkey centromeric satellite. *J. Mol. Biol.* **179**, 629–49.

McGinnis, W., A. W. Shermoen and S. K. Beckendorf 1983. A transposable element inserted just S' to a *Drosophila* glue protein gene alters gene expression and chromatin structure. *Cell* **34**, 75–84.

McGinnis, W., R. L. Barber, J. Wirz, A. Kuroiwa and W. J. Gehring 1984. A homologous protein-coding sequence in *Drosophila* homeotic genes and its conservation in other metazoans. *Cell* **37**, 403–8.

Miklos, G. L. G. 1982. Sequencing and manipulating highly repeated DNA. In *Genome evolution*, G. A. Dover and R. B. Flavell (eds), pp. 41–68. New York: Academic Press.

Miklos, G. L. G. 1985. Localized highly repetitive DNA sequences in vertebrate and invertebrate genomes. In *Evolutionary biology*, R. J. MacIntyre (ed.), New York: Plenum Press.

Miller, C. A. and S. Benzer 1983. Monoclonal antibody cross-reactions between *Drosophila* and human brain (neural antigens/species homology/molecular anatomy/immunohistology). *Proc. Natl Acad. Sci. USA* **80**, 7375–702.

Milner, R. J. and J. G. Sutcliffe 1983. Gene expression in rat brain. *Nucl. Acids Res.* **11**, 5497–520.

Milner, R. J., F. E. Bloom, C. Lai, R. A. Lerner and J. G. Sutcliffe 1984. Brain-specific genes have identifier sequences in their introns. *Proc. Natl Acad. Sci. USA* **81**, 713–17.

Muller, M. M., A. E. Carrasco and E. M. De Robertis 1984. A homeo-box containing gene expressed during oogenesis in *Xenopus*. *Cell* **39**, 157–62.

North, G. 1984a. How to make a fruitfly. *Nature* **311**, 214–16.

North G. 1984b. Multiple levels of gene control in eukaryotic cells. *Nature* **312**, 308–9.

Nusslein-Volhard, C. and E. Wieschaus 1980. Mutations affecting segment number and polarity in *Drosophila*. *Nature* **287**, 795–801.

Ohno, S. 1982. The common ancestry of genes and spacers in the euchromatic region. *Omnis Ordinis Hereditarium a Ordinis Priscum Minutum*. *Cytogenet. Cell Genet.* **34**, 102–11.

Oster, G. and P. Alberch 1982. Evolution and bifurcation of developmental programs. *Evolution* **36**, 444–59.

Raff, E. C. and R. A. Raff 1985. Possible functions of the homeo box. *Nature*, **313**, 185.

Raj, N. B. K. and P. M. Pitha 1983. Two levels of regulation of β-interferon gene expression in human cells. *Proc. Natl Acad. Sci. USA* **80**, 3923–7.

Rubin, G. M. 1983. Dispersed repetitive DNAs in *Drosophila*. In *Mobile genetic elements*, J. A. Shapiro, (ed.), pp. 329–61. New York: Academic Press.

Saigo, K., W. Kugimiya, Y. Matsuo, S. Inouye, K. Yoshioka and S. Yuki 1984. Identification of the coding sequence for a reverse transcriptas-like enzyme in a transposable genetic element in *Drosophila melanogaster*. *Nature*, **312**, 659–61.

Scott, M. P. and A. J. Weiner 1984. Structural relationships among genes that control development: sequence homology between the Antennapedia, ultrabithorax and *fushi tarazu* loci of *Drosophila*. *Proc. Natl Acad. Sci. USA* **81**, 4115–9.

Singer, M. F. 1982. Highly repeated sequences in mammalian genomes. *Int. Rev. Cytol.* 76, 67–112.

Singer, M. F., R. E. Thayer, G. Grimaldi, M. I. Lerman and T. G. Fanning 1983. Homology between the *Kpnl* primate and *Bam* H1 (MIF-1) rodent families of long interspersed repeated sequences. *Nucl Acids Res.* 11, 5739–45.

Slack, J. 1984. A Rosetta stone for pattern formation in animals? *Nature* 310, 364–5.

Spradling, A. C. and G. M. Rubin 1981. *Drosophila* genome organization: conserved and dynamic aspects. *Ann. Rev. Genet.* 15, 219–64.

Strachan, T., E. Coen, D. Webb and G. Dover 1982. Modes and rates of change of complex DNA families of *Drosophila*. *J. Mol. Biol.* 158, 37–54.

Sutcliffe, J. G., R. J. Milner and F. E. Bloom 1983. Cellular localization and function of the proteins encoded by brain-specific mRNAs. *Cold Spring Harbour Symp. Quant. Biol.* 48, 477–84.

Tamkun, J. W., J. E. Schwarzbauer and R. O. Hynes 1984. A single rat fibronectin gene generates three different mRNAs by alternative splicing of a complex exon. *Proc. Natl Acad. Sci. USA* 81, 5140–4.

Thomas, J. B., M. J. Bastiani, M. Bate and C. S. Goodman, 1984. From grasshopper to *Drosophila*: a common plan for neuronal development. *Nature* 312, 203–7.

Ullu, E. and C. Tschudi 1984. *Alu* sequences are processed 7SL RNA genes. *Nature* 312, 171–2.

Weiner, A. J., M. P. Scott and T. C. Kaufman 1984. A molecular analysis of *fushi tarazu*, a gene in *Drosophila melanogaster* that encodes a product affecting embryonic segment number and cell fate. *Cell* 37, 843–51.

White, R. A. H. and M. Wilcox 1984. Protein products of the Bithorax complex in *Drosophila*. *Cell* 39, 163–71.

Wyman, R. J. and J. B. Thomas 1983. What genes are necessary to make an identified synapse? *Cold Spring Harbour Symp. Quant. Biol.* 48, 641–52.

16

The new gene and its evolution

JOHN H. CAMPBELL

ABSTRACT

Discoveries of the elaborate structure of genes raise evolutionary questions that lie outside the scope of traditional theory. Darwinism rationalises how structures evolve, but does not address the more significant issue of the rôles that specialised gene structures play in the process of evolution itself. I suggest that molecular genetics can be integrated into evolutionary theory only through the paradigm that underlies all biological explanation; the structure–function principle. Organisms carry out evolutionary processes as biological functions. To do so, species evolve special structures with *evolutionary functions* just as they evolve other phenotypic structures to carry out various adaptive functions. I shall discuss the evolutionary functions of a variety of structures, ways that these functional capacities evolve, and some implications of evolution being a process that organisms become specialised to carry out instead of just change imposed on the species from the external environment.

INTRODUCTION

Heredity lies at the heart of all theories of evolution. Indeed, evolution is the time dimension of genetics. Darwin's greatest handicap in explaining evolution was a complete ignorance of genetics. This forced him to accept two interwoven causal processes in evolutionary change. Natural selection drove evolution by survival of the fittest but an internal hereditary process generated the individual variability on which selection acts. Darwin was unable to evaluate the relative importance of these two agencies. As time went on, he admitted an increasing rôle for heredity in directing the process of evolution through some sort of Lamarckian mechanism.

The discovery of Mendelian genetics rescued natural selection. Recombination of particulate genes, and ultimately blind mutation, produced continuing variation. Since both processes could be completely random and spontaneous, natural selection no longer had to be diluted with organismal hereditary activities. The stark objectivity of the new Darwinism allowed it to overwhelm its vitalistically tainted competitors. Its causal simplicity also bedazzled the 'Neo-Darwinist' into believing that the simplest possible way that evolution could take place must also be the way that evolution actually did proceed.

This misconception might have been exposed by the growing doubts from the fossil record as to whether Neo-Darwinism addresses the key features of the actual history of life on the planet. In fact, it was insight into heredity that again proved pivotal. The revolution of molecular biology has made it possible to examine directly the structure and physiology of genes. Genetic determinants turn out to be very different from the ineffectual abstractions upon which Neo-Darwinism was based. We still cannot foresee their full rôle in evolution, but it is probably substantial. I suggest that evolution is primarily an internal genetic process. Natural selection, as we know it, occurs and is essential, but it is the junior partner to heredity. Not only the mechanisms but also the goals of evolution are defined by the genetic message. To describe this interpretation I will first sketch the main characteristics of genes as functional structures, then describe the observed evolutionary dynamics of genes, and finally suggest that the phenotype evolves through these same categories of changes.

THE NATURE OF THE GENE

The study of genes as *molecules* has given us the overwhelming flood of information described in preceding chapters of this volume. It has shown genes to have four fundamental characteristics that could not be appreciated earlier.

Genes are units of molecular structure

The modern geneticist extracts genes as molecules from cells, looks at them with the electron microscope, measures their physical properties, such as length, and describes their chemical composition. This is entirely different from the approach that prevailed when Neo-Darwinism was being formulated. T. H. Morgan observed in his 1933 Nobel lecture that: 'There is no consensus of opinion amongst geneticists as to what the genes are – whether they are real or purely fictitious – because at the level at which the genetic experiments lie it does not make the slightest difference whether the gene is a hypothetical unit, of whether the gene is a material particle' (Morgan 1933). Since then, the main concern of genetics has shifted from inheritance of genes to their structure. This approach is fruitful because genes are surprisingly elaborate in structure and this structure is the basis for their function.

Genes are highly individual

Until this decade, the quest of genetics was for universal truths about inheritance – Mendel's laws, Fisher's fundamental law of evolution, Crick's central dogma. This made the simplest genes the ones of choice for study. The resulting lowest common denominator orientation reduced genes to no more than runs of codons in DNA molecules, distinctive only in their sequences. Such forms of genes do exist, and probably are abundant, but they seem to be used only for elementary coding functions. Typically, genes with complex information and active evolutionary dynamics are elaborate, and their significant capacities stem directly from their individualistic structure. This realisation has a profound evolutionary implication. To understand how important phenotypic traits evolve we must describe the actual dynamics of the complex genes involved, and not merely the stereotyped behaviour of simple genes.

For example, it once seemed all but impossible that the mammalian genome could code for hundreds of millions of different antibody molecules that encompass every possible foreign antigen. Yet vertebrates are able to do so and even make antibodies against artificial molecules to which their species had never been exposed in its entire evolutionary history. Both the capacity to code for the immune response and the ability to evolve this information are made possible only by the very special and unique structure of the immune genes (Hood *et al.* 1975).

Even lowly bacteria rely on highly distinctive gene structures. Although the run of the mill genes in bacteria have a stable configuration on the chromosome, those genes that are actively evolving tend to develop into configurations such as transposons. These structures can move around in the genome and can even hitch-hike from one cell to another by associating with other specialised DNA structures called transmissible plasmids. By exchanging transmissible genes among other species in the ecosystem instead of relying on mutations, bacteria are able to evolve far more rapidly and effectively than would be possible through simple mutation and selection.

It took decades for bacteriologists finally to accept the empirical data that bacteria evolve resistances to antibiotics far too fast for a process of mutation and selection (Davies 1971) and it has taken evolutionists even longer to appreciate how micro-organisms actually do evolve. As late as 1970, a treatise on evolution still insisted that 'Microorganisms . . . have the capacity to develop strains resistant to antibiotics and other drugs. This resistance results from the selection of a few resistant mutations of gene combinations exactly as in higher organisms' (Mayr 1970). With the growth of modern genetics we now know that this is a secondary mechanism for bacterial adaptation. Adaptive evolution in bacteria does not centre on mutation and selection. It is foremost a process of restructuring genes into transposable and transmissible forms and then moving them about the cell and the ecosystem (Anderson 1965, Campbell 1981). Thus, for example, the important information for evaluating the potential of a bacterial pathogen to evolve resistance to an antibiotic is not the mutation rate or the frequency of resistant individuals, but the *structure* of the genes that provide that characteristic.

Genes are working molecules of the cell

The third cornerstone of modern genetics is what I call the 'profane' side of the DNA molecule (Campbell 1985). Classical genetics was predicated on the genotype having a sacred status. Genes were inviolable messages from ancient ancestors passed on faithfully from generation to generation, except for rare mutational corruptions. They were transcendental to the activities and metabolism of the organism carrying them. We now realise that this simply is not true. Instead cells actively manipulate the structure of their DNA molecules for both physiological and evolutionary reasons. They engineer changes in their genes with an impressive array of 'gene-processing enzymes' (Campbell 1983). Table 16.1 lists some examples. Cells seem to have enzymes to catalyse almost every imaginable sort of change in the structure of the DNA molecule. Topoisomerase unties knots in DNA strands (Gellert 1981), recA protein inserts single strands of DNA into homologous double-stranded helices (Cox & Lehman 1981), acquisitionase grabs bits of DNA and intercalates them into chromosomes (Miller *et al.* 1981) and so on. These enzymes for manipulating DNA molecules are not only diverse but also numerous. For example, there are thought to be at least 20 different sorts of DNA glycosylases (Alberts *et al.* 1983).

Complex genes, such as multigene families, owe much of their distinctiveness to special enzymic pathways that manipulate their DNA sequences (Campbell 1983). For example, the antibody and T-cell receptor multigene families have highly specialised enzymic pathways to rearrange their DNA in particular somatic cells and the duplicated genes in the haemoglobin families are enzymically 'corrected' one against another in germ line cells during evolution (Slightom *et al.* 1980, Liebhaber *et al.* 1981, 1984).

The ultimate authority over the information content of its DNA would be for the cell actually to write gene segments from scratch, and to edit the result in order to achieve a useful product. Remarkably, even this capacity

Table 16.1   Some gene-processing enzymes.

DNA polymerase	invertase
DNA ligase	translocase
DNA gyrase	transposase
DNA topoisomerase	acquisitionase
DNA methylase	integrase
glycosylase	insertase
back transcriptase	excisase
restriction nuclease	recombinase
helicase	spligase
primase	replicase
	mutase

is realised! Mammals are able to polymerise *de novo* sequences of nucleotides at crucial interior sites in the genes for their antibodies (Siu *et al*. 1984, Kavaler *et al*. 1984). Those physiologically written segments code for particular portions of antibody molecules that recognise their antigens. The creations that are useful are selected from others by an impressively complex multicellular system. A decade ago, geneticists would have considered this capacity unimportant for evolution because it is confined to a small number of highly specialised genes. It is not a general property of genes. However, we now realise that every important gene is probably exceptional in one way or another, and that it is these unique genetic mechanisms that have allowed the important accomplishments of life to evolve.

The full contribution of gene-processing enzymes to evolution remains uncertain. One obvious rôle is to produce mutants. Probably the majority of mutations important in evolution result from the activities of enzymes. This is significant because enzymes are deliberate and precise, the antithesis of the supposed 'randomness' of mutation. Indeed, enzymically induced changes in the structure of multigene families often are referred to as gene 'switching', 'correction', 'rectification', 'excision', and so on, rather than 'mutation'. I also find it suggestive that geneticists have chosen names for gene-processing enzymes that could be used almost as well for the keys on a word processor.

Genes are aware of their surroundings

Genes require information about the organism and its environment in order to express their message appropriately. To meet this demand, complex bacterial genes (operons) have evolved special sensory proteins called 'repressors', which inform them of particular relevant conditions (Jacob & Monod 1961). A typical repressor has two binding sites, one specific for some meaningful metabolite in the cell and a second for a recognition site in the operon. The two binding capacities are mutually exclusive, so that the repressor tells the operon about the concentration of the metabolite by binding to the gene's DNA. Sensor proteins of bacteria are versatile. Some activate rather than repress genes, and some inform genes about hormonal signals generated by the bacterium, or about the status of other genes in the genome (Lewin 1974). Also, some operons receive information from several sensory systems simultaneously.

Eukaryotic genes are more complex and less well understood than those of bacteria. They are tucked away in a membrane-bound nucleus, and germ line cells are isolated in gonads from the soma. Despite these specialisations, receptor proteins still carry hormone molecules directly to sites on the chromosome within the nucleus (Yamamoto & Alberts 1976, Payvar *et al*. 1981). Even germinal cells of vertebrates produce receptors on their cell surfaces for a variety of hormones, growth factors, neuromodulators and so forth (Campbell & Zimmermann 1982). Higher organisms presumably have a far greater potential than bacteria for informing their genes but we still know almost nothing of its evolutionary significance.

TYPES OF EVOLUTIONARY CHANGE IN GENES

In addition to revising our ideas about the fundamental nature of the gene, molecular geneticists also have discovered the types of change that occur in genes during their evolution. Genes undergo four distinct categories of change as listed in Table 16.2. Each of these has been chosen as a definition for 'evolution' by one school or another.

Table 16.2   Dynamic processes of genes.

(1)   fluctuation in frequencies
(2)   change in structure
(3)   adaptation
(4)   advancement

Fluctuations in gene frequencies

Fifty years ago it was known that genes were located on chromosomes and could mutate to alternative states. Their structure was otherwise mysterious. The only information about genes that could be interpreted (other than map positions) concerned the mathematical *frequencies* of their detectable allelic forms in populations. The obvious basis for a genetic theory of evolution was change in this measurable quantity. Neo-Darwinists therefore defined evolution as simply a change in frequencies of gene alleles in a population from one generation to the next (Mayr 1963). Such quantitative change is fully reversible for a gene, and the frequencies of individual alleles can shift back and forth to adapt organisms in the population to their changing environment. New alleles occasionally arise by mutation or become 'fixed' by completely replacing their alternative alleles but these processes are very slow and irrelevant to immediate adaptation. In fact, species evolve numerous and diverse mechanisms especially to minimise both the loss of variation by gene fixation and the formation of new mutations, most of which are deleterious (Mayr 1963). The Neo-Darwinian process concerns fluctuations in the frequencies, but not the forms, of genes.

Change in gene structure

When it became technically possible to read the amino acid sequences of primary gene products (and later the nucleotide sequences of the genes themselves), a new paradigm for molecular evolution became the comparison of homologous sequences taken from related species. Species were found to diverge from one another, and from their ancestors, by the accumulation of small substitutions in the molecular structure of their genes. Irreversible permanent change in DNA structure replaced population dynamics as the definition of gene evolution.

'Evolutionary distance' is now measured as the number of substitutions accumulated in the structure of a gene or protein with time (Kimura 1983). Curiously, this yardstick gives anomalous readings of relatedness among species. It measures the length of time since two species diverged from a common ancestor (Zuckerkandl & Pauling 1965) rather than how much their biology has changed during that time. This is because most base substitutions that accumulate in genes have little effect on phenotype. They are selectively neutral, or nearly so, and driven inexorably into the gene by mutation pressure. Comparative biochemists recognise the distinctiveness of this predominating process of relentless gene change by calling it 'non-Darwinian evolution' (King & Jukes 1969). This implies that Darwinian adaptation through natural selection is a separate, and quantitatively lesser, process of change. The main debate in molecular evolution today is about the relative contributions of neutral and adaptive processes to change in gene structure (Kimura 1983).

Adaptation

There have been two main approaches to the study of adaptive changes in gene structure. One is to expose laboratory populations of micro-organisms to artificial selection pressures and then to analyse the changes by which they adapt. The other is to identify the functionally important parts of genes or proteins and to describe how these vary among related species.

In pioneering the former method, John Langridge spread large numbers of *Escherichia coli* on agar that contained lactobionic acid as the only energy source (Langridge 1969). This sugar is a derivative of lactose, a disaccharide that *E. coli* metabolises using the enzyme β-galactosidase. However, the enzyme cannot hydrolyse the analogue sugar effectively. Mutant colonies were readily obtained and genetic analysis showed that many of them had genetic alterations in their β-galactosidase gene.

Langridge's major discovery was the striking variety of *alternative* ways in which the genes were able to acquire this new function. The adapted state is not a unique one, and when two populations or species adapt to a common environment they can be expected to do so in different ways. This raises a key unresolved question. What determines how a gene will evolve when multiple alternative pathways for adaptation are available (see Bock 1959)? The factors that decide the choice between alternatives probably are as important in determining what will evolve as are the environmental stresses to which the species is adapting (Campbell 1985). So far, few inquiries about gene evolution have reached the degree of sophistication necessary to address this question.

The other approach, of examining the particular parts of macromolecules responsible for functional properties, takes advantage of molecular biology's strong emphasis on structure–function relationships in macromolecules. It is now possible to picture protein molecules in three dimensions and to identify the local regions responsible for substrate binding, catalysis, subunit interaction and so on. Comparing these particular parts of a protein between species shows that adaptation of genes is distinct from just change in structure. Perhaps the most significant revelation from these studies is that the more functionally important a segment of gene is, the slower it changes (Dickerson 1971). In fact, the principal way to locate the functional elements in a gene or protein is to search for regions that have not changed during evolution. This means that selection acts mainly *against change* – a defensive rôle quite distinct from the Neo-Darwinian view of selection as the driving force *for change* in gene frequencies. Natural selection has been likened to an editor whose main activity is to proofread for errors rather than to promote change (King & Jukes 1969).

Advancement

Genes do slowly acquire adaptive changes, but it is clear that not all such changes are of equal importance. Although the majority of selected mutations optimise structural details for the exact functional demands of the environment, a smaller number *advance* the organisation of the gene. They create functional capacity instead of adjusting it, and the gene or protein is developed, not just adapted.

Haemoglobin illuminates this process. Most adaptive alterations in this protein probably adjust its oxygen-binding curve to compensate for changes in the organism's size, habitat and so on (Perutz 1983). In addition, a smaller number of entirely different sorts of changes have *advanced* the haemoglobin molecule from a primitive monomer to its derived tetrameric form. This oxygen-binding protein originally consisted of a single polypeptide chain with a haeme group, a form that still persists in myoglobin (Ingram 1963). When the chains evolved the capacity to aggregate, they acquired the possibility for allosteric interactions between subunits. This allowed oxygen-loading curves to become sigmoidal with a threshold and a narrowed range between the oxygen tensions for loading and unloading. Allostery also permitted oxygen loading to be regulated by concentrations of other ligands, especially hydrogen ion and diphosphoglycerate (Bunn *et al.* 1974). Allosteric regulation is essential to the function of modern haemoglobin, and to enzymes in general. Indeed, nearly all enzymes with sophisticated functions have evolved into oligomers. In the evolutionary history of haemoglobin, the emergence of the capacity for allostery was a landmark advance distinct from its subsequent adaptive fine-tuning.

Several other developments further advanced haemoglobin. One was the evolution of two different types of subunits, alpha and beta, which join together to form the molecule. Heterodimerism allows subunits individually to evolve specialised functions. The importance of this capacity again is evident from the large number of other proteins which also have dissimilar subunits.

Another profound advancement was the organisation of the haemoglobin genes into multigene families. Mammals have tandem arrays of genes for the alpha-type and beta-type subunits (Edgell *et al.* 1983). The gene copies in these families are differentially expressed during ontogeny to match the haemoglobin molecule to the individual demands of the embryo, foetus and adult. Some species have developed additional haemoglobin gene copies for various other special purposes (e.g. see Kitchen *et al.* 1968). Clearly, the vertebrates have actively utilised the opportunities for adaptation opened up by the organisation of haemoglobin genes into multigene families. However, the significant evolutionary step was that which created this special adaptability in the first place.

The distinction between advancement and adaptation stands out in many genes, especially in those with demanding biological functions. Significantly, the two processes depend on different kinds of mutations. Genes adapt mainly through base substitutions and occasional small deletions. In contrast, they advance by a wide variety of rearrangements and additions as well as by point mutations. For example, the genes for antibodies acquired their advanced structure by capturing transposons, shuffling exon gene segments, developing special enzymes for rearranging and mutating gene sequences and repeatedly duplicating segments of genes, whole genes and entire multigene families (Sakano *et al.* 1979, Hood *et al.* 1983, Siu *et al.* 1984). Molecular geneticists interested in structure–function relationships repeatedly have found it important to distinguish the process that originates and develops structure from the lesser process of its subsequent adaptation. Adaptation obviously plays a rôle in the origin of novelty, but the two processes are categorically different.

PRINCIPLES OF EVOLUTIONARY ADVANCEMENT

We have a reasonable understanding of how genes fluctuate in allele frequency, change in structure, and adapt. If progression of form is distinct from these processes it is important to determine just how it occurs. Four extra principles seem to be emerging from studies of gene structure. They are discussed here in order of increasing causal complexity.

The tautology of advancement

Structures that are the most advanced are those that were able to advance the most. This really says no more for evolution than it does for a handful of peas spilled on the ground. Some peas will roll further than others and there must be a reason for the variation. Three can be imagined: (1) some peas might be intrinsically better able to roll, perhaps due to a rounder shape or smoother skin; (2) the surface of the ground may vary from place to place, blocking some peas with bumps and rough spots and facilitating others with depressions or grooves that channel their velocity; (3) chance variation as peas hit the ground and jostle one another will move some peas further than others. The analogous factors in evolution are (1) the genetically determined organisation of the species, (2) the selective environment, and (3) whim.

Obviously it is important to evaluate which of these factors have caused some forms of life to advance so much further than others during their

3×10^9 or so years of existence. However, the interactions between phenotype and ecology of higher organisms seem to be vastly too complex for resolving this issue. Fortunately, advancement at the level of gene structure is more tractable. Here, an important observation is that genes that have evolved conspicuously in a particular direction often have identifiable physiological mechanisms that enhance the ability to change in that direction. This association is strong enough for molecular geneticists to rely on it to dicover significant physiological mechanisms. For example, a number of multigene families with outstandingly large numbers of gene copies are associated in some way with special mechanisms for reproducing gene copies. These include 5S RNA genes (Dancis & Holmquist 1979), rRNA genes (Brown & Dawid 1968), pseudogenes (Denison & Weiner 1982), IAP sequences of mice and other retrovirus-like genes (Kuff et al. 1983), transposons (Shiba & Saigo 1983, Reis et al. 1983), VAT genes of trypanosomes (Hoeijmakers et al. 1980), possibly satellite DNAs (John & Miklos 1979), and unique sequences intercalated into satellite DNA (Bedbrook et al. 1980).

As other examples, the extraordinary uniformity of gene copies in some multigene families imply special enzymic mechanisms for removing mutational variation, and one can predict that specialised genes that occur in similar form among unrelated species of bacteria probably are organised as transposons or associated with transmissible plasmids to aid their dissemination. A species also may progress notably in a particular evolutionary direction because it has overcome some factor that constrains such evolution in other species. This represents another way by which a gene or species can specialise its phenotype for a particular kind of evolution. There is a good deal of speculation about structural constraints on the evolution of macromolecules but neither clear examples nor a comprehensive analysis of this issue has yet appeared.

Preadaptation

A particularly important problem for evolutionary advancement concerns the origin of new form. Once a structure advances enough to be functional it can develop further by selection; but how does it initially develop enough to begin to function? This problem worried Darwin and has yet to be resolved. Two principles are probably important. One relates to the well accepted concept of preadaptation (Mayr 1963). A structure that had evolved originally for one function subsequently acquires a second function. Selection for the original function elaborates the structure until its side effects happened to become useful in other ways. The swim bladder, from which the lung arose, and the reptilian foreleg, which evolved into the avian wing, are familiar preadaptations. The term preadaptation today represents only the opportunism of the evolutionary process without any of the teleological overtones implied by certain earlier evolutionists (Mayr 1982, Langridge, this volume).

Genes are particularly favourable as preadaptations because they can be duplicated by a single mutation. One copy can preserve the original function, while the other is unhindered to acquire a new function. Moreover, a supernumerary gene can become silent and drift to a new structure unimpeded by selective constraints. Gene duplication and divergence have been dominating processes for evolving new proteins (Ohno 1970). Most proteins of eukaryotes are members of families that arose from single ancestral genes (Dayhoff *et al.* 1975). These genes of common descent are homologous in the same sense that the wing of the bird is homologous to the leg of the reptile (Schopf 1982). Of course, not all duplicated genes acquire a new function. The genome of higher organisms is rife with redundant gene copies that have failed to acquire a new function and are degenerating back to random sequence as 'pseudogenes' (Li 1983).

It is significant, but not surprising, that when a duplicated gene acquires a new function that new usage is usually related to the original one. Some groups of homologous enzymes span a wide range of catalytic activities, but most related genes have related functions (for an extraordinary example, see Williams 1984, Hood *et al.* 1985). This suggests that when genes for a function proliferate, the preadaptive potential for further advances of that general function may increase as well. Some genes have spawned many more new functional derivatives and pseudogenes than others, but it is not clear whether this reflects structural features of the genes or just the functions that they serve.

Autopoiesis

A second mechanism for the origin of novelty involves the physical principle of autopoiesis (Zeleny 1980). This is the spontaneous emergence of organisation out of nothing – or, more precisely, out of noise (Prigogine & Stengers 1984).Closed physical systems and open ones near equilibrium gravitate towards definite simple states of maximal entropy and minimal rate of entropy formation, respectively. Dissipative open systems far from equilibrium have entirely different behaviour. They can spontaneously became *more* organised with time. A system that is highly dissipative gives fleeting existence to an astronomical number of random or chaotic local microstructures. Any of these twinkling organisational states has the opportunity to stabilise itself from decay and to extend its organisation catalytically past its borders if it is able. Just a microscopic seed of such 'autopoietic' organisation in an energy dissipating system may grow to macroscopic dimensions, typically as a periodic structure. For example, a thin layer of liquid traversed by a large heat flux characteristically develops a regular periodic pattern of convection that is stable whenever the temperature gradient is applied (Prigogine & Stengers 1984). Analogous self-forming organisation also develops in appropriate dissipative chemical systems and is conjectured to emerge spontaneously even in economic and social systems (see in Zeleny 1980).

Autopoiesis seems particularly applicable to evolution, which generates new self-propagating organisation in a highly dissipative system (Zeleny 1980). A major difference between biological and physical processes is that biological systems are extremely organised to begin with in contrast to the typical system for a physics experiment, such as a container of liquid, on a heat source. The physicist's demonstrations are important to show that dissipating energy does have the capacity to organise matter spontaneously but the appropriate examples to clarify the rôle of autopoiesis in evolution must come from biology and not thermodynamic paradigms.

Autopoiesis plays an essential generative rôle in various forms of genetic organisation. For example, multigene families easily evolve once they get started because after a gene duplicates it has a propensity to vary further in copy number due to non-reciprocal crossing over. Moreover, the larger a family grows, the more intrinsically variable it becomes. If every additional copy in a multigene family were as difficult to achieve as the first or second, one never would expect to see huge multigene families. However, immense families are common and evolve extraordinarily rapidly (Fry & Salser 1977). These satellite DNAs give every indication of being products of a runaway autopoietic process.

Autopoiesis has an even more obvious rôle in the evolution of transposons. A large number of genes scattered along chromosomes and bathed in nucleoplasm rich in enzymes represents the substantial level of organisation that characterises a genome. For example, most cells code for transposases that can synthesise extra copies of a gene and plant them elsewhere in the genome. Brewer's yeast uses a transposase to change sex by copying a silent gene for one or the other sex into a special location where it is expressed (Hicks *et al.* 1979). The two sex-determining genes can be copied because they are flanked by special short DNA sequences that the transposase enzyme recognises as its substrate. Other organisms use transposases for other specialised functions (e.g. Hoeijmakers *et al.* 1980). As long as a transposase, or other gene-processing enzyme, is unrelated to the genes that it acts upon, its activities are causally simple. However, it is possible for the continual shuffling of genes along the DNA molecule to sandwich a transposase gene between recognition sequences for its own enzyme. The enzyme now acts *self-referently* on its own genetic determinant and, in the case of a transposase, a transposon is born. The transposon propagates autonomously by copying itself. Moreover, since its components are genetically specified, the unit can mutate and evolve. Variant copies accumulated in the cell or gene pool can compete with each other for resources, differential fecundity, or *lebensraum* with the 'fitter' form supplanting the others. This evolution is quite distinct from that of other genes because it is *selfish* (Doolittle & Sapienza 1980). Self-reference can allow a gene to evolve for its own sake without dependence on the environment of the organism (Orgel & Crick 1980). Transposons evolve both by modifying the genes they already have and by acquiring new genes.

'Selfish evolution' has been likened to a Darwinian process, with the transposon enjoying the status of an individual 'organism' (Dawkins 1976). Actually the process is considerably more powerful because the 'environment' for the gene is not the inanimate physical world but a milieu determined by a genetic programme. We generally believe that organisms adapt to their environment. However, since transposons can (and actively do) accumulate relevant genes of the host, they are able to adapt their environ-

Table 16.3 Types of genes that accumulate on transposable genes.

transposase genes
transposase target sites
other genes for transposition functions
repressors
initiators and terminators of transcription
other replication systems
genes of selective value to the host
other transposable elements

ment to suit themselves. For example, if the amount or specificity of an enzyme in the cell is suboptimal for transposition, a transposon need not adapt itself to that restrictive condition. It can pick up the gene for that characteristic of the 'environment' and selfishly evolve it. Once the gene becomes part of a selfish element it will automatically evolve to optimise its contribution to the fitness of that element.

The transposons that have been characterised (mainly from bacteria) include elaborate forms, attesting to the creative power of their autopoietic evolution (Kleckner 1981). Table 16.3 lists some of the classes of genes found on transposable units. Accumulating such genes does not merely complicate the transposons. It also gives them new emergent properties. Most notably, transposable elements in bacteria have an inherent tendency to become increasingly autonomous of the host. Table 16.4 lists several stages of independence well represented among bacterial genes. It is believed that genes shuffle actively from one level of autonomy to another as an important process in evolution of bacteria (Luria 1969, Botstein 1980, Lida *et al.* 1980) and possibly of eukaryotic genes as well (Temin 1980). Thus autopoiesis can create new levels of functional activity and can transfer the determinants of evolution from the external environment to the internal evolving structure.

Evolutionary function

Some changes in genes are so deliberate that they can only be understood through the biological concept of function. We are well acquainted with *adaptive functions*, those deliberate biological activities that are carried out

Table 16.4 Progressive levels of autonomy of transposable DNA.

free chromosomal location	(transposable DNA)
freedom from host transposase	(transposon)
self control of replication	
extrachromosomal existence	(plasmid)
independence of various host enzymes	
independence from a single host species	(transfer factor)
extracellular existence	(phage)

to aid survival. In addition, various enzymes that operate on genes have *evolutionary functions* (Campbell 1985). They do not enhance the fitness or survival of the organism, but instead they facilitate its evolution.

An example of an evolutionary function is the rectification of gene copies in a multigene family. Some families contain dozens or hundreds of gene copies with the same function. It is impossible for the individual copies to evolve directly by natural selection because the redundancy makes the contribution of any one copy totally dispensable (Hood *et al.* 1975, Edelman & Gally 1970). For example, only the most dominantly deleterious mutation in a 5S RNA gene of *Xenopus* could possibly be sensitive to selection with 10 000 sister gene copies in the family. In order to expose sequences in a multigene family to natural selection, organisms must change large groups of the genes to a common structure. If they rectify a block with variant genes to wild type, mutations are purged. If they expand a mutant configuration, natural selection can then favour or eliminate that variant allele of the family within the species.

Some multigene families maintain extraordinary levels of homogeneity. Hundreds or thousands of gene copies are kept almost identical even though they readily evolve in concert. Obviously such uniformity can result only from a very effective rectification mechanism, of which eukaryotes have several. A gene copy can be directly corrected to the sequence of another in some multigene families (Slightom *et al.* 1980). Saltatory expansion and contraction, sister chromatid exchange, and extrachromosomal replication and integration probably occur in others (Weiner & Denison 1982). These mechanisms are prerequisites for using multigene families as genetic determinants. Without them not only would a multigene family be unable to evolve its functions, but it would inexorably erode into individual genes. The important point is that rectifying a multigene family promotes its evolution but does not increase fitness. (In fact, the process generates genetic load.) The function of the process is an evolutionary one and not adaptive.

The transposases of bacterial transposons are another set of enzymes with important evolutionary functions. Moving transposons around in a cell does not enhance the cell's fitness at all. In fact, it is probably deleterious on the whole because occasionally a transposon gets inserted into an essential gene. However, transposition is essential to bacterial evolution.

Enzymes are not the only structures with evolutionary functions. Sensory machinery also can regulate cellular processes for evolutionary ends. Repressor systems direct most transposase genes to be active only when appropriate for evolution. Some bacteria and higher organisms have evolved very elaborate sensory systems to induce generalised heritable alterations in their genomes in response to stresses (Echols 1981, McClintock 1984, Cullis 1977).

Since structures with evolutionary functions have entirely different rôles in evolution from those with adaptive functions it is useful to have a specific name for them. Elsewhere (Campbell 1985) I have suggested the term 'evolutionary driver' as the counterpart of the noun 'adaptation'. An evolutionary driver is a genetically determined structure with a function of propelling or steering the evolution of a gene, phenotypic trait or species.

Enzyme pathways that rectify multigene families, transposases and their repressor systems, the recA system for inducible mutagenesis in bacteria and so forth, are examples of evolutionary directors which participate in the evolution of genes. If trends continue, we will probably discover a diversity of other evolutionary functions during the next few years. Most of the special organisational features of eukaryotic genes that have surprised geneticists during the past several decades are suspected of having evolutionary functions (see Table 16.5).

Although the concept of evolutionary function extends some of our ideas about evolution, it retains the conventional notion of 'function'. If the function of the enzyme maltase is to hydrolyse maltose, then surely the function of transposase must be to transpose segments of DNA, and that of acquisitionase is to intercalate bits of foreign bits of DNA into the chromosome. We would call an effect of a structure its 'function' if it provided part or all of the selective advantage for the development or maintenance of that structure. Organisms can evolve structures with evolutionary functions through the same mechanism of *selection by consequence* (Skinner 1981) that underlies the evolution of adaptive structures. All selection is based on the consequences of a structure rather than the structure itself. Species with greater capacity to adapt to their changing environment will benefit from the consequences of that capacity and can eventually prevail (Campbell 1985).

As discussed above, preadaptation and autopoiesis greatly increase the ability to evolve new adaptive functional capacities. Most evolutionary drivers probably evolved from adaptive precursors that acted as preadaptations. The repressor on transposon Tn917 is a case in point (Tomich *et al.* 1978). Its adaptive function is to repress the erythromycin resistance gene of the transposon when erythromycin is absent from the environment. The repressor also has acquired, probably secondarily, control over the expression of the transposase gene. As a result, the presence of erythromycin induces the erythromycin-resistance transposon to move in the cell in addition

Table 16.5 Some eukaryotic genetic structures with possible evolutionary functions.

introns
satellite DNA
pseudogenes
spacer sequences
transposable elements
retroviruses
multigene families
chi sites
mechanisms for:
gene correction
sister chromatid exchange
DNA methylation
back transcription

to expressing a drug-resistant phenotype. Also, many enzymes used for DNA repair have additional functions in recombination and mutagenesis (Kimball 1978, Deserres 1980).

Perhaps the foremost discovery about evolution to come from modern genetics is that organisms carry out evolutionary activities as biological functions. Just how many genes have evolutionary functions is unclear, but the number may be large. Table 16.5 suggest that eukaryotes may actually have a greater amount of DNA concerned with evolution than with phenotypic fitness! A wide variety of structures with evolutionary functions should come as no surprise considering the wealth of adaptations that even the simplest species have evolved. Any adapted structure must be considered a possible potential preadaptation for an evolutionary driver. Thus the more a species or gene advances, the more opportunities it has for recruiting adaptive structures into evolutionary rôles.

EVOLUTION OF PHENOTYPE

As described above, genes change in four distinct ways with time. The frequencies of variant forms fluctuate reversibly in populations, structural alterations become fixed, genes acquire adaptive modifications and they advance in their structure. It is significant that each of these dynamics has its counterpart in the evolution of phenotype.

For example, the distribution of body sizes of European moles fluctuates from one year to another depending on the weather, although there is no evidence for a lasting change. Harsh winters selectively eliminate larger individuals (Stein 1951), but milder years allow them to reaccumulate in the population. Presumably body size, like most quantitative traits in animals, is maintained as a long-term balance between intermittent pressures for increase and for decrease. The overwhelming preponderance of phenotypic changes rapid enough to be seen from generation to generation almost certainly are of this nature and leave no mark on evolution over geological time. They are just 'noise' in the long-term process. Most selection on gene frequencies is probably of the stabilising variety that leads to long-term constancy instead of change in phenotype.

Permanent change in phenotype also does occur, of course. It can be inferred from comparative biology and in some cases can be followed directly in the fossil record. The difficulty is not in detecting change, but in distinguishing changes with no selective importance from adaptations. Obviously changes may be adaptive even though we may not be able to figure out their usefulness. Conversely, and more disconcerting, the selectionist's exercise of devising possible adaptive explanation for a change does not prove that that or any other selective factor caused its evolution. A variety of mechanisms including allometry, mutation pressure, founder effects, fabricational noise, pleiotrophy, drift of various sorts, and gene introgression can cause a phenotypic character to change neutrally or even deleteriously. In recent years, important evolutionists have cautioned against the mistake of assuming that any existing character or phenotypic change automatically must be adaptive (Gould & Lewontin 1979). Also, some palaeontologists believe that most morphological change occurs as part of the process of speciation. The rôle of selection in the formation of new species is poorly understood.

Probably the best guide for picking out the adaptations among evolutionary changes is common sense. The change in the colour of the pepper moth (*Biston betularia*) in England to a darker hue where industrial soot has blackened the tree trunks on which they hide is reasonably inferred to be adaptive (Ford 1964). The red colour of blood probably is not. The colour patterns of mollusc shells are more problematical and may include both selective and non-selective factors (Seilacher 1972).

It is easier to document selection for changes over short periods of time than long ones. However, the gulf between the 'ecological time' of the Neo-Darwinist and 'geological time' is so great that there is little reason to believe that the proportion of changes that are adaptive will be the same for the two. It is not even clear which time scale of change is the more constrained by adaptation. One reason for the current interest in the evolution of gene and protein structure is that it should be so much easier to distinguish adaptive from neutral change where relationships between structure and function are simpler. Yet, even here, selectionists and neutralists bitterly disagree.

Progressive advancement to higher forms is also apparent in phenotypes of organisms. A dominant pattern in the long-term history of life is the appearance of successively higher and higher forms. This does not mean that all lineages inexorably advance. They decidedly do not. The pattern is for most species to get trapped in stasis and for only a minority to continue to advance. Nevertheless, by almost any criterion for 'advancement' that one might choose, life has repeatedly progressed from one record height to another during evolutionary history with the highest forms existing today. Such advancement of phenotype probably involves all four of the principles discussed above for genes.

The tautology of advancement

The truism that the most advanced forms of life come from the lineages that were able to advance the fastest and farthest suggests that higher organisms evolve by different, and more effective, sorts of changes than do more primitive forms. This is indeed the case. Higher forms evolve in ways that were difficult or impossible for earlier forms. A striking progression has been for life to adapt through modifications first in enzyme catalysis, then in morphology, and ultimately in behaviour. Primitive organisms still evolve largely by changing their biochemical pathways. The main adaptive strategy of bacteria, for example, is to meet changes in their environment with new enzymic activities. To this end they have developed the complex system of transposons and plasmids discussed above. Eukaryotes, in contrast, evolve most notably in their morphology, generating new relationship between cells instead of new types of enzymes. Higher animals continue to evolve the occasional new enzyme pathway, for example, a mammalian lactose synthetase for making milk sugar, but a striking discovery from comparative biochemistry is that *Escherchia coli* has essentially the entire repertoire of basic enzyme activities of human cells. By and large, vertebrates, mammals, man and insects are reflections of special developmental pathways. Morphology offers vastly greater scope for change than do enzymes and, from what we know, is easier and quicker to evolve.

Behaviour allows even more varied and rapid adaptation. The most advanced animals evolve largely through their behaviour. For example, ants are counted among the higher arthropods. Their morphology is not conspicuously advanced but their behaviour makes them outstanding. Ants build large architectural structures, conduct agriculture with both plants and animals, co-operate in societies based on altruism, and take slaves. These functional capabilities emerged from advances in behaviour, although with support from specialised morphology and biochemistry.

Another progression is in the source of variation on which selection can operate. Early forms of life depended on spontaneous mutation. Their advanced descendents developed special enzymic pathways for editing alterations in the DNA. Bacteria evolved the more effective capacity to steal new genes from neighbouring species. Eukaryotes advanced in a different direction, evolving a diploid sexual cycle that generates continual quantitative variation by recombination.

Thus, in various ways, the forms of life that have advanced the farthest in phenotype are commensurately endowed with the greatest capacities for evolving.

Preadaptation

This concept is so familiar in classical evolution that it requires little comment here. Most new functional capacities of higher organisms probably originated opportunistically from pre-existing phenotypic structures.

Autopoiesis

Autopoiesis fits less comfortably with current evolutionary thought because it ascribes the cause of change to endogenous structure instead of interferences from the external world. This is a mode of explanation that Neo-Darwinists have explicitly rejected (although it has gained some recent advocacy from followers of so-called non-equilibrium evolutionary theory (Wiley & Brooks 1982, Wicken 1983)). Neo-Darwinism arose in large measure as a repudiation of competing interpretations that evolution was directed from within, such as aristogenesis, orthogenesis, evolution by law, mutationism and so forth (see Langridge, this volume). Nevertheless, autopoiesis and self-reference are integral both to the general process of evolution and to specific cases. In fact, the very origin of life was an autopoietic event, with its chicken-and-egg conundrum that faithful replication of the nucleic acid molecule is a prerequisite for the evolution of genetic information, but information for polymerase activity could evolve only after nucleic acids could be faithfully replicated.

Sexual selection is a more specialised type of autopoietic evolution (Fisher 1930). If female peacocks tend to choose males with large tails as mates, then the genes promoting this attractive masculine character will enjoy a selective advantage. As a consequence, genes that enhance the attraction of females to large-tailed males also will be at an advantage. They will prompt females to mate with the favoured males and hence produce male offspring that are more desirable. Circuits of genes that favour themselves through sexual, social or adaptive processes can erupt into runaway evolution unrelated to economy and ecology. Nevertheless, the directions of such evolution obviously are not random. They must depend in complex ways upon which organisational features of the species are potentially most sensitive to snowballing autopoiesis, and upon the availability of triggers to set the self-reinforcing cycles off. It is suggestive that certain aspects of morphology, such as wing bars, tails and head crests, seem especially prone to exaggeration as sexual adornments of some birds and it is interesting to consider the cause of this pattern.

Far more important self-reference emerges from the action of the organism on its own environment. Neo-Darwinism depends on a Newtonian-like distinction between species on the one hand, and its surroundings as an independent agency on the other. Yet, as Lewontin (1982, 1983) has stressed, this is not realistic. The characteristics of an environment are not independent of the species that inhabit it, but are intimately shaped by the occupants. Probably the most important selective feature of the environment for any higher species is the species itself. Thus, selection pressures are not just dictates of the external world but emerge in large part as the manifestations of the genetic programme of the species itself. This means that the genotype plays two rôles in evolution. It responds with adaptive change and it also generates the cause of that change. What a genotype adapts to is largely its own expression. This self-referent cause and effect increasingly dominates evolution as organisms become more competent to manipulate their environments.

Since even the most doctrinaire evolutionist cannot ignore this self-reference completely, Neo-Darwinism has developed special branches of 'frequency-dependent' and 'density-dependent' selection (Ayala & Campbell 1974). These have given rise to some really exciting insights into evolution, such as the 'Red Queen paradox' (Van Valen 1973) and 'evolutionary stable strategies' (Maynard Smith 1974). Yet, the most conspicuous feature of deterministic mathematical sojourns into self-reference is their overwhelming complexity. Predictability, optimum and determinism become subordinate to instabilities, alternatives, critical points, feedback and oscillations. If there is one overriding lesson to be learned from the evolution of transposable genes it is that the process of evolution has access to dynamic relationships far beyond the trivia of Newtonian cause and effect. The tautology of advancement implies that significant evolutionary progress is a flag for the more powerful forms of evolutionary cause. I suggest that the important issue is not how the phenotype changes due to the effect of evolution, but how it has become specialised to cause evolution.

Evolutionary function

An organism's most powerful mode of action is to carry out important activities as biological functions. Recent discoveries about the physiology of genes prove that evolution is able to produce structures to enhance the evolutionary process as well as to aid immediate survival. Much of the phenotype of higher organisms probably functions as evolutionary drivers. For example, the half of eukaryotic organisation concerned with sexuality, recombination and mutation did not evolve to enhance the survivability or fecundity of the individual, but for its evolutionary function.

The evolution of sexuality poses a problem for the strict adaptationist because it is a burden on the fitness of the individual. We know that it can be dispensed with because occasional species ranging from plants to insects and vertebrates have become asexual. Scrapping sexuality for asexual reproduction probably aided short-term fitness. Nevertheless, higher organisms have preserved the burden of sexuality except for these sporadic cases because asexuality probably leads to *evolutionary* blind alleys.

Recognising that phenotypic structures can have evolutionary functions is a separate issue from identifying the ones that actually are evolutionary drivers. I have already discussed the difficulty of sorting out adaptations from selectively neutral changes. Compounding this difficulty is the third possibility of structures evolving to perform evolutionary functions. Even at the level of genotype it is difficult to distinguish among these three possible reasons for a structure's existence. This difficulty is reflected in the fact that geneticists have proposed all three sorts of explanations for each of the organisational features of eukaryotic genes listed in Table 16.5. The problem of assigning functions to structures becomes further complicated when adaptive structures secondarily acquire rôles in evolution. They can then serve both sorts of functions simultaneously. An example cited above of a protein with such double function is the repressor of transposon Tn917 which simultaneously regulates both the erythromycin-resistance gene and the transposase gene of the transposon.

A critical task before us now is to assess the extent to which phenotypic traits of higher organisms have acquired evolutionary functions. For example, how widely have canalisation systems for reducing noise in development been recruited to guide evolution (see Maynard Smith *et al.* 1985)? A second essential question is how many features of evolution that we take for granted as 'just the way evolution is' actually are results of selection for that mode of evolution? An immediate case is punctuated equilibrium. Is this a specially evolved mode of evolution? Have species developed the capacity or propensity occasionally to reorganise their genomes on a major scale in order to break out of the stasis that otherwise 'straightjackets' established species? Do species evolve special populational, morphological, behavioural or genetic organisation to throw off new species? If so, can we explain how and why those features evolved? Also, what about stasis: is the pronounced tendency of many established species to persist with virtually no change an attribute that evolved from selection?

Relationship of advancement to adaptation and change

Both gene structure and phenotype show the four dynamics listed in order of their dependence on each other in Table 16.2. Each type depends on change at some underlying level, but may occur with or without producing any higher level dynamics. For example, adaptation presupposes change either in the structure of genes or in the frequencies of gene alleles in the gene pool. However, neither a fluctuation in the composition of the gene pool nor a base substitution in a gene implies that adaptation has taken place.

Advancement is also predicated upon some sort of change occuring but clearly does not inevitably accompany it. Superficially, advancement might seem to require adaptive change but this seems not to be so. For example, coupling a multigene family to an enzyme system that actively corrects one gene copy against another could be a critical advance in the organisation of the family even though it had no adaptive importance. Probably complex interactions between the ecology and the structure of organisms determine how fast and in which ways species advance as they change. It is possible that the environment is more important for phenotypic advancement than for advancement of gene structure, but we have yet to understand what situations lead the species to advance instead of simply shifting 'sideways' from one niche to another.

RATES OF EVOLUTION

Descriptions of the molecular evolution of genes, and their extrapolation to phenotype, have two implications for rates of evolutionary change. Most notably, *biological structures* play active rôles in evolution and must be major determinants of how fast or slow genes and species evolve. For example, their elaborate structures for rearranging and transmitting genes allow bacteria to evolve antibiotic resistance much faster than otherwise would be possible. Genetic and phenotypic organisation probably also contribute to slowing down evolution in other cases, although we currently know much less about the mechanisms for stasis. The significance of these genetically determined factors is that they are evolved traits instead of just uncontrolled dictates of the outside environment. To the extent that biological mechanisms participate in evolutionary change they allow its rate to be programmed internally as an attribute of the species.

Secondly, molecular studies show that genes, and presumably phenotype, change in at least four entirely different ways during evolution. These types of change are substantially disconnected from one another and can proceed at independent rates. Analyses of evolutionary rates must begin with a clear description of the sort of change being considered and then proceed to the mechanisms that produce that form of change.

ENDOGENOUS VERSUS EXOGENOUS CAUSE FOR EVOLUTION

The discovery that genes are inherently dynamic instead of passive structures undermines the basic Darwinian proposition that evolution is change forced on the species by the outside environment. The null hypothesis for Neo-Darwinism is that if a species were boxed into an environment so favourable that it exerted no selection pressure, evolutionary change would cease. This is unrealistic. No matter how much a species was environmentally pampered, its genes would evolve. Species are powerfully influenced by their environment, but environmental forces are superimposed on endogenous engines for change and advancement.

The relative importance of endogenous organisation and external environment is perhaps the central question in evolution today. It does not have a simple quantitative answer. One reason is that the species influences its surroundings as discussed above. A more important reason is that the evolutionary rôle of the genetic message varies from one species to another. In particular, genetic constitution becomes increasingly important as the species advances to a higher form.

Consider the most primitive forms at the dawn of life. Their genes had virtually no capacity to 'do' anything and were maximally exposed to environmental conditions. Before enzyme systems arose to detect and repair lesions and to catalyse change in DNA structure, the mutation spectrum was dictated by physical principles and agents. Evolution probably approached the Darwinian model more closely then than at any time since. Subsequent species evolved increasingly elaborate ways to control mutation, suppressing deleterious types and promoting more productive types. They also developed phenotypic wraps to direct the way that environmental insults forced change on their genes. Higher organisms modified their environment and clustered into aggregates in order to be surrounded by a genetically regulated micro-environment instead of raw nature. Ever more sophisticated adaptations became potential preadaptations for developing ever more sophisticated evolutionary functions. Life asserted greater and greater control over its own evolution as it advanced; the forms progressing the fastest and farthest being those that optimised the factors important to evolutionary advancement. The increasing influence of the species in evolution not only results from advancement but also provides the mechanism for the process of advancement. Evolutionary progress means to gain command over one's evolutionary destiny.

This perspective, that the essence of evolution concerns the capacity to evolve rather than survival of the individual, resolves the enigma of 'higher' forms of life. Theories of evolution restricted to survival of the fittest have to dismiss the whole concept of advancement as illusionary. We might call ourselves and our relatives 'higher' than the protista or prokaryotes but for the survivalist this is just an arbitrary value judgement. By the goal of survival, bacteria are just as fit as we are – perhaps even fitter since they maintain far higher populations and have a phenotype that has survived without significant correction far longer than ours. Certainly we are more *complex*, but is it not just an anthropocentric bias to equate complexity with advancement? Complexity seems to be as irrelevant to fitness as any other general characteristic that varies among extant organisms such as size, lifespan, metabolic rate, or degree of specialisation.

In contrast, complexity is intimately related to the capacity to evolve. Organisation and complexity are measures of both the cause and the effect of evolutionary advancement. An increased capacity to evolve effectively should automatically manifest itself as a more sophisticated phenotype. Reciprocally, phenotypic organisation which expands the potential of a species to evolve must advance that process. This is not just a truism because some phenotypic elaborations can adapt a species exquisitely to a particular environmental factor but impede, instead of facilitate, its subsequent evolution. They decrease evolutionary status even though they make the species more specialised. There is nothing arbitrary or anthropocentric about viewing advancement as a relationship of the organism to its evolution instead of to its external environment. Moreover, it turns the problem of evaluating how advanced a particular species is into an evolutionary question demanding an understanding of evolutionary behaviour, instead of just an ecological issue.

Defining evolution as the process of creating capacities for further evolution has a degree of circularity that is entirely appropriate. The ultimate significance of evolving life is not that it can change and adapt, but that it continually achieves new orders of function. Progressive advancement corresponds more closely to the literal meaning of the word evolution, 'to unfold', than does simple change. What is unfolding, however, is not the organism's fitness for Darwinian survival, but its command over its own evolutionary destiny.

ACKNOWLEDGEMENTS

I wish to thank John Langridge for many stimulating discussions and Bob Taylor for suggestions on this manuscript. This work was supported by NSF grant PCM-8120923.

REFERENCES

Alberts, B., D. Bray, J. Lewis, M. Raff, K. Roberts and J. D. Watson 1983. *Molecular biology of the cell.* New York: Garland.

Anderson, E. S. 1965. The ecology of transferable drug resistance in Enterobactereacae. *Ann. Rev. Microbiol.* 22, 131–80.

Ayala, F. J. and C. A. Campbell 1974. Frequency-dependent selection. *Ann. Rev. Ecol. Syst.* 5, 115–38.

Bedbrook, J. R., M. O'Dell and R. B. Flavell 1980. Amplification of rearranged repeated DNA sequences in cereal plants. *Nature* 288, 133–7.

Bock, W. J. 1959. Preadaption and multiple evolutionary pathways. *Evolution* 13, 194–211.

Botstein, D. 1980. A theory of modular evolution for bacteriophages. *Ann. NY Acad. Sci.* **354**, 484–91.

Brown, D. D. and I. B. Dawid 1968. Specific gene amplification in oocytes. *Science* **160**, 272–80.

Bunn, H. F., U. S. Seal and A. F. Scott 1974. The role of 2, 3 diphosphoglycerate in mediating hemoglobin function of mammalian red cells. *Ann. NY Acad. Sci.* **231**, 498–512.

Campbell, A. 1981. Evolutionary significance of accessory DNA elements in bacteria. *Ann. Rev. Microbiol.* **35**, 55–83.

Campbell, J. H. 1983. Evolving concepts of multigene families. *Curr. Top. Biol. Med. Res.* **10**, 401–17.

Campbell, J. H. 1985. An organizational interpretation of evolution. In *Evolution at a crossroads: the new biology and the new philosophy of science*, D. J. Depew and B. H. Weber (eds), pp. 133–67. Cambridge, Mass.: MIT Press.

Campbell, J. H. and E. G. Zimmermann 1982. Automodulation of genes: A mechanism for persisting effects of drugs and hormones in mammals. *Neurobehav. Tox. Terat.* **4**, 435–9.

Cox, M. M. and I. R. Lehman 1981. RecA protein of *Escherichia coli* promotes branch migration, a kinetically distinct phase of DNA strand exchange. *Proc. Natl Acad. Sci.* **78**, 3433–7.

Cullis, C. A. 1977. Molecular aspects of the environmental induction of heritable changes in flax. *Heredity* **38**, 129–54.

Dancis, B. M. and G. P. Holmquist 1979. Telomere replication and fusion in eukaryotes. *J. Theor. Biol.* **78**, 211–24.

Davies, B. D. 1971. Foreword. In *Transferable drug resistance factor R*, S. Mitsuhashi, (ed.), pp. v–vi. Baltimore: University Park Press.

Dawkins, R. 1976. *The selfish gene*. Oxford: Oxford University Press.

Dayhoff, M. O., P. J. McLaughlin, W. C. Barker and L. T. Hunt 1975. Evolution of sequences within protein superfamilies. *Naturwissenschaften* **62**, 154–61.

Denison, R. A. and A. M. Weiner 1982. Human U1 RNA pseudogenes may be generated by both DNA- and RNA-mediated mechanisms. *Mol. Cell. Biol.* **2**, 815.

Deserres, F. J. (ed.) 1980. *DNA repair and mutagenesis in eukaryotes*. New York: Plenum Press.

Dickerson, R. E. 1971. The structure of cytochrome *c* and the rates of molecular evolution. *J. Mol. Evol.* **1**, 26–45.

Doolittle, W. F. and C. Sapienza 1980. Selfish genes, the phenotype paradigm and genomic evolution. *Nature* **284**, 601–3.

Echols, H. 1981. SOS functions, cancer and inducible evolution. *Cell* **25**, 1–2.

Edelman, G. M. and J. A. Gally 1970. Arrangement and evolution of eukaryotic genes. In *Neurosciences: second study program*, F. O. Schmitt (ed.). New York: Rockefeller University Press.

Edgell, M. H., S. C. Hardies, B. Brown, C. Voliva, A. Hill, S. Phillips, M. Comer, F. Burton, S. Weaver and C. A. Hutchison 1983. Evolution of the mouse alpha-globin complex locus. In *Evolution of genes and proteins*, M. Nei and R. K. Koehn (eds), pp. 1–13. Sunderland, Mass.: Sinauer.

Fisher, R. A. 1930. *The genetical theory of natural selection*. Oxford: Clarendon Press.

Ford, E. B. 1964. *Ecological genetics*. London: Methuen.

Fry, D. and W. Salser 1977. Nucleotide sequences of HS-alpha satellite DNA from kangaroo rat *Dipodomys ordii* and characterization of similar sequences in other rodents. *Cell* **12**, 1069–84.

Gellert, M. 1981. DNA topoisomerases. *Ann. Rev. Biochem.* **50**, 879–910.

Gould, S. J. and R. C. Lewontin 1979. The spandrels of San Marco and the Panglossian paradigm: a critique of the adaptionist programme. *Proc. R. Soc. Lond. B.* **205**, 581–98.

Hicks, J., N. Strathern and A. J. S. Klar 1979. Transposable mating-type genes in *Saccharomyces cerevisiae. Nature* **282**, 478–83.

Hoeijmakers, J. H. J., A. C. C. Frasch, A. Bernards, P. Borst and G. A. M. Cross 1980. Novel expression-linked copies of the genes for variant surface antigens in trypanosomes. *Nature* **284**, 78–80.

Hood, L., J. H. Campbell, and S. C. R. Elgin 1975. The organization, expression and evolution of antibody genes and other multigene families. *Ann. Rev. Genet.* **9**, 305–53.

Hood, L. E., T. Hunkapiller and E. Kraig 1983. Stratigies for gene organization and information expression. *Mod. Cell, Biol.* **2**, 305–28.

Hood, L., M. Kronenberg and T. Hunkapiller 1985. T-cell antigen receptors and the immunoglobin supergene family. *Cell* **40**, 225–9.

Ingram, V. M. 1963. *The hemoglobins in genetics and evolution.* New York: Columbia University Press.

Jacob, F. and J. Monod 1961. Genetic regulatory mechanisms in the synthesis of proteins. *J. Mol. Biol.* **3**, 318–56.

John, B. and G. L. G. Miklos 1979. Functional aspects of satellite DNA and heterochromatin. *Int. Rev. Cytol.* **58**, 1–114.

Kavaler, J., M. M. Davis and Yueh-hsiu Chien 1984. Localisation of a T-cell receptor diversity-region element. *Nature* **310**, 421–3.

Kimball, R. F. 1978. The relation of repair phenomena to mutation induction in bacteria. *Mutation Res.* **55**, 85–120.

Kimura, M. 1983. The neutral theory of molecular evolution. In *Evolution of genes and proteins*, M. Nei and R. K. Koehn (eds), pp. 208–33. Sunderland, Mass.: Sinauer.

King, J. L. and T. H. Jukes 1969. Non-Darwinian evolution: random fixation of selectively neutral mutations. *Science* **164**, 788–98.

Kitchen, H., J. W. Eaton and W. J. Taylor 1968. Rapid production of a hemoglobin induced by hemolysis in sheep: hemoglobin C. *Am. J. Vet. Res.* **29**, 2–8.

Kleckner, N. 1981. Transposable elements in prokaryotes. *Ann. Rev. Genet.* **15**, 341–404.

Kuff, E. L., A. Feenstra, D. Lueders, L. Smith, R. Hawley, N. Hozumi and M. Shulman 1983. Intercisternal A-particle genes as movable elements in the mouse genome. *Proc. Natl Acad. Sci.* **80**, 1992–6.

Langridge, J. 1969. Mutations conferring quantitative and qualitative increases in beta-galactosidase activity in *Escherichia coli. Mol. Gen. Genet.* **105**, 74–83.

Lewin, B. 1974. *Gene expression*, vol. 1. New York: John Wiley.

Lewontin, R. C. 1982. Prospectives, perspectives and retrospectives. *Paleobiology* **8**, 309–13.

Lewontin, R. C. 1983. The corpse in the elevator. *New York Rev.* Jan. 20, 34–7.

Li, W. -H. 1983. Evolution and duplication of genes and pseudogenes. In *Evolution of genes and proteins*, M. Nei and R. K. Koehn (eds), pp. 14–37. Sunderland, Mass.: Sinauer.

Lida, S., J. Meyer and W. Arber 1980. Genesis and natural history of IS-mediated transposons. *Cold Spring Harbor Symp. Quant. Biol.* **45**, 27–43.

Liebhaber, S. A., M. Goossens and Y. W. Kan 1981. Homology and concerted evolution at the alpha 1 and alpha 2 loci of human alpha-globin. *Nature* **290**, 26–9.

Liebhaber, S. A., E. F. Rappaport, F. E. Cash, S. D. Ballas, D. Schwartz and S. Surrey 1984. Hemoglobin I mutation encoded at both alpha-globin loci on the same chromosome: concerted evolution in the human genome. *Science* **226**, 1449–51.

Luria, S. E. 1969. Directed genetic change: perspectives from molecular genetics. In *The control of human heredity and evolution*, T. M. Sonneborn (ed.), pp. 1–19. New York: MacMillan Press.

Maynard Smith, J. 1974. The theory of games and the evolution of animal conflicts. *J. Theor. Biol.* **47**, 209–21.

Maynard Smith, J., R. Burian, S. Kaufman, P. Albrech, J. H. Campbell, B. Goodwin, R. Lande, D. Raup and L. Wolpert 1985. Developmental constraints and evolution. *Quant. Rev. Biol.* **60**, 265–87.

Mayr, E. 1963. *Animal species and evolution*. Cambridge, Mass.: Harvard University Press.

Mayr, E. 1970. *Populations, species and evolution*. Cambridge, Mass.: Harvard University Press.

Mayr, E. 1982. *The growth of biological thought*. Cambridge, Mass.: Harvard University Press.

McClintock, B. 1984. The significance of responses of the genome to challenge. *Science* **226**, 792–801.

Miller, H., M. Kirk and H. Echols 1981. SOS induction and autoregulation of the *himA* gene for site specific recombination in *Escherichia coli*. *Proc. Natl Acad. Sci.* **78**, 6754–8.

Morgan, T. H. 1933. The relation of genetics to physiology and medicine. In *Les Pres Nobel en 1938*. Stockholm: Norstedt & Soner.

Ohno, S. 1970. *Evolution by gene duplication*. Berlin: Springer-Verlag.

Orgel, L. F. and F. H. C. Crick 1980. Selfish DNA: the ultimate parasite. *Nature* **284**, 604–7.

Payvar, F., O. Wrange, J. Carlstedt-Duke, S. Okret, J. A. Gustafsson and K. R. Yamamoto 1981. Purified glucocorticoid receptors bind selectively *in vitro* to a cloned DNA fragment whose transcription is regulated by glucocorticoids *in vivo*. *Proc. Natl Acad. Sci.* **78**, 6628–32.

Perutz, M. F. 1983. Species adaptation in a protein molecule. *Mol. Biol. Evol.* **1**, 1–28.

Prigogine, I. and I. Stengers 1984. *Order out of chaos*. New York: Bantam.

Reis, R. J. S., C. K. Lumpkin, Jr, R. J. McGill, K. T. Riabowol and S. Goldstein 1983. Extrachromosomal circular copies of an 'inter-Alu' unstable sequence in human DNA are amplified during *in vitro* and *in vivo* aging. *Nature* **301**, 394–8.

Sakano, H., K. Huppi, G. Heinrich and W. Tonegawa 1979. Sequences at the somatic recombination sites of immunoglobulin light-chain genes. *Nature* **280**, 288–94.

Schopf, T. J. M. 1982. DNA structures: the fourth approach to comparable biology. *Cold Spring Harbor Symp. Quant. Biol.* **47**, 1159–64.

Seilacher, A. 1972. Divaricate patterns in pelecypod shells. *Lethaia* **5**, 325–43.

Shiba, T. and K. Saigo 1983. Retrovirus-like particles containing RNA homologous to the tranposable element *copia* in *Drosophila melanogaster*. *Nature* **302**, 119–24.

Siu, G., M. Kronenberg, E. Straus, R. Haars, T. W. Mak and L. Hood 1984. The structure, rearrangement and expression of D (beta) gene segments of the murine T-cell antigen receptor. *Nature* **311**, 344–9.

Skinner, B. F. 1981. Selection by consequences. *Science* **213**, 501–4.

Slightom, J. L., A. E. Blechl and O. Smithies 1980. Human fetal G gamma- and A gamma-globin genes: complete nucleotide sequences suggest that DNA can be exchanged between these duplicated genes. *Cell* **21**, 627–38.

Stein, G. H. W. 1951. Populationsanalytische Untersuchungen am europäischen Maulwurf. II. Über zeitliche Grössenswankungen. *Zool. Jahrb. (Syst.)* **79**, 567–90.

Temin, H. M. 1980. Origin of retroviruses from cellular movable genetic elements. *Cell* **21**, 599–600.

Tomich, P. K., F. Y. An and D. B. Clewell 1978. A transposon (Tn917) of *Streptococcus faecalis* that exhibits enhanced induction of drug resistance. *Cold Spring Harbor Symp. Quant. Biol.* **43**, 1217–21.

Van Valen, L. M. 1973. A new evolutionary law. *Evol. Theory* **1**, 1–30.

Weiner, A. M. and R. A. Denison 1982. Either gene amplification or gene conversion may maintain the homogeneity of the multigene family encoding human Ul small nuclear RNA. *Cold Spring Harbor Symp. Quant. Biol.* **47**, 1141–9.

Wicken, J. S. 1983. Entrophy, information and nonequilibrium. *Syst. Zool.* **32**, 438–43.

Wiley, E. O. and D. R. Brooks 1982. Victims of history – a nonequilibrium approach to evolution. *Syst. Zool.* **31**, 1–24.

Williams, A. F. 1984. The immunoglobulin supergene family takes shape. *Nature* **308**, 12.

Yamamoto, K. R. and B. M. Alberts 1976. Steroid receptors: elements for modulation of eukaryotic transcription. *Ann. Rev. Biochem.* **45**, 721–46.

Zeleny, M. (ed.) 1980. *Autopoiesis, dissipative structures and spontaneous social orders*. Boulder, Col.: Westview Press.

Zuckerkandl, E. and L. Pauling 1965. Evolutionary divergence and convergence in proteins. In *Evolving genes and proteins*, V. Bryson and H. J. Vogel (eds), pp. 97–166. New York: Academic Press.

Index